AF686460

Novel Approaches in Contaminant Hydrology and Groundwater Remediation

Novel Approaches in Contaminant Hydrology and Groundwater Remediation

Guest Editor

Meysam Vadiati

MDPI

Basel • Beijing • Wuhan • Barcelona • Belgrade • Novi Sad • Cluj • Manchester

Guest Editor
Meysam Vadiati
Fellowship Program
Global Affairs
University of California, Davis
CA
USA

Editorial Office
MDPI AG
Grosspeteranlage 5
4052 Basel, Switzerland

This is a reprint of the Special Issue, published open access by the journal *Hydrology* (ISSN 2306-5338), freely accessible at: https://www.mdpi.com/journal/hydrology/special_issues/02699A5C2V.

For citation purposes, cite each article independently as indicated on the article page online and as indicated below:

Lastname, A.A.; Lastname, B.B. Article Title. *Journal Name* **Year**, *Volume Number*, Page Range.

ISBN 978-3-7258-3967-4 (Hbk)
ISBN 978-3-7258-3968-1 (PDF)
https://doi.org/10.3390/books978-3-7258-3968-1

© 2025 by the authors. Articles in this book are Open Access and distributed under the Creative Commons Attribution (CC BY) license. The book as a whole is distributed by MDPI under the terms and conditions of the Creative Commons Attribution-NonCommercial-NoDerivs (CC BY-NC-ND) license (https://creativecommons.org/licenses/by-nc-nd/4.0/).

Contents

hydrology MDPI

Article

Biogeochemical Permeable Barrier Based on Zeolite and Expanded Clay for Immobilization of Metals in Groundwater

Nadezhda Popova [1], Grigoriy Artemiev [1], Inga Zinicovscaia [2,3,4], Nikita Yushin [2], Ludmila Demina [1], Kirill Boldyrev [5], Denis Sobolev [5] and Alexey Safonov [1,*]

1 A.N. Frumkin Institute of Physical Chemistry and Electrochemistry, Russian Academy of Sciences, 31-4, Leninsky Prospect, 199071 Moscow, Russia
2 Joint Institute for Nuclear Research, 6 Joliot-Curie Str., 141980 Dubna, Russia
3 Horia Hulubei National Institute for R&D in Physics and Nuclear Engineering, 30 Reactorului Str. MG-6, 077125 Magurele, Romania
4 The Institute of Chemistry, 3 Academiei Str., 2028 Chisinau, Moldova
5 The Nuclear Safety Institute of the Russian Academy of Sciences 52, Bolshaya Tulskaya Street, 115191 Moscow, Russia
* Correspondence: alexeysafonof@gmail.com

Abstract: Groundwater samples contaminated with potentially toxic elements (PTE), including metals and nitrate ions, were collected at a depth of 8–10 m from the Siberian Chemical Plant multicomponent waste storage. The possibility of developing a permeable biogeochemical barrier with zeolite and lightweight expanded clay aggregate (LECA) was investigated. The mass fraction and properties of several metals (Mn, Fe, Co, Ni, Cu, Zn, Cd, Hg and Pb) were determined to investigate their fixation on the chosen materials at the given experimental conditions. It was established that metals in sulfide or phosphate forms can be effectively immobilized via biomineralization on LECA, whereas metals from the non-chalcogen group are primarily retained in the form of phosphates. The formation of biogenic deposits of iron sulfide, which serve as a sorption–precipitation phase during the immobilization of the majority of metals, is an important aspect of the LECA loading process. The use of LECA and zeolite in the form of a two-component barrier is feasible based on the data obtained. It is assumed that metal immobilization processes occur due to sorption mechanisms in the zone of zeolite loading. Microbial nitrate removal and the formation of iron sulfide phases under reducing conditions, which form a geochemical barrier for metals, are expected in the LECA zone.

Keywords: mesoporous zeolite; lightweight expanded clay aggregate (LECA); aquifer contamination; permeable barrier; potentially toxic elements (PTE); sorption; biomineralization

Citation: Popova, N.; Artemiev, G.; Zinicovscaia, I.; Yushin, N.; Demina, L.; Boldyrev, K.; Sobolev, D.; Safonov, A. Biogeochemical Permeable Barrier Based on Zeolite and Expanded Clay for Immobilization of Metals in Groundwater. *Hydrology* **2023**, *10*, 4. https://doi.org/10.3390/hydrology10010004

Academic Editor: Meysam Vadiati

Received: 7 December 2022
Revised: 15 December 2022
Accepted: 17 December 2022
Published: 24 December 2022

Copyright: © 2022 by the authors. Licensee MDPI, Basel, Switzerland. This article is an open access article distributed under the terms and conditions of the Creative Commons Attribution (CC BY) license (https://creativecommons.org/licenses/by/4.0/).

1. Introduction

Electroplating enterprises, solid waste landfills, as well as mining activities, ore processing, chemical and metalworking industries contribute to the contamination of both underground and surface waters with potentially toxic elements (PTE: heavy metals and acid anions). The most pronounced detrimental effect on the environment can be caused by industrial effluents stored in the form of dumps or sludge, and in surface storage pools [1–3]. Notwithstanding the fact that many techniques for remediation of contaminated soils and surface water bodies have been created and are continuously improved, the treatment of contaminated groundwater continues to be an expensive and difficult task.

Various impermeable (e.g., cut-off walls in the ground) or permeable engineering barriers have traditionally been used to prevent pollutant migration in groundwater [4]. Permeable barriers are considered to be a better long-term solution, as they do not disrupt groundwater movement and do not induce blockages in the geologic horizon [5–7]. A variety of natural materials, including zeolites [8–10], limestone [6,11], apatite [12], artificially created materials such as cement-based filter media (CBFM) [13], waste products (e.g., fly ash [14]), as well as numerous organic materials and their compositions with minerals [15], can be used to create these barriers. The formation of in situ geochemical barriers using reducing agents, such as zero-valent iron [16,17] and other additives, is another method of treating groundwater. Their injection into the groundwater formation creates local zones for metal immobilization.

One of the promising remediation strategies for combating multicomponent contamination is the application of in situ biogeochemical barriers based on sorbents and highly porous material, as well as microorganism growth stimulation by organic compounds. Numerous studies have employed this strategy [18,19]. The development of microbial biofilms that protect microorganisms from the toxic effects of the contaminated environment and allow the microbial community to effectively cope with it is critical for the successful operation of such barriers [20]. Shortcomings of this approach include the necessity of extra nutrition for the microorganisms as well as biofouling of highly effective sorptive materials causing deterioration of their properties. Biofilms, on the other hand, can improve material sorption capacity since their polysaccharide matrix contains a significant number of functional groups, which can participate in metal sorption [21–24]. Furthermore, microbial processes can result in the formation of secondary mineral phases, such as sulfide or ferruginous phases, which additionally contribute to metal immobilization [25–27].

Previously, we have studied the possibility of using zeolite, apatite, expanded clay, vermiculite, and other materials along with organic stimulation as a groundwater permeable barrier near a storage facility for the immobilization of radionuclides, such as cesium, strontium, uranium, and technetium, from radioactive waste [28]. The effects of natural groundwater microflora on the sorption properties of materials were studied. It was demonstrated that microbiological effects did not significantly alter the sorption properties of the examined materials under the operating conditions of the barrier. In another study, the possibility of using vermiculite, lightweight expanded clay aggregate (LECA), perlite, zeolite, and shungite as filtration barrier in the aquifer near a solid domestic waste landfill for Cd and Cr (VI) immobilization was investigated. Based on the results obtained by authors of the present study and other researchers [4], the most optimal approach in case of purification of groundwater with multi-component contamination is development of permeable barriers consisting of several materials that create different conditions for the immobilization of various contaminants.

The aim of the present study was to assess the possibility of using zeolite and LECA as a permeable biogeochemical engineering barrier in groundwater with high nitrate and sulfate contamination for metal ion immobilization, taking as an example, groundwater collected near the multicomponent waste storage of the Siberian Chemical Plant (Tomsk region, Russia).

A field test to create a groundwater biogeochemical barrier was conducted there previously [29]. In less than one month, as a result of a single injection of organic matter, the studied area was cleared of nitrate ions, but the effect was transient; in one year, concentrations of contaminants comparable to the initial ones were observed in the stonecrop zone. In this case, enhancing the stable development of microorganisms in biofilms in contaminated groundwater could be a critical solution for in situ nitrate, radionuclide, and various metal removals.

2. Materials and Methods

In the present study, a groundwater sample was collected from a depth of 8–10 m in the area of the basin of the Siberian Chemical Plant storage of polycomponent wastes. Water samples were taken after pumping one and a half well volumes in sterile, 2 L plastic bottles, hermetically sealed and stored at a temperature of +4 °C in a refrigerator. The sample chemical composition and microbiological properties can be found in [29]. It contained high concentrations of major components, nitrate ions and calcium, as well as PTE—Mn, Fe, Ni, Cu, Sr, and Zn (Table 1).

Table 1. Characteristics of the analyzed groundwater sample.

Major Elements *, mg/L		Trace Elements **, µg/L					
pH	6.34 ± 0.2	Al	145 ± 4.4	As	0.6 ± 0.1	Cs	0.01 ± 0.001
salt concentration	6670 ± 200	Si	9893 ± 297	Br	128 ± 0.2	Ba	612 ± 18.4
Na^+	647 ± 19	Sc	2 ± 0.1	Se	2.7 ± 0.1	La	0.74 ± 0.02
K^+	8.9 ± 0.3	Ti	3.1 ± 0.1	Rb	0.25 ± 0.2	Ce	1.16 ± 0.04
Ca^{2+}	762.3 ± 22.8	V	0.9 ± 0.02	Sr	993 ± 30	W	0.3 ± 0.01
Mg^{2+}	139.7 ± 4.2	Cr	5.06 ± 0.2	Zr	0.1 ± 0.01	Pb	0.92 ± 0.03
$NH_4{}^+$	12.3 ± 0.03	Mn	4482 ± 135	Nb	0.3 ± 0.01	Th	0.04 ± 0.001
$NO_3{}^-$	3849 ± 115	Fe	17563 ± 527	Mo	2.24 ± 0.06	U	0.72 ± 0.02
$SO_4{}^{2-}$	467 ± 14	Co	16.5 ± 0.5	Ru	0.03 ± 0.001		
Cl^-	6.3 ± 0.2	Ni	145 ± 4.4	Rh	2.03 ± 0.06		
$HCO_3{}^-$	305.1 ± 9.2	Cu	60 ± 1.8	Pd	0.02 ± 0.001		
P_{tot}	32.5 ± 1	Zn	150 ± 4.5	Cd	11.3 ± 0.34		

* Detection limit = 100 µg/L, with the exceptions of K, Cl = 10 µg/L, and P = 25 µg/L. ** Detection limit: As, Nb, Ba = 0.1 µg/L; Br = 16 µg/L; Se = 1.6 µg/L; Zr, Nb, Ru, Pd, Cd, W = 0.01 µg/L; Mo, Pb = 0.02 µg/L, Cs = 0.001 µg/L; La = 0.003 µg/L; Ce = 0.004 µg/L; Th, U = 0.002 µg/L.

2.1. Carriers

Natural zeolite (Clinoptilolite type) from the Chola deposit (Transbaikalia, Russia) was purchased from the "Zeolite-Trade" company (http://www.zeolite.spb.ru/ accessed on 17 January 2021). The zeolite "Trade" consists of isometric aggregates of 3–5 mm and microaggregates of micron size with thin isometric pores and elongate channels. This structure ensures simultaneous high filtration and sorption properties. The density of the "Trade" zeolite is 2.2–2.6 g/cm^3, the specific surface area is 10.1 m^2/g, and the bulk weight is 1.02–1.2 g/cm. Natural zeolite was ground up and sieved, and the fraction with a size 300–100 µm was used for further experiments.

Lightweight expanded clay aggregate (**LECA**) produced by the PJSC "Keramzit" (Serpuhov, Russia, https://zao-keramzit.com) (accessed on 26 January 2021). Is a mixture of clay minerals (smectite, beydelite) heated at 1200 °C. It is composed of highly porous aggregates up to 5 mm in size and pores ranging from several nanometers to 0.5 mm. Despite its high porosity and specific surface, the LECA has a low chemical activity due to the highest and most stable oxidation state of its components during burning, when water and organic matter are completely removed.

2.2. Experimental Design

2.2.1. Sorption before and after Biofouling

In the first stage, the experiments were performed on materials with biofilm formed by the groundwater microbial community. The biofouling was performed in aerobic conditions for 14 days in Adkins media inoculated with 10% of groundwater sample. The medium contained NH_4Cl—1.0; KH_2PO_4—0.75; K_2HPO_4—1.5; $NaNO_3$—1.0; NaCl—0.8; Na_2SO_4—0.1; $MgSO_4 \cdot 7H_2O$—0.1; KCl—0.1, yeast extract—0.5; glucose—1.0; CH_3COONa—1.0, pH = 7. Glucose and sodium acetate Sigma Aldrich (Darmstadt, Germany) (https://www.sigmaaldrich.com/ accessed on 1 March 2021.) in concentrations of 1 g/L were used as carbon sources and electron donors. Filtration and freeze drying were used to separate the biofilm-containing materials from the cultivation medium.

Sorption experiments were conducted in 100 mL flasks for 24 h at vigorous agitation with the same groundwater. The solution volume was 50 mL, and the sorbent dosage was 0.5 g. Metals were added from nitrate solutions (Sigma Aldrich (Darmstadt, Germany)) at a concentration of 10 mg /L (per metal). All experiments were conducted in triplicate, and the average values were used for further calculations.

The metal uptake q (mg/g sorbent) was calculated using the following equation:

$$q = \frac{V(C_i - C_f)}{m} \quad (1)$$

and sorption removal efficiency, R (%), from the equation:

$$R = \frac{C_i - C_f}{C_i} * 100 \quad (2)$$

where q is the amount of metal ions adsorbed on the sorbent in mg/g; V is the volume of solution in ml; C_i is the initial concentration of a metal in mg/L, C_f is the final metal concentration in the solution in mg/L, and m is the mass of sorbent in g.

Desorption was carried out with low mineralized model water ($NaHCO_3$—25.2, $MgSO_4 \times 7H_2O$—36.6, $CaCl_2 \times 6H_2O$—223.9, $MgCO_3$—3.2, $NaNO_3$—1000, pH 7.1) In vials, the material was stirred at 120 rpm for 2 h at room temperature. All chemicals were high purity grade from Sigma Aldrich (Darmstadt, Germany).

2.2.2. Metal Immobilization during Microbial Growth

In the second stage, an experiment on accumulation and biomineralization of mineral carriers with simultaneous growth of microorganisms from the solution of underground water was carried out. Sodium acetate and glucose high purity grade (Sigma Aldrich, Darmstadt, Germany) at a concentration of 1 g/L were used as a carbon source and electron donors. The experiment was carried out in hermetically sealed vials for one month to achieve anaerobic mineralization of iron and sulfur. Metals at the same concentrations as in the sorption experiment were added to the medium in the beginning of the experiment.

The medium in the first stage of the experiment contained phosphates from groundwater samples; in the second variant, phosphates at a concentration of 500 mg/L in the form of potassium phosphate high purity grade (Sigma Aldrich, Darmstadt, Germany) were added. Desorption was carried out according to the procedure described for sorption experiment.

2.3. Methods

The chemical composition of the water samples was analyzed immediately after sample collection and filtration through a 0.45 μm glass filter by inductively coupled plasma–mass spectrometry (ICP-MS) on mass spectrometer Xseries II ICP-MS (Thermo Fisher Scientific, Waltham, MA, USA) and ICP-OES on ICP-OES CID Spectrometer); iCAP 6500 (Thermo Fisher Scientific, Waltham, MA, USA, https://www.fishersci.com/shop/products/icap-6500duoview-icp-oes-spect/NC1982295 accessed on 19 January 2022).

The determination of Eh and pH values was carried out using an ANION-4100 pH meter/ionomer (Russia) with an electrode combination. Anion and cation concentrations were measured by a CGE capillary electrophoresis system (Capel-105M, LUMEX Instruments, Saint Peterburg, Russia, https://www.lumexinstruments.com/catalog/capillary-electrophoresis/capel-105m.php accessed on 10 January 2020).

Copper, Cd, and Pb concentrations in the solutions were determined by AAS (Thermo Scientific iCE 3400 series, Waltham, MA, USA, https://www.thermofisher.com/order/catalog/product/942350023411) (accessed on 6 December 2022) with electrothermal atomization. Calibration solutions were prepared from a 1 g/L stock solution (AAS standard solution; Merck, Darmstadt, Germany).

The mass fraction of other elements was determined using neutron activation analysis at the pulsed fast reactor IBR-2 (Frank Laboratory of Neutron Physics, Joint Institute for Nuclear Research, Dubna, Russia). The concentration of Mn was determined by irradiation for 3 min at a thermal neutron flux of 1.2×10^{12} n cm^{-2} s^{-1}, and measurement time was 15 min. To determine the mass fraction of elements with long-lived isotopes: Cr, Co, Zn, Sr, Ba, and Hg samples were irradiated for 4 days at a neutron flux of 1.1×10^{11} cm^{-2} s^{-1}. Gamma spectra of induced activity were obtained after 4 and 20 days using three Canberra HPGe detectors with an efficiency of 40–55% and resolution of 1.8–2.0 keV at 1332 keV ^{60}Co total-absorption peak. The analysis of the spectra was performed using the Genie2000 software by Canberra (https://www.mirion.com/products/genie-2000-basic-spectroscopy-software) (accessed on 6 December 2022), with peak fitting verification in interactive mode.

The calculation of the concentrations was carried out using the software "Concentration" developed in FLNP [30].

Biofilm development was detected using confocal scanning microscopy. The samples were washed with distillate water to remove planktonic cells prior to storing in a 96% alcohol solution for biofilm fixation. ConA (lectin conjugated with the fluorescent dye Alexa Fluor 488 (C11252, ThermoFisher) in phosphate buffer at a dilution of 1:500) and SYBR Green II (S7564, ThermoFisher), which binds to nucleic acids (primarily RNA), were used to stain the samples. ConA binds to bacterial wall monosaccharides and EPS, SYBR Green II to nucleic acid. Staining was performed in the dark for 30 min on a shaker at room temperature. The samples were analyzed using a Zeiss LSM880 confocal microscope (Zeiss, Germany). The images were acquired with x20 and x40 objectives and argon lasers with wavelengths of 488 nm for detecting ConA fluorescence and 543 nm for detecting SYBR Green II. The Nomarski contrast method was applied to detect uncolored particles (LECA and zeolite). The obtained images were analyzed using the ImageJ software package with the plugin BioFormats 5.8.2 (https://docs.openmicroscopy.org/bio-formats/5.8.2/about/index.html) (accessed on 6 December 2022) and BioFilmAnalyzer v.1.0 [31].

Organic carbon was determined using a total organic carbon analyzer: Shimadzu TOC-V CSN (Kyoto, Japan).

Respiration activity was determined using the MTT test under oxic and anoxic conditions [32]. Before spectrophotometry of the oxidized formazan complex, the samples were centrifuged at 7000 g to remove the clay suspension.

Materials surface analysis before and after sorption was performed using a S3400N scanning electron microscope (Hitachi, Santa Clara, CA, USA). Analysis samples were removed from the liquid medium by filtration and dried at room temperature in a nitrogen glove box to a constant weight. For SEM analysis, the samples were placed on an aluminum holder using electrically conductive tape, and vacuum carbon deposition (Q150T E Plus) was carried out (vacuum 4–3, current 50 A). The samples were analyzed in two modes, SE and BSE, at a voltage of 20 kV.

Fourier-transform infrared (FT-IR) spectroscopy was used to confirm the presence of the functional groups in the microbial samples. Infrared spectra were recorded in the 4000–550 cm^{-1} region using a Thermo Nicolet Nexus 4700 FT IR Spectrometer (Thermo Fisher Scientific, Waltham, MA, USA).

The speciation of metals in solution was assessed by thermodynamic modeling in the PhreeqC 2.1 software with the llnl.dat thermodynamic database [33]. The saturation indices (*SI*) were determined as follows:

$$SI = logIAP - logK_s \tag{3}$$

where *IAP* is the product of activities of the relevant ions, and K_s is the equilibrium constant. At *SI* > 0, formation of the studied phase is predicted.

3. Results

3.1. Zeolite and LECA Biofilm Characterization

The microbial biofilm formation and the accumulation of organic carbon on the materials occurred after a single stimulation with glucose. Organic carbon is predominantly present as the biofilm exopolysaccharide matrix. The maximum carbon accumulation on zeolite was observed on days 20–30, and on LECA on days 15–20 (Table 2). LECA had a higher total carbon mass fraction, reaching 12.9 mg/g. A gradual biofilm degradation on both materials was observed after 40 days. After 60 days, the carbon mass fraction on zeolite decreased to the initial values of biofilm development; for LECA, the decrease was 15% from the maximum.

Table 2. Kinetics of total organic carbon accumulation on the studied materials (mg/g) *.

Sample	Time, Days								
	0	5	7	15	20	30	40	50	60
Zeolite	0.12 ± 0.004	3.8 ± 0.13	4.6 ± 0.15	7.1 ± 0.22	8.5 ± 0.3	8.4 ± 0.3	4.5 ± 0.15	3.9 ± 0.13	3.5 ± 0.11
LECA	0.26 ± 0.01	4.5 ± 0.15	7.9 ± 0.3	12.8 ± 0.45	12.9 ± 0.4	12.4 ± 0.43	11.5 ± 0.38	11.0 ± 0.35	10.9 ± 0.33

* Uncertainty of the TOC result is less than 5%, according to the Shimadzu TOC Measurement Manual.

The morphology of the samples was visualized using confocal laser scanning microscopy (Figure 1). It was discovered that the LECA coverage by biofilm was more even than biofouling on zeolite. On the 20th day, the total area of polysaccharides on LECA was 75 ± 3.8%, taking into account that the total area of fouling was 89 ± 4.3%. The area covered by polysaccharides on zeolite was 54 ± 2.6%, the total fouling area being 59 ± 2.9%.

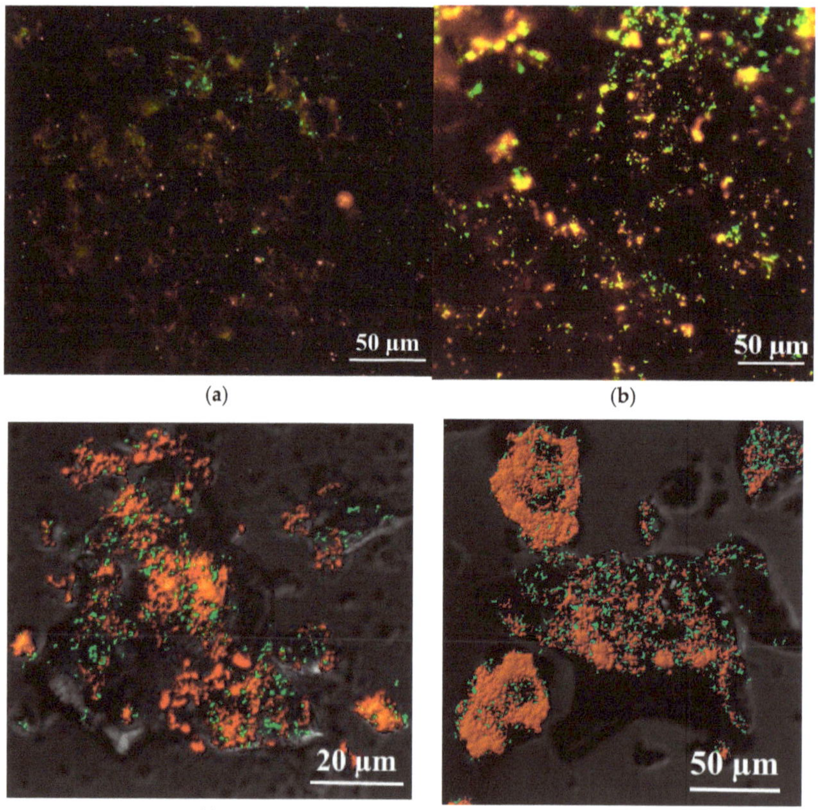

(**a**) (**b**) (**c**) (**d**)

Figure 1. Micrographs of materials on the 20th day: zeolite–bio (**a**), zeolite–biomineralization (**b**), LECA–bio (**c**) and LECA–biomineralization (**d**). Red channel is polysaccharides, and green is nucleic acids.

The area covered by cells (according to nucleic acid staining) on LECA was on average two times greater than that on zeolite (Table 3, which also contains data on the formation of biofilms in the biomineralization experiment). A similar trend was observed with the total covered area, in the case of zeolite, it was 53 ± 2.6% and for LECA, 84 ± 4.1%. The high biofouling of LECA is primarily due to its porous structure and larger surface area compared to zeolite.

Table 3. Topological parameters of biofilm on analyzed materials (after 20 days of growth) studied by confocal scanning laser microscopy.

Sample	Nucleic Acid, %	Polysaccharides, %	Total Area of Fouling, %
Zeolite–bio	5 ± 0.2	54 ± 2.6	59 ± 2.9
LECA–bio	14 ± 0.7	75 ± 3.8	89 ± 4.3
Zeolite–biomineralization	8 ± 0.4	45 ± 2.2	53 ± 2.6
LECA–biomineralization	13 ± 0.6	71 ± 3.6	84 ± 4.1

3.2. Metal Immobilization on Materials before and after Biofouling (Sorption Experiment)

The results of metal accumulation on raw and biofilm-coated materials are presented in Figure 2. The efficiency of metal sorption on LECA did not exceed 20%, with the exception of mercury, when the sorption efficiency was slightly above 50%. Sorption efficiency on zeolite was significantly higher. Thus, for cadmium, strontium, mercury, and manganese, it was higher than 90%, and for nickel, zinc, copper, and lead, it was in the range of 80–90%. Formation of biofilm on the analyzed materials had a multidimensional effect on the efficiency of sorption. For LECA, the efficiency of all metals' immobilization increased. The highest efficiency (60–80%) was observed for chromium, zinc, cadmium, and copper. For nickel, cobalt, strontium, and barium, the increase in the efficiency of immobilization was less pronounced. In the case of zeolite, the formation of biofilm resulted in the decrease in manganese, cobalt, nickel, copper, zinc, strontium, cadmium and barium immobilization. Moreover, for cobalt, copper, strontium and barium, the decrease in the sorption efficiency was significant (20–50%). For lead and mercury, the efficiency of immobilization was almost unaffected by the biofilm. Thus, the formation of biofilms on materials with high immobilization characteristics inhibits metal accumulation.

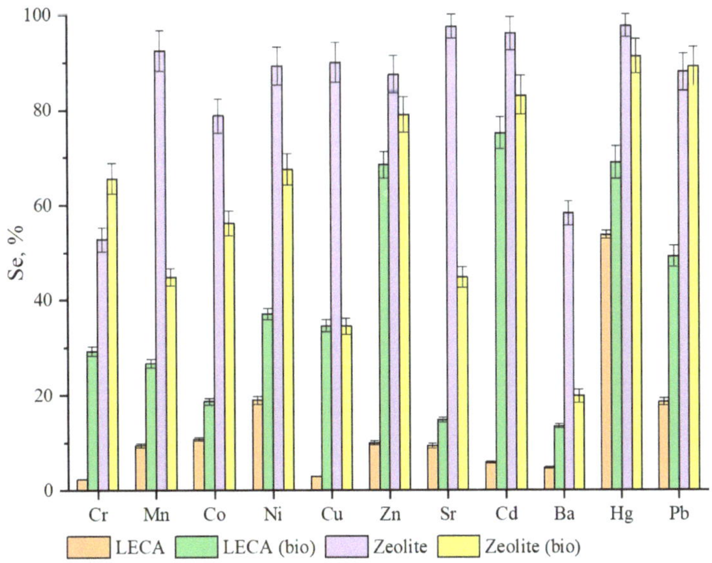

Figure 2. Metal immobilization on untreated and biofilm-coated materials.

3.3. Metal Immobilization on Materials during Biofouling (Bioaccumulation and Biomineralization Experiment)

The values of the efficiency of metal immobilization on the studied materials during biomass growth with and without the addition of the excess of phosphates in the medium are presented in Table 4. Metal immobilization on materials, particularly for LECA, increased in the biomineralization experiments. Immobilization on LECA was higher than on zeolite, which can be attributed to the greater surface biofouling. The addition of phosphates contributed to the significant increase in strontium and barium immobilization and resulted in 100% immobilization of mercury.

Table 4. Efficiency (%) of metal immobilization on materials in biomineralization experiments with phosphates (bmp) and without phosphates (bm).

Metal	LECA (bio)	LECA (bm)	LECA (bmp)	Zeolite (bio)	Zeolite (bm)	Zeolite (bmp)
Cr	29.3 ± 1.0	97.5 ± 4.4	98.7 ± 4.3	65.5 ± 1.6	90.7 ± 4.0	91.5 ± 3.9
Mn	26.7 ± 1.0	98.7 ± 4.2	99.4 ± 4.5	44.8 ± 0.9	89.4 ± 3.7	90.8 ± 3.7
Co	18.6 ± 0.8	99.1 ± 4.5	98.1 ± 4.3	56.1 ± 1.9	82.3 ± 3.9	84.7 ± 2.9
Ni	37.1 ± 1.7	99.6 ± 4.5	99.5 ± 4.4	67.4 ± 2.0	82.32.6	99.5 ± 3.9
Cu	34.6 ± 1.4	92.5 ± 4.0	89.1 ± 3.0	34.5 ± 0.8	85.4 ± 2.4	88.1 ± 2.6
Zn	68.4 ± 3.4	98.9 ± 4.4	99.8 ± 4.7	78.9 ± 3.6	92.1 ± 4.0	90.4 ± 3.7
Sr	14.7 ± 0.7	31.2 ± 1.2	100 ± 5.0	44.7 ± 1.2	57.9 ± 1.9	100 ± 4.5
Cd	75 ± 3.0	98.1 ± 4.7	99.2 ± 4.3	83 ± 3.0	98.4 ± 3.8	99.1 ± 4.4
Ba	13.4 ± 0.3	16.2 ± 0.7	89.7 ± 3.8	19.7 ± 0.7	31.2 ± 1.5	98.5 ± 4.2
Hg	68.9 ± 2.5	100 ± 4.4	100 ± 4.5	91.2 ± 2.6	100 ± 4.8	100 ± 4.5
Pb	49.1 ± 1.8	97.9 ± 4.2	98.6 ± 3.9	89.1 ± 2.4	99.8 ± 4.8	99.3 ± 4.3

3.4. *Evaluation of the Binding Strength of Immobilized Forms of Metals on Analyzed Materials*

Table 5 report the data related to the efficiency of metal desorption using groundwater as a desorbing agent (2 h of mixing). According to the results, the strength of metal binding on zeolite was higher despite its lower fouling. In the experiment with LECA coated with biofilm, the highest efficiency of desorption was obtained for cobalt, chromium, copper, barium, and manganese, while for cadmium and zinc, it was very low. On zeolite coated with biofilms, the maximum desorption was observed for chromium and the minimum for cadmium. Thus, in sorption experiments, the majority of metals was not strongly immobilized on organic microbial biofilms.

Table 5. Efficiency of metal desorption from the LECA and zeolite using groundwater as a desorbing agent.

Metal	LECA (bio)	LECA (bm)	LECA (bmp)	Zeolite (bio)	Zeolite (bm)	Zeolite (bmp)
Cr	64.9 ± 2.9	2.2 ± 0.1	2.11 ± 0.01	41.9 ± 2.1	2.1 ± 0.01	1.7 ± 0.03
Mn	72.8 ± 3.5	1.3 ± 0.07	2.11 ± 0.01	32.3 ± 1.5	1.5 ± 0.01	1.9 ± 0.03
Co	81.6 ± 3.9	3.9 ± 0.09	4.1 ± 0.1	30.4 ± 1.4	3.1 ± 0.02	2.9 ± 0.4
Ni	49.4 ± 2.5	1.1 ± 0.03	1.8 ± 0.01	39.4 ± 1.9	0.9 ± 0.01	1.3 ± 0.01
Cu	54.3 ± 2.8	1.5 ± 0.05	1.71 ± 0.01	39.8 ± 1.7	1.9 ± 0.2	1.8 ± 0.03
Zn	43.4 ± 2.1	1.2 ± 0.06	0.9 ± 0.003	27.4 ± 1.2	3.9 ± 0.1	1.5 ± 0.03
Sr	52.9 ± 2.7	34.1 ± 1.4	1.91 ± 0.03	38.9 ± 1.7	28.3 ± 0.08	0.9 ± 0.02
Cd	43.6 ± 2.0	1.6 ± 0.05	1.5 ± 0.02	17.4 ± 0.8	1.3 ± 0.01	1.7 ± 0.01
Ba	77.7 ± 3.1	42.4 ± 2.1	2.11 ± 0.5	29.2 ± 1.0	20.1 ± 1.0	1.1 ± 0.02
Hg	12.8 ± 0.6	4.5 ± 0.0	3.1 ± 0.7	5.8 ± 0.01	1.1 ± 0.05	0.3 ± 0.001
Pb	49.2 ± 2.2	1.5 ± 0.5	0.34 ± 0.01	23.7 ± 1.2	1.1 ± 0.05	0.84 ± 0.02

The metal binding strength on both materials increased significantly during the biomineralization experiment. The highest efficiency of desorption for both materials was observed for Ba and Sr, and for other elements, it was less than 5%. At the addition of phosphates, the efficiency of elements desorption on LECA was less than 5% and on zeolite less than 3%.

4. Discussion

The formation of biofilm on the studied materials occurred differently. The adhesion of biofilms on LECA was strengthened due to the more developed surface macrostructure and high roughness. However, since mesopores are inaccessible for microorganisms, the zeolite surface was not affected by biofouling. As a result, it can be assumed that biofouling and biofilm formation will influence metal immobilization only on LECA. Metal binding on zeolite can be influenced by biofilms as well as by the material's surface. The experiments revealed that the surfaces of materials, biofilms, and mineral phases formed during microorganism growth and contributed to metal immobilization. Furthermore, biomineralization was mainly responsible for metal immobilization. The mechanisms of metal fixation by microbial biofilms have been thoroughly investigated. They include physical and physicochemical adsorption, such as ion exchange or formation of complexes on biofilm sorption centers [34].

4.1. The Role of Biofouling in Metals Immobilization

It is known that bacterial biofilms consist of a matrix with up to 90–95% polysaccharides based on β-glucuronic acid. The sorption sites of biofilms include hydroxyl (alcohols, carbohydrates), carboxyl (fatty acids, proteins, organic acid residues), amino groups (proteins and nucleic acids), esters (lipids), sulfhydryl groups (cysteine residues, proteins), aldehyde groups (aldehydes and polysaccharides), internal carbonyl groups (ketones and polysaccharides), and phosphate groups [35].

IR spectra recorded before and after the biofouling processes confirmed the materials' biogenic fouling. The spectrum of LECA after biofilm formation differed significantly from the spectrum of the raw material. Aluminosilicate-like bands were observed prior to biofilm formation: 1035, 799, and 775 cm^{-1}. Two maxima, 993 and 918 cm^{-1}, which are indicative of the stretching vibrations of the C-O and C-C groups, were identified on the band in the range of 1200–900 cm^{-1} after biofouling (Figure 3a). The spectrum of the zeolite sample showed the appearance of ν(OH) stretching vibrations in the range of 3600–3300 cm^{-1}, as well as the appearance of a band at 1414 cm^{-1}, which can be attributed to the δ(COH) vibration. In addition, the spectrum of the bio-treated zeolite contained water vibration bands: $\nu(H_2O)$ = 3528, 3381 cm^{-1} and $\delta(H_2O)$ = 1640 cm^{-1}, which along with the hydroxy group, can be part of the polysaccharide matrix.

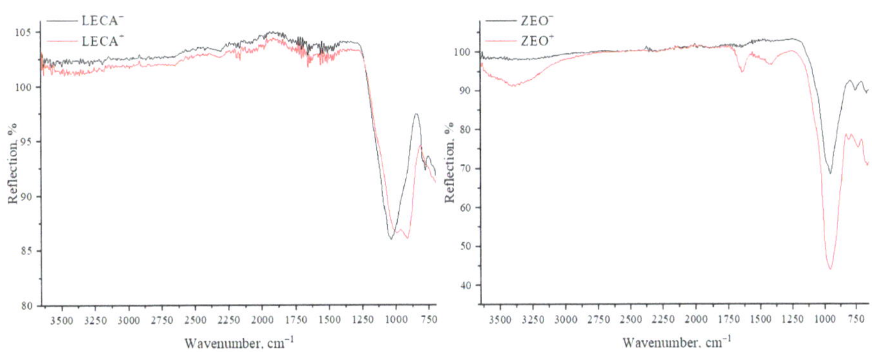

(**a**) (**b**)

Figure 3. IR spectra of materials studied before and after biofouling: LECA (**a**) and zeolite (**b**).

According to the literature (Table A1), the most common complexing agents capable of chelating almost all metals used in the current study are carboxyl, hydroxyl, and thiol groups. Amino (and amide) groups are able to form compounds with Cr, Zn, Ni, Cu, Mn, Cd, Sr, and Hg. Phosphoryl groups have the lowest affinity for Mn, Co, and Sr. Although carbonyl groups can bind Cr, Ni, Cu, Hg, and Pb, their chelate formation is much less active.

Microbial biofilms can be considered as polyfunctional adsorbents for the majority of metals. At the same time the overlapping of the material's specific surface by the biological matrix can lead to the decrease in the material sorption capacity. This may explain the decrease in zeolite with biofilms sorption capacity toward Mn, Co, Ni, Cu, Sr, and Ba. It can be concluded that functional groups of zeolite play a dominant role in the binding of the mentioned metals.

4.2. The Role of Biomineralization in Metal Immobilization

Previously, a microbial community capable of reducing iron, sulfur, and nitrogen compounds metabolically was discovered in water samples. In the present work, the effect of a single addition of organic matter on physico-chemical conditions was investigated experimentally (Figure 4). Significant shifts in the medium's redox potential toward the reduction region, as well as the reduction of nitrate ions, were observed up to day 30. Following the establishment of strongly reducing conditions, the concentration of sulfate ions decreased as a result of the sulfate reduction process, which led to sulfide reduction. The microbial processes were more active in the presence of LECA.

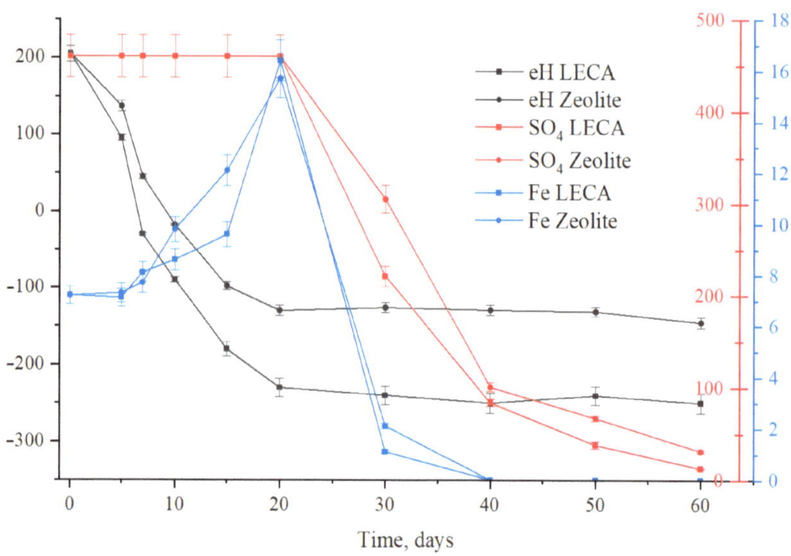

Figure 4. Kinetics of eH and Fe and SO_4^{-2} concentration change in solutions in biomineralization experiment.

Optional conditions for the solid mineral phases formation were determined according to calculation in PhreeqC code (the thermodynamic database llnl.dat was used) [33]. Sulfate-reductive conditions, a shift in the redox potential to the reduction side, and an increase in carbonate ion concentration are required. Carbonate mineral precipitation is primarily caused by anoxic microbial respiration. Sulfide and phosphates were formed in the system as a result of the sulfate reduction. Table 6 and Figure 5 show the results of the SI calculations of solid mineral phases under sulfate reduction conditions. For zinc phosphate (Hopeite, $Zn_3(PO_4)_2{:}4H_2O$), the SI was also greater than 0 (SI = 4.12).

Table 6. Thermodynamic modeling (PhreeqC 2.1 software with the llnl.dat) of mineral phases formation at saturation indices (SI) > 0.

Phase	Without Additions, pH 6.5	Sulfates Addition 500 mg, pH 8	Sulfates Addition 500 mg, Phosphates Addition, pH 8	Phosphates Addition, 500 mg	Formula
			Carbonates		
Aragonite	−0.8	1.7	1.6	1.6	$CaCO_3$
Calcite	−0.7	1.9	1.7	1.7	$CaCO_3$
Cerussite	2.1	2.2	2.1	2.1	$PbCO_3$
Dolomite	−0.7	4.6	4.2	4.2	$CaMg(CO_3)_2$
Dolomite-dis	−2.3	3.0	2.5	2.5	$CaMg(CO_3)_2$
Dolomite-ord	−0.7	4.6	4.2	4.2	$CaMg(CO_3)_2$
Huntite	−7.5	3.3	2.3	2.3	$CaMg_3(CO_3)_4$
Hydrocerussite	3.5	3.7	3.4	3.4	$Pb_3(CO_3)_2(OH)_2$
Monohydrocalcite	3.5	3.7	3.4	3.4	$CaCO_3{:}H_2O$
Magnesite	−1.6	1.0	0.7	0.7	$MgCO_3$
Otavite	1.0	3.4	3.3	3.3	$CdCO_3$
Rhodochrosite	−0.4	1.6	1.4	1.4	$MnCO_3$
Siderite	−0.7	0.8	0.7	0.7	$FeCO_3$
Smithsonite	−1.2	1.1	0.5	0.5	$ZnCO_3$
Strontianite	−0.3	2.4	2.4	2.4	$SrCO_3$
$ZnCO_3{:}H_2O$	−0.6	1.7	1.1	1.1	$ZnCO_3{:}H_2O$
			Sulfides		
Alabandite	−1.6	0.9	0.6	0.7	MnS
Bornite	84.1	96.5	96.4	90.8	Cu_5FeS_4
Cattierite	9.2	13.3	13.3	12.4	CoS_2
CdS	14.4	17.3	17.2	15.8	CdS
Chalcocite	30.8	34.2	34.2	32.8	Cu_2S
Chalcopyrite	22.1	27.7	27.6	24.8	$CuFeS_2$
CoS	3.8	4.3	4.3	4.8	CoS
Covellite	14.2	17.7	17.7	16.3	CuS
Galena	13.0	13.6	13.4	12.1	PbS
Metacinnabar	18.8	22.2	22.2	20.8	HgS
Millerite	6.3	9.6	9.6	8.2	NiS
Pyrite	6.7	12.4	12.3	9.5	FeS_2
Pyrrhotite	1.6	3.7	3.5	2.1	FeS
Troilite	1.7	3.8	3.6	2.3	FeS
Vaesite	9.1	16.1	16.0	13.2	NiS_2
Wurtzite	7.4	10.3	9.76	8.3	ZnS
			Phosphates		
Hopeite			4.67	4.6	$Zn_3(PO_4)_2{:}4H_2O$
$Pb_4O(PO_4)_2$			6.7	6.7	$Pb_4O(PO_4)_2$
$PbHPO_4$			5.0	5.0	$Pb_4O(PO_4)_2$

Table 6. *Cont.*

Phase	Without Additions, pH 6.5	Sulfates Addition 500 mg, pH 8	Sulfates Addition 500 mg, Phosphates Addition, pH 8	Phosphates Addition, 500 mg	Formula
			Oxyhydroxide		
Delafossite	4.9	9.4	9.3	9.3	$CuFeO_2$

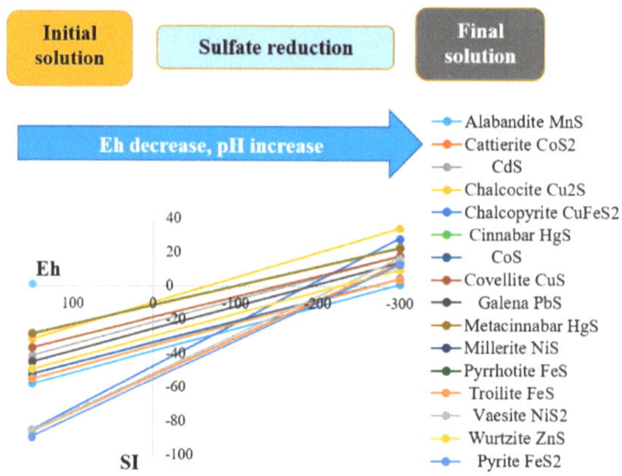

Figure 5. The results of solid mineral phases saturation index (SI) calculation under sulfate reduction conditions.

As a result of biogenic mineralization, according to BTC llnl.dat, the formation of the following metal phases was predicted: Mn in the form of $MnCO_3$ (Rhodochrosite) and in the form of MnS phases (Akabandite), Co as CoS_2 (Cattierite), Cd as $CdCO_3$ (Otavite) and CdS, Cu as Cu_2S (Chalcocite) and CuS (Covellite), Hg as HgS (Cinnabar), Ni as NiS (Mullerite) and NiS_2 (Vaesite), Zn as $ZnCO_3$ (Smithsonite) ZnS (Wurtzite), Pb in the form of $Pb_3(CO_3)_2(OH)_2$ (Hydrocerussite) and $PbCO_3$ (Cerussite), as well as PbS (Galena), Fe in the form of $FeCO_3$ (Siderite) and various sulfide phases such as FeS_2 (Pyrite) and FeS (Troilite, Pirrhotite). The addition of phosphates, as well as the production of biogenic carbonate during decomposition of organic carbon, led to a significant decrease in the desorption of strontium and barium, while for other metals, the effect was less pronounced. This can be explained by the formation of calcium phosphate and biogenic calcite (the SI for monohydrocalcite was 0.42, for aragonite > 1, calcite > 1.2, and dolomite > 4), which with high probability, participated in the coprecipitation of strontium (including strontianite) and barium. The addition of phosphates could also lead to the formation of zinc phosphates of the $Zn_3(PO_4)_2{:}4H_2O$ (Hopeite) type.

Iron phases under reducing conditions are the most valuable in the process of biogenic minerals formation. Although the formation of iron oxyhydroxides +3 was not anticipated, sulfide phases were predicted to form because of the highly reducing environment. The elemental maps (S, Fe) acquired using electron microscopy provided further evidence (Figure 6). The accumulation of sulfur and iron phases on LECA (L) and zeolite (Z) after microbial transformation (2) was observed. No substantial iron sulfide crystals were discovered by scanning electron microscopy since the experiment's time period was insufficient for their formation. The observed sulfide–iron formations were most likely associated with a microbial biofilm.

It is known that ferriferous phases (siderite, pyrite, and other iron sulfides) are active sorbents of metals. The resulting sulfide reacts with chalcophile metals [36] such as copper, iron and zinc [37–39]. In a study by Jong et al. [40], it was found that biogenic iron sulfide is a highly effective adsorbent for a wide range of metals and non-metals, including Pb(II),

Cu(II), Cd(II), Zn(II), Ni(II), Fe(II), and As(V). In addition to sorption, the formation of sulfide phases can lead to reduction and stabilization of the reduced phases of metals with varying degrees of oxidation, forming a reductive barrier on the material's surface.

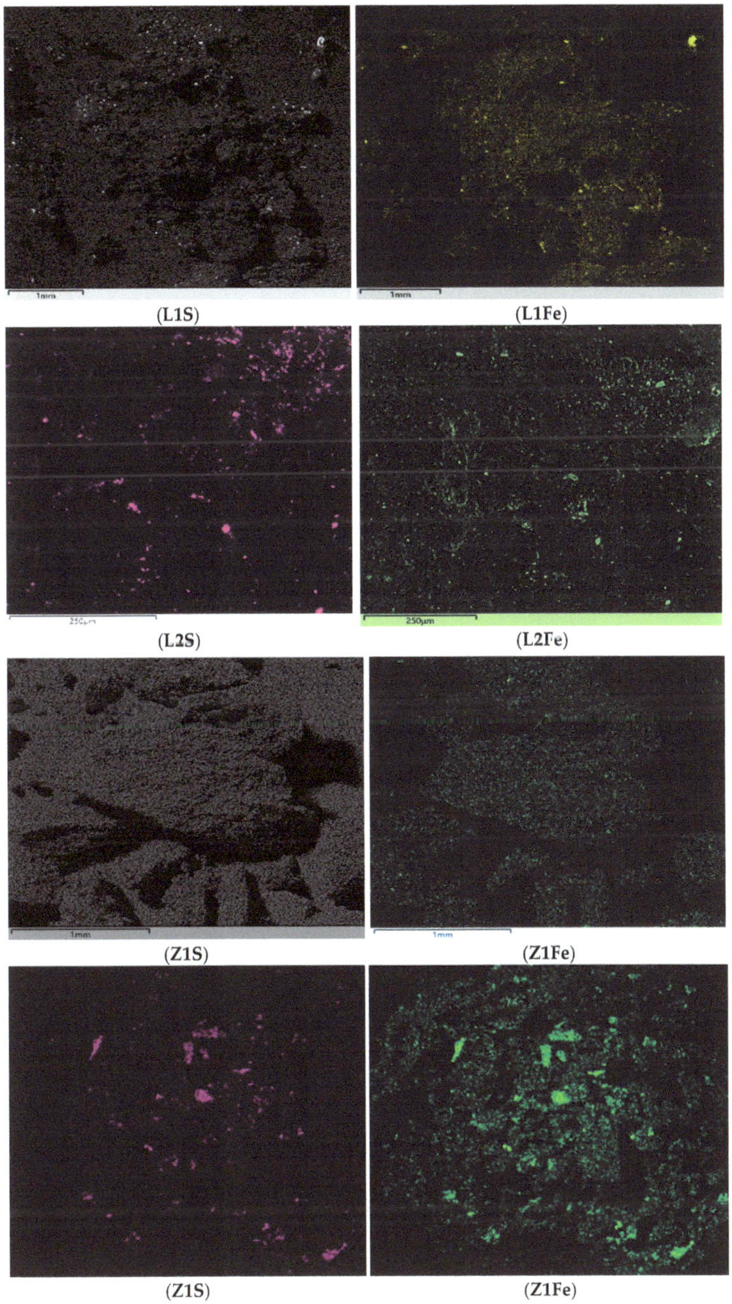

Figure 6. Microphotographs of the analyzed samples. Color mapping: purple for sulfur; yellow-green for iron.

5. Conclusions

The parameters of Cr, Mn, Co, Ni, Cu, Zn, Sr, Cd, Ba, Hg, and Pb immobilization on LECA and zeolite were established taking as an example the permeable biogeochemical barrier near the Siberian Chemical Plant multicomponent waste storage. The geochemical modeling approach predicted the formation of solid mineral phases of iron sulfides and other metals.

It has been established that microbial biofouling does not always promote metal immobilization on the mineral base of the barrier. Mesoporous materials with a high initial sorption capacity were affected by biofilm fouling, which reduced their sorption efficiency. However, biofouling had a beneficial effect on materials with a high surface area of macropores and a characteristically low capacity of metal sorption, significantly increasing their sorption capacity.

The non-uniformity of biofouling on zeolite and LECA demonstrated in this study suggested the feasibility of developing a permeable barrier for the purification of contaminated groundwater with specific functionality. The first component of the barrier can be zeolite, which is less susceptible to fouling and effectively immobilizes metals. LECA can be used as the second part of the barrier to remove metals, nitrates, and sulfates. It was discovered that the biomineralization process on LECA effectively retained metals in sulfide and phosphate forms. For metals of the non-chalcogen group, immobilization is possible with the addition of phosphates. The formation of biogenic iron sulfide precipitate during expanded clay loading is crucial, as it provides a sorption–precipitation phase for the immobilization of the majority of metals.

Author Contributions: Conceptualization, Investigation, Writing—original draft, A.S.; Formal analysis, Investigation, Writing—original draft, N.P.; Formal analysis, Investigation., L.D.; Investigation. N.Y.; Formal analysis K.B.; Investigation D.S.; Investigation G.A.; Investigation, Writing—original draft I.Z. All authors have read and agreed to the published version of the manuscript.

Funding: This work was supported by state assignments from The Ministry of Science and Higher Education of the Russian Federation (#AAAA-A16-11611091001) and was performed using the equipment of the Core Facilities Center of IPCE RAS (CKP FMI IPCE RAS).

Institutional Review Board Statement: Not applicable.

Informed Consent Statement: Not applicable.

Data Availability Statement: Not applicable.

Conflicts of Interest: The authors declare no conflict of interest. The authors declare that they have no known competing financial interests or personal relationships that could have appeared to influence the work reported in this paper.

Appendix A

Table A1. The role of organic groups in metal immobilization according to literature data.

	Carboxyl	Carbonyl	Hydroxyl	Amino	Phosphoryl	Thiol (SH)
Cr	[41,42]	[43]	[44,45]	[44]	[44]	[41,45]
Mn	[46]	-	[46,47]	[46]	-	[46]
Co	[48,49]	[48]	[49,50]	[48]	-	[51]
Ni	[52,53]	[52,53]	[53,54]	[53]		-
Cu	[53,55,56]	[53,57]	[57]	[57]	[53,55,57]	[55]
Zn	[58–60]	-	[58]	[58]	[58,59]	[51,58,61,62]
Sr	[63]	-		[64]	-	[63]
Cd	[65–67]	-	[53,57]	[65,67]	[41]	[46]
Ba	[63]	-	-	-	[68]	[63]
Hg	[69–71]	[69]	[70,71]	[70,71]	[71]	[62,72–74]
Pb	[53,75]	[53]	[76,77]	-	[53,75–78]	[62,72–74]

References

1. Safonov, A.V.; Boguslavsky, A.E.; Boldyrev, K.A.; Gaskova, O.L.; Naimushina, O.S.; Popova, N.M. Geochemical Modeling of the Uranium Behavior in Groundwater near the Sludge Storages during Bioremediation. *Geochem. Int.* **2021**, *59*, 56–65. [CrossRef]
2. Boguslavskii, A.E.; Gas'kova, O.L.; Shemelina, O.V. Geochemical Model of the Environmental Impact of Low-Level Radioactive Sludge Repositories in the Course of Their Decommissioning. *Radiochemistry* **2016**, *58*, 323–328. [CrossRef]
3. Safonov, A.V.; Boguslavsky, A.E.; Gaskova, O.L.; Boldyrev, K.A.; Shvartseva, O.S.; Khvashchevskaya, A.A.; Popova, N.M. Biogeochemical Modelling of Uranium Immobilization and Aquifer Remediation Strategies near Nccp Sludge Storage Facilities. *Appl. Sci.* **2021**, *11*, 2875. [CrossRef]
4. Conca, J.; Strietelmeier, E.; Lu, N.; Ware, S.D.; Taylor, T.P.; Kaszuba, J.; Wright, J. Treatability Study of Reactive Materials to Remediate Groundwater Contaminated with Radionuclides, Metals, and Nitrates in a Four-Component Permeable Reactive Barrier. In *Handbook of Groundwater Remediation Using Permeable Reactive Barriers*; Elsevier: Amsterdam, The Netherlands, 2003; pp. 221–252.
5. Benner, S.G.; Blowes, D.W.; Gould, W.D.; Herbert, R.B.; Ptacek, C.J. Geochemistry of a Permeable Reactive Barrier for Metals and Acid Mine Drainage. *Environ. Sci. Technol.* **1999**, *33*, 2793–2799. [CrossRef]
6. Ludwig, R.D.; McGregor, R.G.; Blowes; Benner, S.G.; Mountjoy, K. A Permeable Reactive Barrier for Treatment of Heavy Metals. *Ground Water* **2002**, *40*, 59–66. [CrossRef]
7. Thiruvenkatachari, R.; Vigneswaran, S.; Naidu, R. Permeable Reactive Barrier for Groundwater Remediation. *J. Ind. Eng. Chem.* **2008**, *14*, 145–156. [CrossRef]
8. Vignola, R.; Bagatin, R.; De Folly D'Auris, A.; Flego, C.; Nalli, M.; Ghisletti, D.; Millini, R.; Sisto, R. Zeolites in a Permeable Reactive Barrier (PRB): One Year of Field Experience in a Refinery Groundwater-Part 1: The Performances. *Chem. Eng. J.* **2011**, *178*, 204–209. [CrossRef]
9. Faisal, A.A.H.; Hmood, Z.A. Groundwater Protection from Cadmium Contamination by Zeolite Permeable Reactive Barrier. *Desalination Water Treat* **2015**, *53*, 1377–1386. [CrossRef]
10. Rocha, L.C.C.; Zuquette, L.V. Evaluation of Zeolite as a Potential Reactive Medium in a Permeable Reactive Barrier (PRB): Batch and Column Studies. *Geosciences* **2020**, *10*, 59. [CrossRef]
11. Skinner, S.J.W.; Schutte, C.F. The Feasibility of a Permeable Reactive Barrier to Treat Acidic Sulphate- and Nitrate-Contaminated Groundwater. *Water SA* **2006**, *32*, 129–135. [CrossRef]
12. Fuller, C.C.; Bargar, J.R.; Davis, J.A. Molecular-Scale Characterization of Uranium Sorption by Bone Apatite Materials for a Permeable Reactive Barrier Demonstration. *Environ. Sci. Technol.* **2003**, *37*, 4642–4649. [CrossRef] [PubMed]
13. Holmes, R.R.; Hart, M.L.; Kevern, J.T. Reuse of Drinking Water Treatment Waste for Remediation of Heavy Metal Contaminated Groundwater. *Groundw. Monit. Remediat.* **2019**, *39*, 69–79. [CrossRef]
14. Taha, G.M. Utilization of Low-Cost Waste Material Bagasse Fly Ash in Removing of Cu^{2+}, Ni^{2+}, Zn^{2+}, and Cr^{3+} from Industrial Waste Water. *Groundw. Monit. Remediat.* **2006**, *26*, 137–141. [CrossRef]
15. Thakur, A.K.; Vithanage, M.; Das, D.B.; Kumar, M. A Review on Design, Material Selection, Mechanism, and Modelling of Permeable Reactive Barrier for Community-Scale Groundwater Treatment. *Environ. Technol. Innov.* **2020**, *19*, 100917. [CrossRef]
16. Cantrell, K.J.; Kaplan, D.I.; Wietsma, T.W. Zero-Valent Iron for the in Situ Remediation of Selected Metals in Groundwater. *J. Hazard. Mater.* **1995**, *42*, 201–212. [CrossRef]
17. Gu, B.; Liang, L.; Dickey, M.J.; Yin, X.; Dai, S. Reductive Precipitation of Uranium(VI) by Zero-Valent Iron. *Environ. Sci. Technol.* **1998**, *32*, 3366–3373. [CrossRef]
18. Gibert, O.; Assal, A.; Devlin, H.; Elliot, T.; Kalin, R.M. Performance of a Field-Scale Biological Permeable Reactive Barrier for in-Situ Remediation of Nitrate-Contaminated Groundwater. *Sci. Total Environ.* **2019**, *659*, 211–220. [CrossRef]
19. Borch, T.; Kretzschmar, R.; Kappler, A.; Van Cappellen, P.; Ginder-Vogel, M.; Voegelin, A.; Campbell, K. Biogeochemical Redox Processes and their Impact on Contaminant Dynamics. *Environ. Sci. Technol.* **2010**, *44*, 15–23. [CrossRef]
20. Silva, G.; Pennafirme, S.; Lopes, R.; Lima, I.; Crapez, M. Imaging Techniques for Monitoring Bacterial Biofilms in Environmental Samples—An Important Tool for Bioremediation Studies. *BAOJ Microbiol.* **2017**, *3*, 1–15.
21. Zinicovscaia, I.; Yushin, N.; Grozdov, D.; Abdusamadzoda, D.; Safonov, A.; Rodlovskaya, E. Zinc-Containing Effluent Treatment Using Shewanella Xiamenensis Biofilm Formed on Zeolite. *Materials* **2021**, *14*, 1760. [CrossRef]
22. Zinicovscaia, I.; Safonov, A.; Boldyrev, K.; Gundorina, S.; Yushin, N.; Petuhov, O.; Popova, N. Selective Metal Removal from Chromium-Containing Synthetic Effluents Using Shewanella Xiamenensis Biofilm Supported on Zeolite. *Environ. Sci. Pollut. Res.* **2020**, *27*, 10495–10505. [CrossRef] [PubMed]
23. Zinicovscaia, I.; Yushin, N.; Grozdov, D.; Vergel, K.; Popova, N.; Artemiev, G.; Safonov, A. Metal Removal from Nickel-Containing Effluents Using Mineral–Organic Hybrid Adsorbent. *Materials* **2020**, *13*, 4462. [CrossRef] [PubMed]
24. Zinicovscaia, I.; Yushin, N.; Grozdov, D.; Safonov, A.; Ostovnaya, T.; Boldyrev, K.; Kryuchkov, D.; Popova, N. Bio-Zeolite Use for Metal Removal from Copper-Containing Synthetic Effluents. *J. Environ. Health Sci. Eng.* **2021**, *19*, 1383–1398. [CrossRef] [PubMed]
25. Kumar, N.; Chaurand, P.; Rose, J.; Diels, L.; Bastiaens, L. Synergistic effects of sulfate reducing bacteria and zero-valent iron on zinc removal and stability in aquifer sediment. *Chem. Eng.* **2015**, *260*, 83–89. [CrossRef]
26. Upadhyay, S.; Sinha, A. Role of microorganisms in permeable reactive bio-barriers (PRBBs) for environmental clean-up: A review. *Global NEST J.* **2018**, *20*, 269–280.

27. He, Y.T.; Wilson, J.T.; Wilkin, R.T. Transformation of Reactive Iron Minerals in a Permeable Reactive Barrier (Biowall) Used to Treat TCE in Groundwater. *Environ. Sci. Technol.* **2008**, *42*, 6690–6696. [CrossRef] [PubMed]
28. Safonov, A.V.; Andryushchenko, N.D.; Ivanov, P.V.; Boldyrev, K.A.; Babich, T.L.; German, K.E.; Zakharova, E.V. Biogenic Factors of Radionuclide Immobilization on Sandy Rocks of Upper Aquifers. *Radiochemistry* **2019**, *61*, 99–108. [CrossRef]
29. Safonov, A.V.; Babich, T.L.; Sokolova, D.S.; Grouzdev, D.S.; Tourova, T.P.; Poltaraus, A.B.; Zakharova, E.V.; Merkel, A.Y.; Novikov, A.P.; Nazina, T.N. Microbial Community and in Situ Bioremediation of Groundwater by Nitrate Removal in the Zone of a Radioactive Waste Surface Repository. *Front. Microbiol.* **2018**, *9*, 1985. [CrossRef]
30. Pavlov, S.S.; Dmitriev, A.Y.; Frontasyeva, M.V. Automation system for neutron activation analysis at the reactor IBR-2, Frank Laboratory of Neutron Physics, Joint Institute for Nuclear Research, Dubna, Russia. *JRNC* **2016**, *309*, 27–38. [CrossRef]
31. Bogachev, M.I.; Volkov, V.Y.; Markelov, O.A.; Trizna, E.Y.; Baydamshina, D.R.; Melnikov, V.; Murtazina, R.R.; Zelenikhin, P.V.; Sharafutdinov, I.S.; Kayumov, A.R. Fast and simple tool for the quantification of biofilm-embedded cells sub-populations from fluorescent microscopic images. *PLoS ONE* **2018**, *13*, e0193267. [CrossRef]
32. Trafny, E.A.; Lewandowski, R.; Zawistowska-Marciniak, I.; Stępińska, M. Use of MTT assay for determination of the biofilm formation capacity of microorganisms in metalworking fluids. *World J. Microbiol. Biotechnol.* **2013**, *29*, 1635–1643. [CrossRef] [PubMed]
33. Parkhurst, D. User's Guide to PHREEQC—A Computer Program for Speciation, Reaction-Path, Advective-Transport, and Inverse Geochemical Calculations. *US Geol. Surv. Water-Resour. Investig. Rep.* **1995**, *143*, 95–4227. [CrossRef]
34. Javanbakht, V.; Alavi, S.A.; Zilouei, H. Mechanisms of Heavy Metal Removal Using Microorganisms as Biosorbent. *Water Sci. Technol.* **2014**, *69*, 1775–1787. [CrossRef] [PubMed]
35. Sag, Y.; Kutsal, T. Recent Trends in the Biosorption of Heavy Metals: A Review. *Biotechnol. Bioprocess Eng.* **2001**, *6*, 376–385. [CrossRef]
36. Viggi, C.C.; Pagnanelli, F.; Toro, L. Sulphate Reduction Processes in Biological Permeable Reactive Barriers: Column Experimentation and Modeling. In *Chemical Engineering Transactions*; Italian Association of Chemical Engineering—AIDIC: Milan, Italy, 2011; Volume 24, pp. 1231–1236.
37. Marius, M.S.; James, P.A.B.; Bahaj, A.S.; Smallman, D.J. Influence of Iron Valency on the Magnetic Susceptibility of a Microbially Produced Iron Sulphide. In *Journal of Physics: Conference Series*; Institute of Physics Publishing: Bristol, UK, 2005; Volume 17, pp. 65–69.
38. Mokone, T.P.; van Hille, R.P.; Lewis, A.E. Metal Sulphides from Wastewater: Assessing the Impact of Supersaturation Control Strategies. *Water Res.* **2012**, *46*, 2088–2100. [CrossRef]
39. Martins, M.; Faleiro, M.L.; Barros, R.J.; Veríssimo, A.R.; Barreiros, M.A.; Costa, M.C. Characterization and Activity Studies of Highly Heavy Metal Resistant Sulphate-Reducing Bacteria to Be Used in Acid Mine Drainage Decontamination. *J. Hazard. Mater.* **2009**, *166*, 706–713. [CrossRef] [PubMed]
40. Jong, T.; Parry, D.L. Adsorption of Pb(II), Cu(II), Cd(II), Zn(II), Ni(II), Fe(II), and As(V) on Bacterially Produced Metal Sulfides. *J. Colloid Interface Sci.* **2004**, *275*, 61–71. [CrossRef]
41. Sobol, Z.; Schiestl, R.H. Intracellular and Extracellular Factors Influencing Cr(VI and Cr(III) Genotoxicity. *Environ. Mol. Mutagen.* **2012**, *53*, 94–100. [CrossRef]
42. Li, S.; Zhou, Z.; Tie, Z.; Wang, B.; Ye, M.; Du, L.; Cui, R.; Liu, W.; Wan, C.; Liu, Q.; et al. Data-Informed Discovery of Hydrolytic Nanozymes. *Nat. Commun.* **2022**, *13*, 1–12. [CrossRef]
43. Jobby, R.; Jha, P.; Yadav, A.K.; Desai, N. Biosorption and Biotransformation of Hexavalent Chromium [Cr(VI)]: A Comprehensive Review. *Chemosphere* **2018**, *207*, 255–266. [CrossRef]
44. An, H.; Tian, T.; Wang, Z.; Jin, R.; Zhou, J. Role of Extracellular Polymeric Substances in the Immobilization of Hexavalent Chromium by Shewanella Putrefaciens CN32 Unsaturated Biofilms. *Sci. Total Environ.* **2022**, *810*, 151184. [CrossRef] [PubMed]
45. Viti, C.; Marchi, E.; Decorosi, F.; Giovannetti, L. Molecular Mechanisms of Cr(VI) Resistance in Bacteria and Fungi. *FEMS Microbiol. Rev.* **2014**, *38*, 633–659. [CrossRef] [PubMed]
46. Huang, H.; Zhao, Y.; Xu, Z.; Ding, Y.; Zhou, X.; Dong, M. A High Mn(II)-Tolerance Strain, Bacillus Thuringiensis HM7, Isolated from Manganese Ore and Its Biosorption Characteristics. *PeerJ* **2020**, *8*, e8589. [CrossRef] [PubMed]
47. Snyder, M.S. *Biological Control of Manganese in Water Supplies in the Presence of Humic Acids*; University of Kentucky: Lexington, KY, USA, 2013.
48. Fawzy, M.A.; Hifney, A.F.; Adam, M.S.; Al-Badaani, A.A. Biosorption of Cobalt and Its Effect on Growth and Metabolites of Synechocystis Pevalekii and Scenedesmus Bernardii: Isothermal Analysis. *Environ. Technol. Innov.* **2020**, *19*, 100953. [CrossRef]
49. Su, Y.; Sun, S.; Liu, Q.; Zhao, C.; Li, L.; Chen, S.; Chen, H.; Wang, Y.; Tang, F. Characterization of the Simultaneous Degradation of Pyrene and Removal of Cr(VI) by a Bacteria Consortium YH. *Sci. Total Environ.* **2022**, *853*, 158388. [CrossRef]
50. Waris, A.; Din, M.; Ali, A.; Afridi, S.; Baset, A.; Khan, A.U.; Ali, M. Green Fabrication of Co and Co_3O_4 nanoparticles and Their Biomedical Applications: A Review. *Open Life Sci.* **2021**, *16*, 14–30. [CrossRef]
51. Dulay, H.; Tabares, M.; Kashefi, K.; Reguera, G. Cobalt Resistance via Detoxification and Mineralization in the Iron-Reducing Bacterium Geobacter Sulfurreducens. *Front. Microbiol.* **2020**, *11*, 600463. [CrossRef]
52. Levett, A.; Gleeson, S.A.; Kallmeyer, J. From Exploration to Remediation: A Microbial Perspective for Innovation in Mining. *Earth Sci. Rev.* **2021**, *216*, 103563. [CrossRef]

53. Haque, M.M.; Mosharaf, M.K.; Haque, M.A.; Tanvir, M.Z.H.; Alam, M.K. Biofilm Formation, Production of Matrix Compounds and Biosorption of Copper, Nickel and Lead by Different Bacterial Strains. *Front. Microbiol.* **2021**, *12*, 1385. [CrossRef]
54. Haider, A.; Ijaz, M.; Ali, S.; Haider, J.; Imran, M.; Majeed, H.; Shahzadi, I.; Ali, M.M.; Khan, J.A.; Ikram, M. Green Synthesized Phytochemically (Zingiber Officinale and Allium Sativum) Reduced Nickel Oxide Nanoparticles Confirmed Bactericidal and Catalytic Potential. *Nanoscale Res. Lett.* **2020**, *15*, 1–11. [CrossRef]
55. Lin, H.; Wang, C.; Zhao, H.; Chen, G.; Chen, X. A Subcellular Level Study of Copper Speciation Reveals the Synergistic Mechanism of Microbial Cells and EPS Involved in Copper Binding in Bacterial Biofilms. *Environ. Pollut.* **2020**, *263*, 114485. [CrossRef] [PubMed]
56. Fang, L.; Yang, S.; Huang, Q.; Xue, A.; Cai, P. Biosorption Mechanisms of Cu(II) by Extracellular Polymeric Substances from Bacillus Subtilis. *Chem. Geol.* **2014**, *386*, 143–151. [CrossRef]
57. Cheng, X.; Xu, W.; Wang, N.; Mu, Y.; Zhu, J.; Luo, J. Adsorption of Cu2+ and Mechanism by Natural Biofilm. *Water Sci. Technol.* **2018**, *78*, 721–731. [CrossRef] [PubMed]
58. Thomas, S.A.; Mishra, B.; Myneni, S.C.B. High Energy Resolution-X-Ray Absorption Near Edge Structure Spectroscopy Reveals Zn Ligation in Whole Cell Bacteria. *J. Phys. Chem. Lett.* **2019**, *10*, 2585–2592. [CrossRef]
59. Li, C.C.; Wang, Y.J.; Du, H.; Cai, P.; Peijnenburg, W.J.G.M.; Zhou, D.M. Influence of Bacterial Extracellular Polymeric Substances on the Sorption of Zn on γ-Alumina: A Combination of FTIR and EXAFS Studies. *Environ. Pollut.* **2017**, *220*, 997–1004. [CrossRef]
60. Basak, G.; Lakshmi, V.; Chandran, P.; Das, N. Removal of Zn(II) from Electroplating Effluent Using Yeast Biofilm Formed on Gravels: Batch and Column Studies. *J. Environ. Health Sci. Eng.* **2014**, *12*, 8. [CrossRef]
61. Desmau, M.; Carboni, A.; Le Bars, M.; Doelsch, E.; Benedetti, M.F.; Auffan, M.; Levard, C.; Gelabert, A. How Microbial Biofilms Control the Environmental Fate of Engineered Nanoparticles? *Front. Environ. Sci.* **2020**, *8*, 82. [CrossRef]
62. Kumar, R.; Umar, A.; Kumar, G.; Nalwa, H.S. Antimicrobial Properties of ZnO Nanomaterials: A Review. *Ceram. Int.* **2017**, *43*, 3940–3961. [CrossRef]
63. Deng, N.; Stack, A.G.; Weber, J.; Cao, B.; De Yoreo, J.J.; Hu, Y. Organic–Mineral Interfacial Chemistry Drives Heterogeneous Nucleation of Sr-Rich (Bax, Sr1−x)SO4 from Undersaturated Solution. *Proc. Natl. Acad. Sci. USA* **2019**, *116*, 13221–13226. [CrossRef]
64. Liu, M.; Dong, F.; Zhang, W.; Nie, X.; Wei, H.; Sun, S.; Zhong, X.; Liu, Y.; Wang, D. Contribution of Surface Functional Groups and Interface Interaction to Biosorption of Strontium Ions by Saccharomyces Cerevisiae under Culture Conditions. *RSC Adv.* **2017**, *7*, 50880–50888. [CrossRef]
65. Zhang, J.; Li, Q.; Zeng, Y.; Zhang, J.; Lu, G.; Dang, Z.; Guo, C. Bioaccumulation and Distribution of Cadmium by Burkholderia Cepacia GYP1 under Oligotrophic Condition and Mechanism Analysis at Proteome Level. *Ecotoxicol. Environ. Saf.* **2019**, *176*, 162–169. [CrossRef] [PubMed]
66. Boyanov, M.I.; Kelly, S.D.; Kemner, K.M.; Bunker, B.A.; Fein, J.B.; Fowle, D.A. Adsorption of Cadmium to Bacillus Subtilis Bacterial Cell Walls: A PH-Dependent X-Ray Absorption Fine Structure Spectroscopy Study. *Geochim. Cosmochim. Acta* **2003**, *67*, 3299–3311. [CrossRef]
67. Xu, S.; Xing, Y.; Liu, S.; Luo, X.; Chen, W.; Huang, Q. Co-Effect of Minerals and Cd(II) Promoted the Formation of Bacterial Biofilm and Consequently Enhanced the Sorption of Cd(II). *Environ. Pollut.* **2020**, *258*, 113774. [CrossRef]
68. Martinez-Ruiz, F.; Jroundi, F.; Paytan, A.; Guerra-Tschuschke, I.; Abad, M.D.M.; González-Muñoz, M.T. Barium Bioaccumulation by Bacterial Biofilms and Implications for Ba Cycling and Use of Ba Proxies. *Nat. Commun.* **2018**, *9*, 1–9. [CrossRef] [PubMed]
69. Bourdineaud, J.P.; Durn, G.; Režun, B.; Manceau, A.; Hrenović, J. The Chemical Species of Mercury Accumulated by Pseudomonas Idrijaensis, a Bacterium from a Rock of the Idrija Mercury Mine, Slovenia. *Chemosphere* **2020**, *248*, 126002. [CrossRef] [PubMed]
70. Fathollahi, A.; Coupe, S.J.; El-Sheikh, A.H.; Sañudo-Fontaneda, L.A. The Biosorption of Mercury by Permeable Pavement Biofilms in Stormwater Attenuation. *Sci. Total Environ.* **2020**, *741*, 140411. [CrossRef]
71. Dash, H.R.; Das, S. Interaction between Mercuric Chloride and Extracellular Polymers of Biofilm-Forming Mercury Resistant Marine Bacterium: Bacillus Thuringiensis PW-05. *RSC Adv.* **2016**, *6*, 109793–109802. [CrossRef]
72. Desmau, M.; Levard, C.; Vidal, V.; Ona-Nguema, G.; Charron, G.; Benedetti, M.F.; Gélabert, A. How Microbial Biofilms Impact the Interactions of Quantum Dots with Mineral Surfaces? *NanoImpact* **2020**, *19*, 100247. [CrossRef]
73. Yu, Q.; Szymanowski, J.; Myneni, S.C.B.; Fein, J.B. Characterization of Sulfhydryl Sites within Bacterial Cell Envelopes Using Selective Site-Blocking and Potentiometric Titrations. *Chem. Geol.* **2014**, *373*, 50–58. [CrossRef]
74. Mishra, B.; Shoenfelt, E.; Yu, Q.; Yee, N.; Fein, J.B.; Myneni, S.C.B. Stoichiometry of Mercury-Thiol Complexes on Bacterial Cell Envelopes. *Chem. Geol.* **2017**, *464*, 137–146. [CrossRef]
75. Templeton, A.S.; Trainor, T.P.; Spormann, A.M.; Newville, M.; Sutton, S.R.; Dohnalkova, A.; Gorby, Y.; Brown, G.E. Sorption versus Biomineralization of Pb(II) within Burkholderia Cepacia Biofilms. *Environ. Sci. Technol.* **2003**, *37*, 300–307. [CrossRef] [PubMed]
76. Sowmya, M.; Mohamed Hatha, A.A. Cadmium and Lead Tolerance Mechanisms in Bacteria and the Role of Halotolerant and Moderately Halophilic Bacteria in Their Remediation. In *Handbook of Metal-Microbe Interactions and Bioremediation*; CRC Press: Boca Raton, FL, USA, 2017; pp. 557–573. ISBN 9781315153353.

77. Kumari, S.; Mangwani, N.; Das, S. Interaction of Pb(II) and Biofilm Associated Extracellular Polymeric Substances of a Marine Bacterium Pseudomonas Pseudoalcaligenes NP103. *Spectrochim Acta A Mol. Biomol. Spectrosc.* **2017**, *173*, 655–665. [CrossRef] [PubMed]
78. Ngwenya, B.T.; Sutherland, I.W.; Kennedy, L. Comparison of the Acid-Base Behaviour and Metal Adsorption Characteristics of a Gram-Negative Bacterium with Other Strains. *Appl. Geochem.* **2003**, *18*, 527–538. [CrossRef]

Disclaimer/Publisher's Note: The statements, opinions and data contained in all publications are solely those of the individual author(s) and contributor(s) and not of MDPI and/or the editor(s). MDPI and/or the editor(s) disclaim responsibility for any injury to people or property resulting from any ideas, methods, instructions or products referred to in the content.

hydrology

Article

Combined Well Multi-Parameter Logs and Low-Flow Purging Data for Soil Permeability Assessment and Related Effects on Groundwater Sampling

Francesco Maria De Filippi * and **Giuseppe Sappa**

Department of Civil, Building and Environmental Engineering (DICEA), Sapienza University of Rome, 00184 Rome, Italy
* Correspondence: francescomaria.defilippi@uniroma1.it

Abstract: Cost-effective remediation is increasingly dependent on high-resolution site characterization (HRSC), which is supposed to be necessary prior to interventions. This paper aims to evaluate the use of low-flow purging and sampling water level data in estimating the horizontal hydraulic conductivity of soils. In a new quali-quantitative view, this procedure can provide much more information and knowledge about the site, reducing time and costs. In case of high heterogeneity along the well screen, the whole procedure, as well as the estimation method, could be less effective and rigorous, with related issues in the purging time. The result showed significant permeability weighted sampling, which could provide different results as the pump position changes along the well screen. The proposed study confirms this phenomenon with field data, demonstrating that the use of multiparameter well logs might be helpful in detecting the behaviour of low-permeability layers and their effects on purging and sampling. A lower correlation between low-flow permeability estimations and LeFranc test results was associated with high heterogeneity along the screen, with a longer purging time. In wells P43, MW08 and MW36, due to the presence of clay layers, results obtained differ for almost one order of magnitude and the purging time increases (by more than 16 min). However, with some precautions prior to the field work, the low-flow purging and sampling procedure could become more representative in a shorter time and provide important hydrogeological parameters such as hydraulic conductivity with many tests and high-resolution related results.

Keywords: groundwater monitoring; water sampling; low-flow; wells; soil permeability

Citation: De Filippi, F.M.; Sappa, G. Combined Well Multi-Parameter Logs and Low-Flow Purging Data for Soil Permeability Assessment and Related Effects on Groundwater Sampling. *Hydrology* **2023**, *10*, 12. https://doi.org/10.3390/hydrology10010012

Academic Editor: Meysam Vadiati

Received: 19 December 2022
Revised: 30 December 2022
Accepted: 30 December 2022
Published: 2 January 2023

Copyright: © 2023 by the authors. Licensee MDPI, Basel, Switzerland. This article is an open access article distributed under the terms and conditions of the Creative Commons Attribution (CC BY) license (https://creativecommons.org/licenses/by/4.0/).

1. Introduction

In groundwater monitoring, site characterization is a complex process composed of many components and activities. It is carried on when a potential contamination threatens specific targets and should be necessary to provide background data on the site sensitivity to anthropogenic impact [1,2]. In situ monitoring plays a key role in groundwater protection and management, as it is the only way to find out reliable aquifer properties not only to assess the contamination but also to understand complex hydrogeological and hydrochemical processes [3–5]. In particular, the assessment of porous aquifer vulnerability to contaminants and the eventual successful remediation measures are closely linked to the results of a well-performed groundwater monitoring, which in turn, depends on a highly structured well network, the field technicians' skills and the financial resources of the involved stakeholders [6,7]. During these activities, well purging is usually a required and mandatory operation, carried out before groundwater sampling. Nowadays, low-flow purging and sampling is a well-known and consolidated methodology in environmental monitoring, consisting of pumping water at low flowrates (in fine-grained soils from 0.1 to 1 L/min) and minimizing the induced stabilized drawdown in the well (usually few tens of cm maximum) [8–11]. These limits are rather unstable for both practical and theoretical issues: stabilized low flowrates are not easy to set due to specific well conditions or pumps' technical limits, and even with the lowest flow rate of the aforementioned range, particular hydrogeological settings with low-permeable soils does not allow in achieving acceptable low drawdowns. In addition, the defined ranges of "low-flow" conditions are strictly connected to the aquifer geometry and properties, which are sometimes unknown.

The misleading common belief is that the low-flow purging is only related to the pumped value, whereas one has to consider the induced groundwater flow to the well, which depends on the aquifer properties too [12–14]. In this sense, the only measured parameter that can give information about the right conditions is the drawdown ΔH, or even better the ratio $\Delta H/H$ (where H is the thickness of the aquifer), which allows to scale the drawdown taking into account the aquifer geometry also [15,16].

Regardless of the foregoing data that are strictly related to the hydraulics, the aim of this technique is mainly to correctly purge the well in order to obtain representative groundwater samples, with a reduced stress on the aquifer and less volumes of groundwater disposal at the same time. However, the word "representative" is still an object of discussion among scientists and practitioners, because representativeness often depends on the focus of the study, driven by specific compounds or related parameters [14,17,18].

The difficulty in collecting high-quality groundwater samples is related both to the specific sampling technique as well as the well construction and soil heterogeneity in terms of hydraulic conductivity. The presence of unknown low permeability layers or lenses in the aquifer may involve a contaminant back diffusion, modifying sample quality values over time [3,19–22]. Hence, collecting a formation water sample is always the main goal, but as the aquifer is heterogeneous and the screen length is sometimes not properly designed, the result of the operation will be always permeability-weighted sampling [23,24]. Consequently, the qualitative and quantitative aspects of low-flow purging and sampling cannot be separated and must be investigated more, in order to maximize the aquifer knowledge as well as the representativeness of the collected groundwater sample. The reduced time-steps of the site characterization process combined with the high resolution of the obtained results is crucial for stakeholders [25–28].

Recent studies demonstrated that this technique might be useful for estimating the horizontal hydraulic conductivity (K_H) of the aquifer and for strengthening the preliminary site assessment, without further and expensive investigations [15,29,30].

The variability of groundwater flow along the well screen and the presence of vertical flows may represent an issue [23,31,32]. This is mainly due to soil heterogeneities along the screen that affects both the evaluated K_H (a depth-weighted value) and the groundwater sample characteristics (permeability weighted). Recent studies show that vertical multiparameter well logs are a useful tool to assess soil heterogeneities and flow exchanges in the well water column [33,34]. For this reason, in this study, multi-parameter vertical logs combined with low-flow purging have been carried out in a landfill-monitoring network (17 monitoring wells), to achieve a more in-depth site characterization. The results obtained show that the use of a combined quali-quantitative approach could be helpful to understand the flow regime better along the screen well, identifying possible outliers of the low-flow proposed methodology for estimating aquifer K_H, due to low-permeability layers.

2. Study Area and Geological Framework

The study area is in Italy, in the landfill of Borgo Montello, few kilometres far from Latina, in the Latium Region (Figure 1). The location is in the southern portion of the Roman countryside and on the southern slopes of the volcanic region of the Alban Hills.

The related hydrographic basin is that of the Astura River, which overall extends for about 400 km^2. Its length, from the northern part, located in the highest area of the Alban Hills, is about 35 km. The Tyrrhenian Sea is about 10 km far from the study area, which is near the so called "Agro Pontino", a swampy wetland area, currently reclaimed with an intense agricultural vocation and related high groundwater impacts [35].

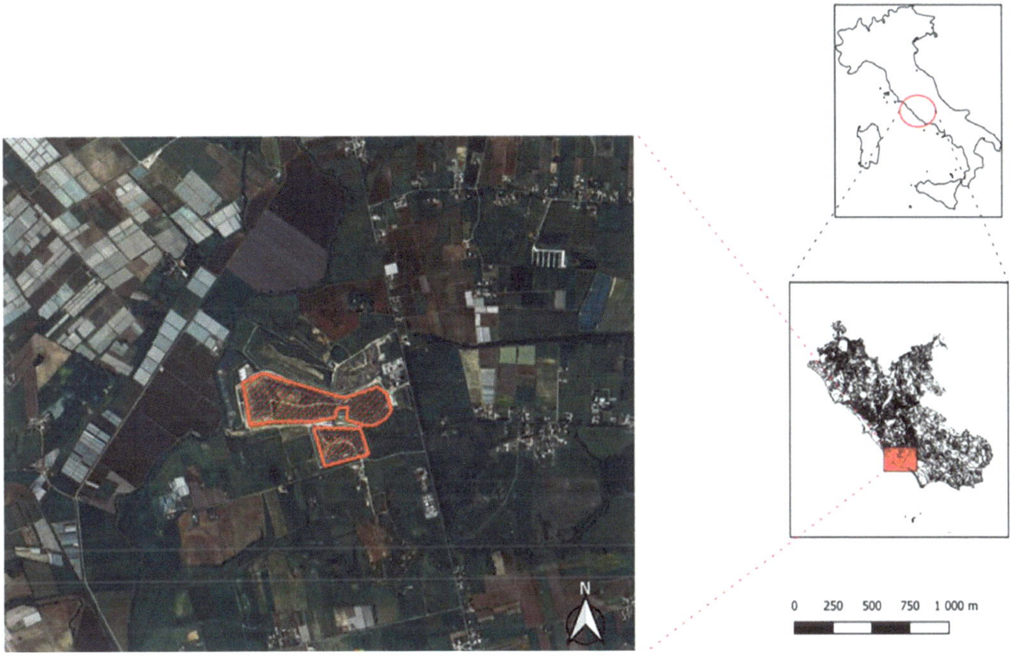

Figure 1. Borgo Montello landfill study area.

The presence of these morphological features is interconnected with a complex regional geology, characterized by the pyroclastic soils and lavas of Alban Hills, whose boundaries are difficult to delimit because they have been covered by more recent alluvial soils towards the external part.

The "Agro Pontino" area represents a sector of the Plio-Pleistocene backdeep from the Volsci Range that characterizes the central part of the Tyrrhenian border of the Italian peninsula and for which important new studies have been performed at the regional tectonic level [36,37].

The intense subsidence allowed the sedimentation of marine deposits, mainly clayey formations, with a thickness of many hundreds of meters. Clayey formations, practically impermeable, can be considered the regional groundwater basement. Hence, pyroclastic rocks and tuffs, locally covered by sands, host the main aquifer of the area (Figure 2).

Their limited outcrop largely depends on the erosion process related to the overlying sands. The presence of silty clayey levels may constitute local aquitards, whereas the base aquiclude is related to the presence of the previously mentioned marine clay formations [38]. The landfill site is located few kilometres far from the small town of Borgo Montello. It is divided in two waste disposal basins, protected by underground hydraulic barriers (black contours in Figure 2) and surrounded by the monitoring well network, consisting of 17 points. The well depths are between 14.5 (minimum) and 41 m (maximum), whereas monitoring point elevations range from 11 to 29.19 m asl. (Table 1). Based on the piezometric surveys previously carried out, groundwater flow is locally directed NE-SW (Figure 2).

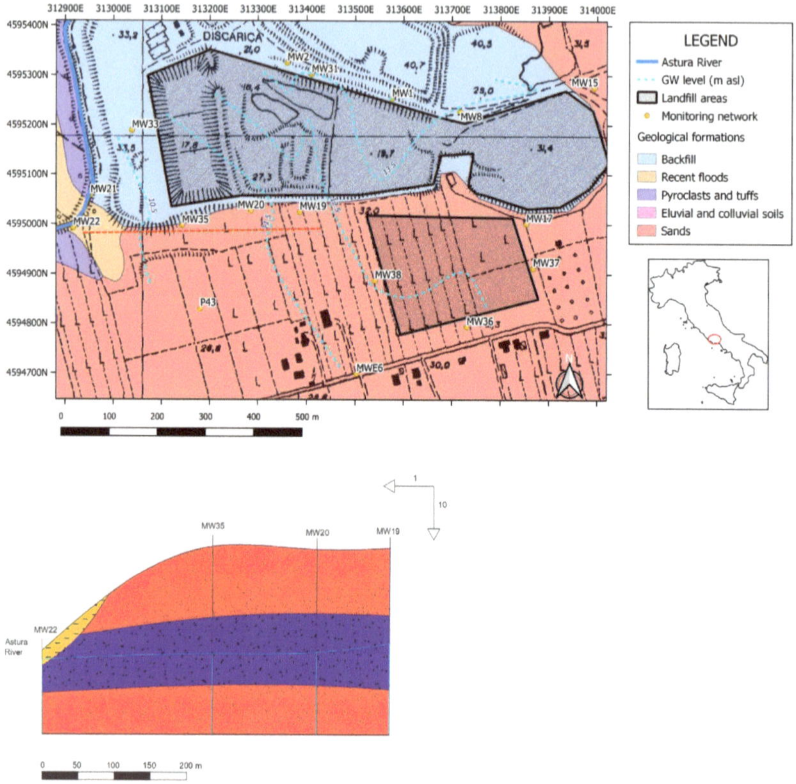

Figure 2. Geological map and cross section of the landfill well monitoring network area.

Table 1. Coordinates (WGS84) and depth of the landfill monitoring wells.

ID	N-WGS84 (°)	E-WGS84 (°)	Depth (m)	Altitude (m asl)	Depth to Water (m)	Water Table (m asl)
MW1	41.4855	12.7661	34.0	28.11	13.7	14.41
MW2	41.4861	12.7635	20.0	22.76	8.8	13.96
MW8	41.4853	12.7678	40.0	28.58	15.8	12.78
MW15	41.4858	12.7711	41.0	28.77	15.6	13.17
MW31	41.4859	12.7641	30.0	23.19	10.4	12.79
MW17	41.4833	12.7695	37.0	22.22	9.3	12.92
MW19	41.4834	12.7627	36.0	29.19	17	12.19
MW20	41.4834	12.7639	25.0	24.86	14	10.86
MW21	41.4836	12.7587	20.0	11.04	1.3	9.74
MW22	41.4830	12.7583	14.5	11.23	1.2	10.03
MW33	41.4848	12.7597	30.0	32.6	21.8	10.8
MW35	41.4831	12.7610	30.0	25.17	14.4	10.77
MW36	41.4814	12.7681	33.0	26.88	14.7	12.18
MW37	41.4825	12.7697	37.0	28.99	16.2	12.74
MW38	41.4822	12.7658	40.0	28.41	16.8	11.56
MWE6	41.4805	12.7654	32.0	25.19	13.8	11.39
P43	41.4816	12.7615	33.5	26.70	15.87	10.83

3. Materials and Methods

Instrumentation and Measurements

Monitoring and field data collection have been carried out in March 2022, from the 28th to the end of the month. Both the low-flow purging technique and the multi-parametric log measurements involved all the 17 monitoring wells around the landfill perimeter. The depth to water was measured before purging started, and it was measured again, during the procedure, at increasing time steps (1, 2, 4, 8, and 16 min) using a water level meter instrument. A low-flow rate was achieved using a 12 v Proactive Hurricane submersible pump with booster, allowing to obtain a minimum flow of 0.1 L/min (Figure 3).

(**a**) (**b**)

Figure 3. Instrumentation used for well purging and groundwater sampling in low-flow conditions: (**a**) 12 v battery and booster for pumping flow regulation; (**b**) water level meter and HANNA HI7609829 multiparameter probe.

Flow-rates values have been manually measured, using a graduated vessel at known volume and measuring the time to fill it. The values generally were within the usual range of 0.1–1 L/min, even if in some cases, it has been possible to obtain stabilization of both chemical-physical parameters and water level at higher flow-rates, due to the high permeability of layers. Water temperature (T), electrical conductivity (EC), pH, dissolved oxygen (DO) and Redox potential (ORP) were the groundwater chemical-physical parameters measured, at the same time steps of the drawdowns. Their stabilization was determined in the field using a HANNA HI7609829 probe and a flow cell, keeping groundwater with no air contact during the reading (Figure 3).

The ranges defined for parameters' stabilization between two successive readings follow the specific thresholds suggested previously [39]. This allowed for the end of purging activities and the beginning of water sampling for laboratory analyses. Regarding the multiparameter logs, they have been executed with a Seba Hydrometrie 5 W MPS D8 Multiparameter probe. The depth measurement frequency was every one meter, waiting at each step for parameter stabilization (Figure 3). The stabilized water level measured in each well during the low-flow purging was used as an input data, as well as the well radius and depth, for assessing aquifer horizontal hydraulic conductivity K_H [15].

This methodology is mainly based on the Dupuit/Thiem theory for unconfined/confined aquifers and its assumptions for steady-state groundwater flow to a fully penetrated well, which are supposed to be better respected in such flow conditions. Due to the unknown value of the radius of influence, an iterative procedure is proposed using the empirical Sichardt's formula. The whole process of calculation is represented as a flow-chart in Figure 4.

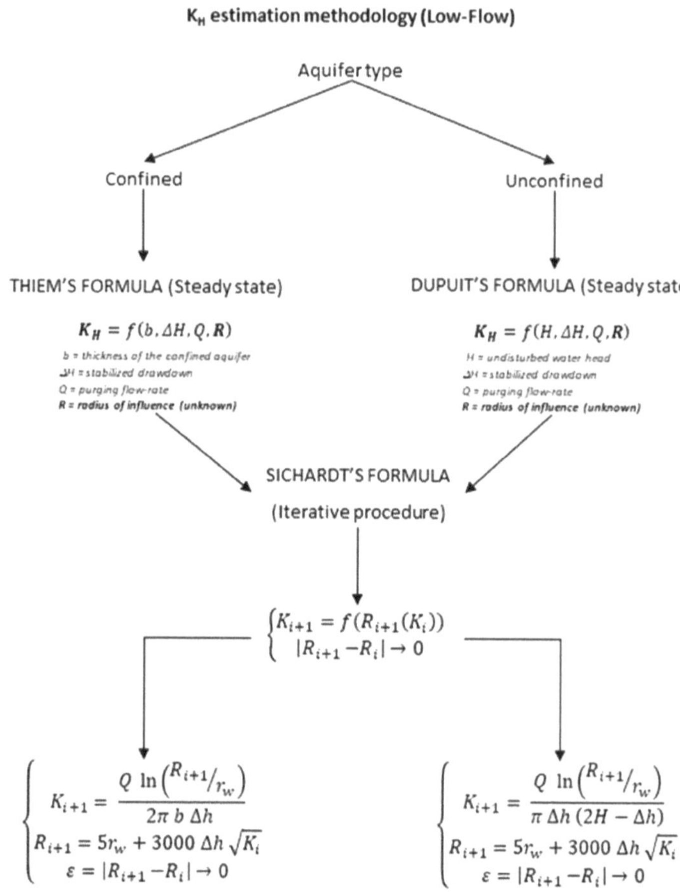

Figure 4. Flow-chart of the proposed methodology for K_H estimation with low-flow purging.

Vertical flows may be considered mostly null compared to the horizontal ones, so, more realistically, water can be assumed to move horizontally through the well screen length. The hypothesis becomes less true in the case of non-homogeneous and/or anisotropic aquifers with a series of overlapping low and high permeability layers, compromising K_H estimation results as well as the water sample characteristics. The aim of this study is to verify and to quantify this deviation of results coming from the proposed method, with the help of multiparameter well log measurements.

In Table 2, the main data referring to stratigraphy (top and bottom of the main geological layers), aquifer thickness and well construction information (well depth, screen length and position) are presented for each monitoring point of the well network. The main geological formations are pyroclastic soils, silty sands and silty clays, whose average K_H values are known from previous LeFranc tests carried on in the same site or from literature (pyroclastic soil: 2×10^{-5} m/s; silty sands: 1×10^{-6} m/s; and silty clays: 1×10^{-8} m/s).

Table 2. Static depth to water (DTW) measurements, screen position and length, and stratigraphic data referred to monitoring points.

Well ID	DTW	Well Screen			Unit	Top	Bottom	Thickness
-	(m)	Z_{TOP} (m)	Z_{BOT} (m)	L (m)	-	(m)	(m)	(m)
MW33	21.82	6	30	24	pyroclastic soil silty sand pyroclastic soil silty sand	3.7 13.4 20 28.8	13.4 20 28.8 30	9.7 6.6 8.8 1.2
MW37	16.18	6	36.5	30.5	pyroclastic soil	6.9	36.5	29.6
MW17	9.3	6	32	26	silty sand pyroclastic soil silty clay	9.2 12 31.1	12 31.1 32	2.8 19.1 0.9
MW15	15.5	20	39	19	silty clay pyroclastic soil silty sand	10.7 20 30	20 30 39	9.3 10 9
MW35	14.34	3	30	27	silty sand pyroclastic soil silty sand	4.6 15.4 25	15.4 25 28.3	10.8 9.6 3.3
MW20	13.92	3	25	22	silty sand pyroclastic soil silty sand	2.1 15 22	15 22 24.3	12.9 7 2.3
MW19	17.58	4	33.5	29.5	silty sand pyroclastic soil silty sand	9.2 17.3 30	17.3 30 33.5	8.1 12.7 3.5
MW31	10.56	2.8	22	19.2	pyroclastic soil silty sand pyroclastic soil	2.8 5 12.9	5 12.9 26.5	2.2 7.9 13.6
MW02	8.78	15	20	5	silty clay pyroclastic soil pyroclastic soil	11.5 13 15	13 15 20	1.5 2 5
MW01	13.6	20	24	4	silty clay pyroclastic soil	8 20	20 24	12 4
MW08	15.75	18	34.6	16.6	silty clay pyroclastic soil silty sand	11.2 19.5 32.2	19.5 32.2 34.6	8.3 12.7 2.4
MW21	1.38	5	16.3	11.3	silty clay silty sand	5.1 10.1	10.1 16.3	5 6.2
MW22	1.19	3	13.5	10.5	silty sand pyroclastic soil silty sand	4 7.8 11.5	7.8 11.5 13	3.8 3.7 1.5
MW36	14.52	10	33	19.5	silty sand pyroclastic soil silty sand	6 16.4 31.5	16.4 31.5 33	10.4 15.1 1.5
P43	15.87	12.7	33.5	20.8	silty sand pyroclastic soil silty sand silty clay	14.7 23 26.4 30	23 26.4 30 33.5	8.3 3.4 3.6 3.5
MWE6	13.61	12.5	32	19.5	silty sand pyroclastic soil silty sand silty clay	11.2 20.4 23.7 28.8	20.4 23.7 28.8 32	9.2 3.3 5.1 3.2
MW38	15.8	10	40	30	silty clay pyroclastic soil silty sand	16.1 20.1 32.7	20.1 32.7 40	4 12.6 7.3

Data presented in Table 2 show that, in the study area, the main aquifer is both in confined and unconfined conditions, requiring a double approach for the procedure of K_H estimation, as represented in Figure 4. The choice of the aquifer type is simply defined by comparing the depth to water (DTW) value and the well screen top (Z_{TOP}).

4. Results

The results obtained for saturated hydraulic conductivity coming from low-flow purging (K_{LOW}) are reported in Table 3. They have been compared with values coming from Le Franc tests and averaged along the depth (K_{LEF}). Both values, in fact, are referred to average values weighted on different layer lengths along the well screen and below the water table, as the total horizontal hydraulic conductivity of the aquifer is expressed by using the following equation:

$$K_H = \frac{\sum_i K_i h_i}{\sum_i h_i}$$

Table 3. Total K_H results obtained for low-flow purging (K_{LOW}) and LeFranc (K_{LEF}) for monitoring wells. The red values are for confined conditions while the black values are for unconfined conditions.

Well ID	K_{LEF} (m/s)	K_{LOW} (m/s)	Q (L/min)	Q (m^3/s)	ΔH (m)	Q/ΔH (m^2/s)	ΔH/H (−)
MW33	1.72×10^{-5}	1.01×10^{-5}	2	3.33×10^{-5}	0.28	1.19×10^{-4}	0.03
MW37	2.00×10^{-5}	1.74×10^{-5}	2	3.33×10^{-5}	0.06	5.56×10^{-4}	0.00
MW17	1.69×10^{-5}	1.22×10^{-5}	2	3.33×10^{-5}	0.06	5.56×10^{-4}	0.00
MW15	1.10×10^{-5}	1.37×10^{-5}	2	3.33×10^{-5}	0.15	2.22×10^{-4}	0.01
MW35	1.41×10^{-5}	8.35×10^{-6}	2	3.33×10^{-5}	0.19	1.75×10^{-4}	0.01
MW20	1.38×10^{-5}	1.55×10^{-5}	2	3.33×10^{-5}	0.12	2.78×10^{-4}	0.01
MW19	1.59×10^{-5}	9.56×10^{-6}	2	3.33×10^{-5}	0.13	2.56×10^{-4}	0.01
MW31	1.82×10^{-5}	1.09×10^{-5}	2	3.33×10^{-5}	0.14	2.38×10^{-4}	0.01
MW02	2.00×10^{-5}	2.98×10^{-5}	2	3.33×10^{-5}	0.15	2.22×10^{-4}	0.03
MW01	2.00×10^{-5}	1.60×10^{-5}	2	3.33×10^{-5}	0.40	8.33×10^{-5}	0.10
MW08	1.70×10^{-5}	5.15×10^{-6}	2	3.33×10^{-5}	0.22	1.52×10^{-4}	0.01
MW21	5.58×10^{-7}	3.94×10^{-7}	0.3	5.00×10^{-6}	0.65	7.69×10^{-6}	0.06
MW22	8.81×10^{-6}	1.46×10^{-5}	1	1.67×10^{-5}	0.05	3.33×10^{-4}	0.00
MW36	1.65×10^{-5}	2.08×10^{-5}	2	3.33×10^{-5}	0.06	5.56×10^{-4}	0.00
P43	4.47×10^{-6}	1.51×10^{-5}	2	3.33×10^{-5}	0.07	4.76×10^{-4}	0.00
MWE6	4.24×10^{-6}	2.49×10^{-5}	2	3.33×10^{-5}	0.04	8.33×10^{-4}	0.00
MW38	1.09×10^{-5}	5.31×10^{-6}	2	3.33×10^{-5}	0.47	7.09×10^{-5}	0.05

Four classes related to different correlation degrees have been defined, to quantify the number of wells in which K values, obtained with the proposed method, better match with the results of the LeFranc test, as follows:

- Low correlation for $|K_{LEF} - K_{LOW}| > 1$ order of magnitude (o.m.);
- Medium correlation for $0.5 < |K_{LEF} - K_{LOW}| < 1$ o.m;
- High correlation for $0.25 < |K_{LEF} - K_{LOW}| < 0.5$ o.m;
- Very high correlation for $|K_{LEF} - K_{LOW}| < 0.25$ o.m.

About 75% of the K values results showed a high correlation with the average LeFranc test results, whereas no results showed a low correlation, i.e., over 1 o.m. different (Figure 5a). In particular, poor results were obtained for MWE6, P43 and MW08 wells (medium class) and, to a lesser extent, for MW38 (high class) (Figure 5b).

Results from well multiparameter logs have been coupled with stratigraphy and low-flow purging data for each chemical-physical parameter measured. As expected, temperature (T) was not so useful to assess the heterogeneities, due to the slight changes of values along the screen as well as the almost immediate stabilization during the purging operation. Regarding dissolved oxygen (D.O.), the measured values were very different between the two different probes used for logs and purging, probably due to incorrect calibration; therefore, these values could not be compared.

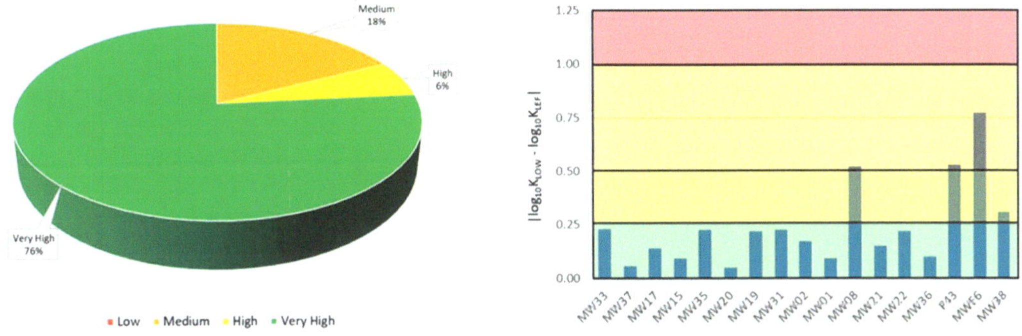

(**a**) (**b**)

Figure 5. Correlation results between permeability values estimated with the low-flow methodology (K_{LOW}) and values obtained with the LeFranc tests (K_{LEF}).

Hence, the base product of results obtained via this approach is presented in Figures 6–8, where coupled logs and purging data are referred to EC, and pH and ORP parameters, respectively, for MWE6, P43, MW08 and MW38 wells. These wells present high soil heterogeneity along the screen. Figures 9–11 present the results of wells with lower soil heterogeneity along the screen (MW19, MW33, MW35 and MW36).

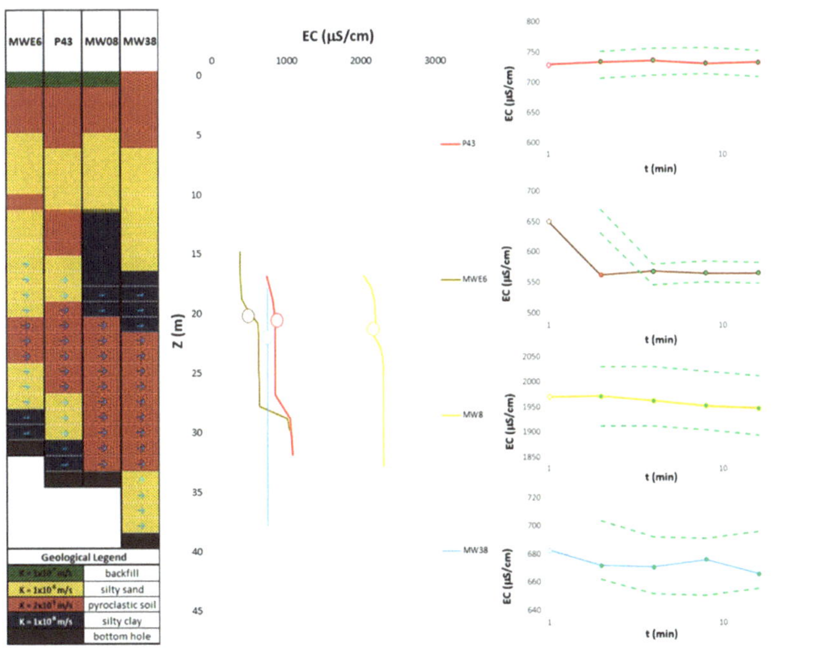

Figure 6. Layout of coupled log-purging (z-t) EC results for high-heterogeneity well types (MWE6, P43, MW8 and MW38).

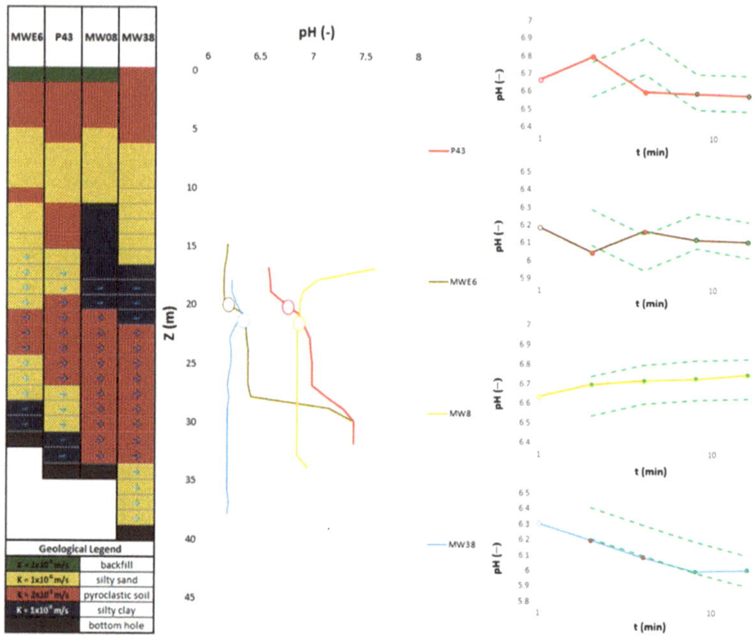

Figure 7. Layout of coupled log-purging (z-t) pH results for high-heterogeneity well types (MWE6, P43, MW8 and MW38).

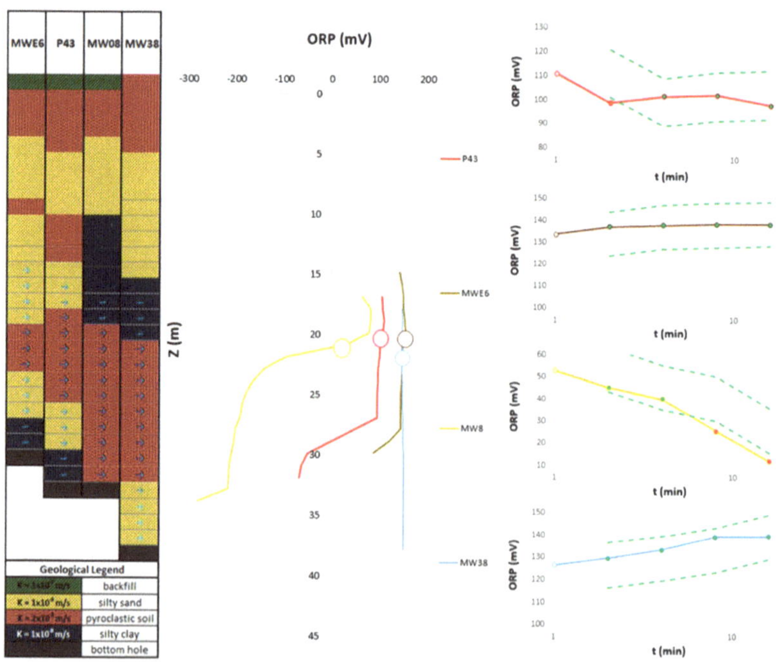

Figure 8. Layout of coupled log-purging (z-t) ORP results for high-heterogeneity well types (MWE6, P43, MW8 and MW38).

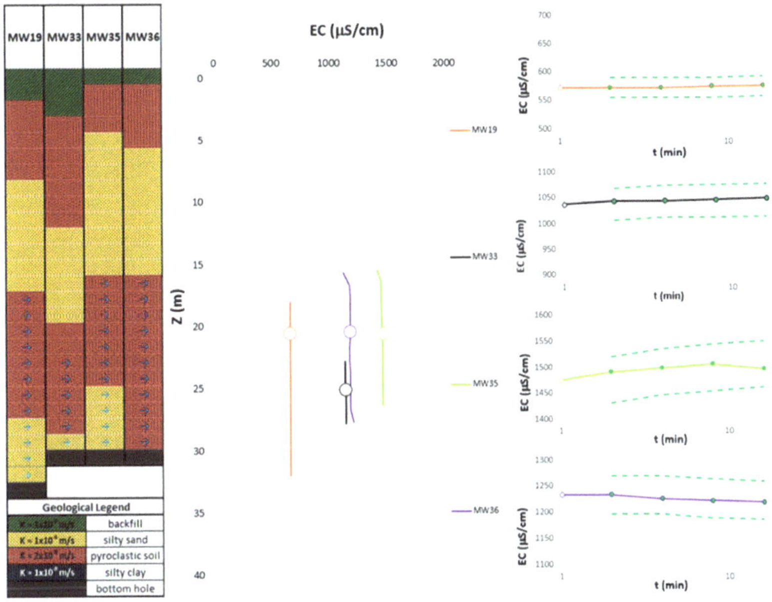

Figure 9. Layout of coupled log-purging (z-t) EC results for low-heterogeneity well types (MW19, MW33, MW35 and MW36).

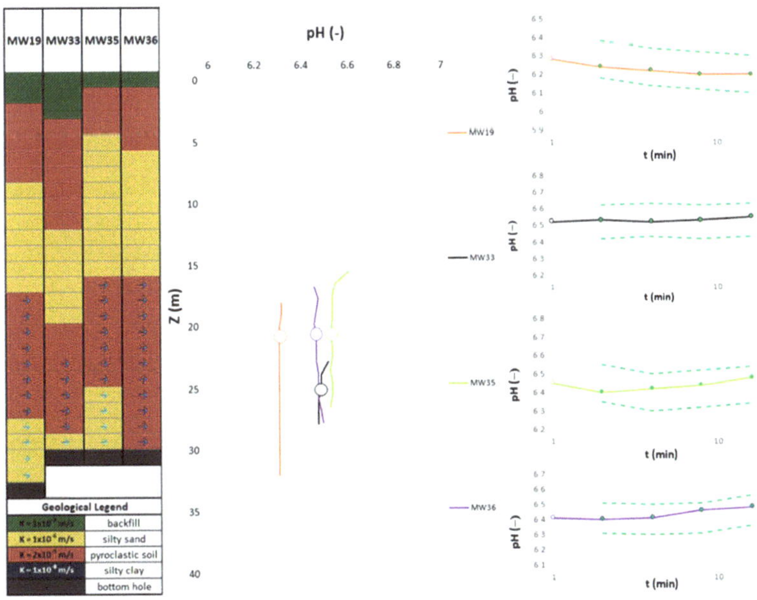

Figure 10. Layout of coupled log-purging (z-t) pH results for low-heterogeneity well types (MW19, MW33, MW35 and MW36).

In the left side of the layout, soil stratigraphy is represented for the different wells considered, as well as the screen length (grey horizontal lines) and the saturated zone (light blue arrows). Below this, a small geological legend contains soil types and permeability values of the geological formation (obtained using LeFranc tests). In the central chart, multiparameter log results are presented for wells taken into account.

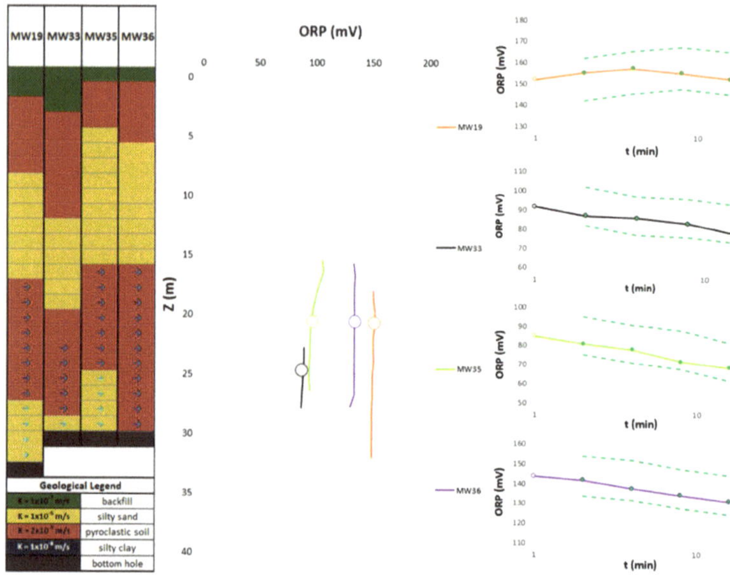

Figure 11. Layout of coupled log-purging (z-t) ORP results for low-heterogeneity well types (MW19, MW33, MW35 and MW36).

The pump depth (coloured circle on the log trend) was about 20 m for each well (except for MW36) and can be considered as the "starting point" for the following step of measuring parameters during low-flow purging over time (right side of Figures). In the purging step, the stabilization of parameters was assumed to be obtained only when differences between two subsequent readings did not exceed the specific thresholds proposed previously [39].

To graphically show this aspect, two dashed green lines, representing the maximum and minimum values of stabilization criterion, outline the range of acceptance. The points falling outside this range are red coloured and indicate no stabilization of the specific parameter considered.

5. Discussion

The obtained results clearly show that low-flow purging piezometric data, collected during the groundwater monitoring activities, can be used for the assessment of horizontal hydraulic conductivity of the aquifer investigated. Almost 75% of the estimations highly match with the permeability values obtained from the LeFranc test results, calculated as a weighting average along the saturated zone of the well screen. Regarding the remaining 25% of values, even if the values showed lower correlations (high and medium), the deviations from the LeFranc test values were never greater than one order of magnitude (o.m.). Lower correlated results of the proposed methodology have been obtained in MWE6, P43, MW08 and MW31 wells, in which soil heterogeneity due to low-permeability layers along the well screen was much more marked. This fact is reflected by a sharp decrease or increase in some physical-chemical parameters at the interface between the low- and high-permeability layers, suggesting that both drawdown and parameter stabilization may become much more difficult in these conditions, especially depending on the pump position. This is confirmed by unstable parameter values measured during purging operations (Figures 6–8), thereby highlighting that, even in high-permeability aquifers, as it is in this case study, the presence of thin low-permeability layers could affect the purging procedure in terms of time and costs. Instead, relatively homogeneous portions of the aquifer (maximum about 1 o.m. from layer to layer along the well screen) presented better results in terms of permeability (K_H) estimation and parameters' stabilization (Figures 9–11), showing almost vertical profiles of multi-parameter log results. Therefore, using these latter values, it is possible to correlate the geological heterogeneity with physical-chemical variation along the well screen, after which the pump position can be chosen to correctly intercept the main aquifer flow, sometimes not involving low-permeability layers and reducing the purging time a priori. In this study, pH was the most useful parameter for detecting layer K_H variations along the well screen, even if D.O. was not considered due to the issues related to instrumentation, as mentioned in Section 4.

Starting from the idea proposed by Harte et al. (2021) [40], who defined a heterogeneity factor (HF) for the use of his purging analyser tool (PAT), a new similar parameter has been defined in this study, considering not only the variability of permeability, but also the number of overlapping layers with different K_H and the saturated screen length. In this way, the heterogeneity is dependent on the monitoring well construction and the undisturbed groundwater level.

The heterogeneity factor along the well screen depends on the number of layers (i) and their permeability values (K), and it is defined as it follows:

$$F_H = \log_{10}\left(\frac{\max K_i}{\min K_i}\right) \cdot \frac{i}{L_{SS}}$$

where L_{SS} is the saturated screen length, calculated as:

$$L_{SS} = \begin{cases} L_T \text{ (Confined)} \\ Z_{BOT} - DTW \text{ (Unconfined)} \end{cases}$$

Hence, the L_{SS} is the total screen length (L_T) in case of confined aquifer conditions, and equal to the difference between the bottom well depth (Z_{BOT}) and depth to water (DTW) in case of unconfined conditions.

The F_H calculated for monitoring wells and the correlation between the results coming from the LeFranc tests and the proposed method was good (Figure 12). Whether the stratigraphy is known, the use of this factor prior to the purging operations, even using approximate K_H values of geological layers, might help to individuate those monitoring wells where permeability estimation and physical-chemical parameter stabilization will be difficult to achieve.

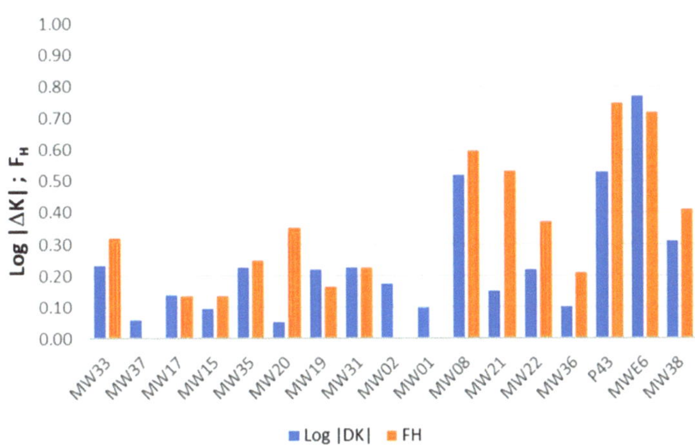

Figure 12. Graphical comparison between horizontal hydraulic conductivity precision and the proposed heterogeneity factor F_H for each monitoring well.

6. Conclusions

Starting from previous research works that demonstrated the reliability of low-flow purging data for the estimation of soil horizontal hydraulic conductivity (K_H), this study aimed to focus on eventual weaknesses of the methodology proposed by De Filippi et al. (2020) [15] in the case of high geological heterogeneity along the well screen. Coupling drawdowns and depth multiparameter values obtained with well logs, the results showed that the precision of the proposed method is somehow related to both the geological heterogeneity and well screen design. In high-permeability aquifers, where also purging rates higher than 1 L/min can be used, the presence of low-permeability layers could affect both quantitative and qualitative results. Therefore, in the case of several overlapping layers with different K_H values, the position of the pump plays a key role in the precision of this parameter estimation, as well as in reducing/raising purging time, with a cascading effect on permeability-weighted samples collected. Multiparameter well logs, carried out before the procedure, could be very helpful to detect the interface between high- and low-permeability layers, usually suggested by a sharp decrease/increase in some specific parameters along the well screen. In this specific study, temperature (T) was not useful for the mentioned purposes, whereas pH was the best parameter for detecting the layers' K_H variations along the well screen.

To support the procedure steps of low-flow purging and sampling, a heterogeneity factor, which takes into account both the soil and the screen characteristics as well as the geological features, has been defined. The use of this factor, prior to field work, can be helpful to know which monitoring wells require more attention, possibly with multiparameter log execution, or by being careful about the pumping depth that should be chosen based on well stratigraphy and screen knowledge.

Future studies will improve the knowledge of the K_H estimation technique using low-flow purging data as well as the parameters' behaviour during the procedure in different hydrogeological contexts, continuing with a quali-quantitative approach. The position of the pump along the well screen and its effects should be furthermore studied and verified with measurements during the field work. The objective will be providing depth profiles of parameters changing over time during the low-flow purging and comparing them with the results of a dedicated FEM model.

Author Contributions: Conceptualization, F.M.D.F. and G.S.; methodology, F.M.D.F.; validation, G.S.; writing—original draft preparation, F.M.D.F.; writing—review and editing, F.M.D.F. and G.S.; supervision, G.S.; project administration, G.S. All authors have read and agreed to the published version of the manuscript.

Funding: This research received no external funding.

Data Availability Statement: The datasets generated during and/or analysed during the current study are available from the corresponding author on reasonable request.

Acknowledgments: The authors want to kindly thank Indeco srl for the opportunity to perform the study on their site and for the useful trials for obtaining the results of this research. The authors are also very grateful to Emilio D'Amato for the support in well logs measurements and Lorenzo Del Grammastro and Davide Bramato for their patience and helpful work during the well purging activities.

Conflicts of Interest: The authors declare no conflict of interest.

References

1. Preziosi, E.; Petrangeli, A.B.; Giuliano, G. Tailoring Groundwater Quality Monitoring to Vulnerability: A GIS Procedure for Network Design. *Environ. Monit. Assess.* **2013**, *185*, 3759–3781. [CrossRef] [PubMed]
2. Chakraborty, A.; Prakash, O. Optimal Monitoring Locations for Identification of Ambivalent Characteristics of Groundwater Pollution Sources. *Environ. Monit. Assess.* **2022**, *194*, 664. [CrossRef]
3. Ding, X.H.; Feng, S.J.; Zheng, Q.T. Forward and Back Diffusion of Reactive Contaminants through Multi-Layer Low Permeability Sediments. *Water Res.* **2022**, *222*, 118925. [CrossRef]

4. Streeter, M.T.; Vogelgesang, J.; Schilling, K.E.; Burras, C.L. Use of High-Resolution Ground Conductivity Measurements for Denitrifying Conservation Practice Placement. *Environ. Monit. Assess.* **2022**, *194*, 784. [CrossRef] [PubMed]
5. Oli, I.C.; Opara, A.I.; Okeke, O.C.; Akaolisa, C.Z.; Akakuru, O.C.; Osi-Okeke, I.; Udeh, H.M. Evaluation of Aquifer Hydraulic Conductivity and Transmissivity of Ezza/Ikwo Area, Southeastern Nigeria, Using Pumping Test and Surficial Resistivity Techniques. *Environ. Monit. Assess.* **2022**, *194*, 719. [CrossRef]
6. Rivard, C.; Bordeleau, G.; Lavoie, D.; Lefebvre, R.; Malet, X. Can Groundwater Sampling Techniques Used in Monitoring Wells Influence Methane Concentrations and Isotopes? *Environ. Monit. Assess.* **2018**, *190*, 191. [CrossRef] [PubMed]
7. Taheri, K.; Missimer, T.M.; Amini, V.; Bahrami, J.; Omidipour, R. A GIS-Expert-Based Approach for Groundwater Quality Monitoring Network Design in an Alluvial Aquifer: A Case Study and a Practical Guide. *Environ. Monit. Assess.* **2020**, *192*, 684. [CrossRef]
8. Barcelona, M.J.; Varljen, M.D.; Puls, R.W.; Kaminski, D. Ground Water Purging and Sampling Methods: History vs. Hysteria. *Gr. Water Monit. Remediat.* **2005**, *25*, 52–62. [CrossRef]
9. Kaminski, D. *Low-Flow Groundwater Sampling: An Update on Proper Application and Use;* Present Organised by EnviroEquip; QED Environmental Systems Inc.: Ann Arbor, MI, USA; Oakland, CA, USA, 2006.
10. Robbins, G.; Higgins, M. Low Flow Sampling and Hydraulic Conductivity Analysis. Available online: https://www.epoc.org/resources/Documents/LowFlow6-5-18/Low%20Flow%20Lecture%20notes%206-5-18.pdf (accessed on 18 December 2022).
11. Weaver, J.M.C.; Cavé, L.; Talma, A.S. *Groundwater Sampling: A Comprehensive Guide for Sampling Methods;* Water Research Commission: Gezina, South Africa, 2007; ISBN 9781770055452.
12. Gomo, M.; Vermeulen, D.; Lourens, P. Groundwater Sampling: Flow-Through Bailer Passive Method Versus Conventional Purge Method. *Nat. Resour. Res.* **2018**, *27*, 51–65. [CrossRef]
13. Qi, S.; Hou, D.; Luo, J. Optimization of Groundwater Sampling Approach under Various Hydrogeological Conditions Using a Numerical Simulation Model. *J. Hydrol.* **2017**, *552*, 505–515. [CrossRef]
14. Wang, Y.; Hou, D.; Qi, S.; O'Connor, D.; Luo, J. High Stress Low-Flow (HSLF) Sampling: A Newly Proposed Groundwater Purge and Sampling Approach. *Sci. Total Environ.* **2019**, *664*, 127–132. [CrossRef] [PubMed]
15. De Filippi, F.M.; Iacurto, S.; Ferranti, F.; Sappa, G. Hydraulic Conductivity Estimation Using Low-Flow Purging Data Elaboration in Contaminated Sites. *Water* **2020**, *12*, 898. [CrossRef]
16. Sappa, G.; Filippi, F.M. De Quali-Quantitative Considerations on Low-Flow Well Purging and Sampling. *Acque Sotter. Ital. J. Groundw.* **2021**, *10*, 9–14. [CrossRef]
17. Molofsky, L.J.; Richardson, S.D.; Gorody, A.W.; Baldassare, F.; Connor, J.A.; McHugh, T.E.; Smith, A.P.; Wylie, A.S.; Wagner, T. Purging and Other Sampling Variables Affecting Dissolved Methane Concentration in Water Supply Wells. *Sci. Total Environ.* **2018**, *618*, 998–1007. [CrossRef] [PubMed]
18. Han, Y.L.; Tom Kuo, M.C.; Fan, K.C.; Lin, C.Y.; Chuang, W.J.; Yao, J.Y. Radon as a Complementary Well-Purging Indicator for Sampling Volatile Organic Compounds in a Petroleum-Contaminated Aquifer. *Gr. Water Monit. Remediat.* **2007**, *27*, 130–134. [CrossRef]
19. You, X.; Liu, S.; Dai, C.; Guo, Y.; Zhong, G.; Duan, Y. Contaminant Occurrence and Migration between High- and Low-Permeability Zones in Groundwater Systems: A Review. *Sci. Total Environ.* **2020**, *743*, 140703. [CrossRef]
20. Brooks, M.C.; Yarney, E.; Huang, J. Strategies for Managing Risk Due to Back Diffusion. *Groundw. Monit. Remediat.* **2021**, *41*, 76–98. [CrossRef]
21. Borden, R.C.; Cha, K.Y. Evaluating the Impact of Back Diffusion on Groundwater Cleanup Time. *J. Contam. Hydrol.* **2021**, *243*, 103889. [CrossRef]
22. Yang, M.; Annable, M.D.; Jawitz, J.W. Back Diffusion from Thin Low Permeability Zones. *Environ. Sci. Technol.* **2015**, *49*, 415–422. [CrossRef]
23. McMillan, L.A.; Rivett, M.O.; Tellam, J.H.; Dumble, P.; Sharp, H. Influence of Vertical Flows in Wells on Groundwater Sampling. *J. Contam. Hydrol.* **2014**, *169*, 50–61. [CrossRef]
24. Mcmillan, L.; Rivett, M.; Tellam, J.; Dumble, P.; Sharp, H. Water Quality Sample Origin in Wells under Ambient Vertical Flow Conditions. In Proceedings of the EGU General Assembly 2013, Vienna, Austria, 7–12 April 2013.
25. Vienken, T.; Kreck, M.; Hausmann, J.; Werban, U.; Dietrich, P. Innovative Strategies for High Resolution Site Characterization: Application to a Flood Plain. *Acque Sotter. Ital. J. Groundw.* **2014**, *3*, 7–14. [CrossRef]
26. Dijkshoorn, P.; Mori, P.; Zaffaroni, M.C. High Resolution Site Characterization as Key Element for Proper Design and Cost Estimation of Groundwater Remediation. *Acque Sotter. Ital. J. Groundw.* **2014**, *3*, 17–27. [CrossRef]
27. Liu, G.; Butler, J.J.; Reboulet, E.; Knobbe, S. Bestimmung von Vertikalprofilen Der Hydraulischen Durchlässigkeit Mit Direct Push-Methoden. *Grundwasser* **2012**, *17*, 19–29. [CrossRef]
28. Liu, G.; Butler, J.J.; Bohling, G.C.; Reboulet, E.; Knobbe, S.; Hyndman, D.W. A New Method for High-Resolution Characterization of Hydraulic Conductivity. *Water Resour. Res.* **2009**, *45*, 1–6. [CrossRef]
29. Aragon-Jose, A.T.; Robbins, G.A. Low-Flow Hydraulic Conductivity Tests at Wells That Cross the Water Table. *Ground Water* **2011**, *49*, 426–431. [CrossRef]
30. Robbins, G.A.; Aragon-Jose, A.T.; Romero, A. Determining Hydraulic Conductivity Using Pumping Data from Low-Flow Sampling. *Ground Water* **2009**, *47*, 271–286. [CrossRef]

31. Harte, P.T. In-Well Time-of-Travel Approach to Evaluate Optimal Purge Duration during Low-Flow Sampling of Monitoring Wells. *Environ. Earth Sci.* **2017**, *76*, 251. [CrossRef]
32. Mcmillan, L.A.; Michael, O.; Tellam, J.H.; Birmingham, B. Groundwater Quality Sampling at Contaminated Sites: The Long and the Short of It. Available online: https://www.envirotech-online.com/article/water-wastewater/9/in-situ-inc/groundwater-quality-sampling-at-contaminated-sites-the-long-and-the-short-of-it/1827 (accessed on 18 December 2022).
33. Sappa, G.; Grelle, G.; Ferranti, F.; De Filippi, F.M. Temperature Logs to Evaluate Groundwater-Surface Water Interaction (Sabato River at Avellino, Campania). *Rend. Online Soc. Geol. Ital.* **2019**, *47*, 108–112. [CrossRef]
34. Vitale, S.A.; Robbins, G.A. Characterizing Groundwater Flow in Monitoring Wells by Altering Dissolved Oxygen. *Groundw. Monit. Remediat.* **2016**, *36*, 59–67. [CrossRef]
35. Sappa, G.; Rossi, M.; Coviello, M. Effetti Ambientali Del Sovrasfruttamento Degli Acquiferi Della Pianura Pontina (Lazio). In Proceedings of the Aquifer Vulnerability and Risk 2nd International Workshop, Parma, Italy, 14–16 September 2005; pp. 1–16.
36. Cardello, G.L.; Vico, G.; Consorti, L.; Sabbatino, M.; Carminati, E.; Doglioni, C. Constraining the Passive to Active Margin Tectonics of the Internal Central Apennines: Insights from Biostratigraphy, Structural, and Seismic Analysis. *Geosciences* **2021**, *11*, 160. [CrossRef]
37. Alessandri, L.; Cardello, G.L.; Attema, P.A.J.; Baiocchi, V.; De Angelis, F.; Del Pizzo, S.; Di Ciaccio, F.; Fiorillo, A.; Gatta, M.; Monti, F.; et al. Reconstructing the Late Pleistocene—Anthropocene Interaction between the Neotectonic and Archaeological Landscape Evolution in the Apennines (La Sassa Cave, Italy). *Quat. Sci. Rev.* **2021**, *265*, 107067. [CrossRef]
38. Boni, C.; Bono, P.; Calderoni, G.; Lombardi, S.; Turi, B. Indagine idrogeologica e geochimica sui rapporti tra ciclo carsico e circuito idrotermale nella pianura pontina (Lazio meridionale). *Geol. App. Idrog.* **1980**, *15*, 203–247.
39. US EPA. Environmental protection agency region i low stress (low flow) purging and sampling procedure for the collection of from monitoring. *Environ. Prot.* **1996**, *1*, 1–13.
40. Harte, P.T.; Perina, T.; Becher, K.; Levine, H.; Rojas-Mickelson, D.; Walther, L.; Brown, A. *Evaluation and Application of the Purge Analyzer Tool (PAT) to Determine In-Well Flow and Purge Criteria for Sampling Monitoring Wells at the Stringfellow Superfund Site in Jurupa Valley, California, in 2017*; U.S. Geological Survey: Reston, VA, USA, 2021. [CrossRef]

Disclaimer/Publisher's Note: The statements, opinions and data contained in all publications are solely those of the individual author(s) and contributor(s) and not of MDPI and/or the editor(s). MDPI and/or the editor(s) disclaim responsibility for any injury to people or property resulting from any ideas, methods, instructions or products referred to in the content.

hydrology

MDPI

Article

Characterizing Trace Metal Contamination and Partitioning in the Rivers and Sediments of Western Europe Watersheds

Aline Grard and Jean-François Deliège *

FOCUS Research Unit—Freshwater and OCeanic Sciences Unit of Research, PeGIRE Laboratory, Aquapôle Research Center, Faculty of Sciences, Department of Biology Ecology and Evolution, University of Liège, Allée de la Découverte 11, Building B53, Quartier Polytech 1, 4000 Liège, Belgium
* Correspondence: jfdeliege@uliege.be; Tel.: +32-4-366-23-53

Abstract: Adsorption and desorption processes occurring on suspended and bed sediments were studied in two datasets from western Europe watersheds (Meuse and Mosel). Copper and zinc dissolved and total concentrations, total suspended sediment concentrations, mass concentrations, and grain sizes were analyzed. Four classes of mineral particle size were determined. Grain size distribution had to be considered in order to assess the trace metal particulate phase in the water column. The partitioning coefficients of trace metals between the dissolved and particulate phases were calculated. The objective of this study was to improve the description of the processes involved in the transportation and fate of trace metals in river aquatic ecosystems. Useful data for future modelling, management and contamination assessment of river sediments were provided. As it is confirmed by a literature review, the copper and zinc partitioning coefficients calculated in this study are reliable. The knowledge related to copper and zinc (e.g., partitioning coefficients) will allow us to begin investigations into environmental modelling. This modelling will allow us to consider new sorption processes and better describe trace metal and sediment fates as well as pressure–impact relationships.

Keywords: trace metals; sediment contamination; watershed; partitioning coefficient; water quality; modelling

Citation: Grard, A.; Deliège, J.-F. Characterizing Trace Metal Contamination and Partitioning in the Rivers and Sediments of Western Europe Watersheds. *Hydrology* **2023**, *10*, 51. https://doi.org/10.3390/hydrology10020051

Academic Editor: Meysam Vadiati

Received: 11 January 2023
Revised: 10 February 2023
Accepted: 13 February 2023
Published: 16 February 2023

Copyright: © 2023 by the authors. Licensee MDPI, Basel, Switzerland. This article is an open access article distributed under the terms and conditions of the Creative Commons Attribution (CC BY) license (https://creativecommons.org/licenses/by/4.0/).

1. Introduction

The sediments in large rivers are often considered as a quantitative issue associated with erosion, hydromorphology, etc. However, these sediments are carrying organic and mineral pollutants, so that they can be sources of pollution [1]. The description of the processes that determine the transportation of particulates and dissolved matter in the hydrographic network is of primary importance to characterize pollution emerging from elements with high speciation. A micropollutant is defined as a substance detectable in the environment at very low concentrations (microgram per liter or even nanogram per liter). Very low concentrations of these substances can cause negative effects on living organisms depending on their toxicity, persistence and/or bioaccumulation [2]. Micropollutants are a major issue in the management of surface water quality around the world [3–5]. The biogeochemical processes and the loads from the watershed that modulate micropollutant concentrations have received considerable attention in recent literature. Mathematical modelling helps to simulate the fate and transportation of micropollutants and supports water quality management decisions [3].

Several international environmental organizations consider sediment contamination as a major risk in aquatic environments because heavy metals remain in the environment as micropollutants of high toxicity [1].

The chemical forms (free ions, bound complexes, etc.) of trace metals are controlled by the physicochemical and biological characteristics of the system. Their mobility, transportation and partitioning in natural river systems (including water columns and sediments)

involve very complex processes that depend on the physicochemical properties of the contaminants, water and sediments [6]. Major variations of these forms and concentrations are due to the biogeochemical processes which drive speciation reactions. The dissolved concentrations of these elements depend on their interactions with solid components in the sediment, trough adsorption/desorption and dissolution/precipitation reactions coupled to complexation, acidification or redox reactions.

The chemical speciation is reported as the distribution of an element amongst defined chemical species in a system [7]. Trace metals are highly persistent in sediments since they cannot be degraded. Once they have entered an aquatic ecosystem, trace metals are involved in biogeochemical processes and distributed under many different chemical forms. Their reactivity is (i) determined by their interaction with a large variety of inorganic and organic compounds and (ii) controlled by interdependent acid–base, redox, complexation, adsorption, and precipitation reactions [8]. Trace metals interact with a broad spectrum of biotic and abiotic components via dynamic interrelated processes. As a result, they exist in different forms such as free uncomplexed ions, bound to small inorganic or organic ligands, particles, or inorganic or organic colloids, with different reactivity and biological availability [8].

This study focused on Cu and Zn (dissolved and inorganic forms) while interacting with water columns and sediments. The characterization of the dissolved and particulate forms is also a challenge [9]. Attention has to be paid to the solid–liquid partitioning coefficient of trace metals [10]. The partitioning of trace metals between the particulate concentration and dissolved concentration in the water column, expressed as the partition coefficient (Kd), depends on the chemical processes affecting the trace metal phases. Most elements have been documented [11].

The form of a trace metal is one of the major factors for the characterization of some specific pollution. It influences trace metal behavior, ability to sediment, and downstream transportation (depending on the flow velocity and other hydrodynamic conditions of the watercourse).

Particles in a river can be divided into two parts, permanently suspended particles and settleable particles, depending on their grain size and density as well as the flow conditions. The bed sediment is an integral part of a river which also constitutes a sink for metals [1]. After sedimentation on the riverbed, the particles can return to suspension in the water column and may lead to the redistribution of trace metals between the particulate and dissolved phases [10]. Particularly, it is believed that fine sediments of an aquatic system, such as silts and clays, are of extreme importance in the trace metal transportation due to their relative high surface area and their physicochemical properties facilitating then their capacity to adsorb substances [1]. Suspended particles in the water column are usually in a temporary state of exchange with the bed sediment reservoir. On one hand, the hydrodynamic processes of sedimentation, resuspension and generation of particles, coupled with their high reactivity, play a key role in the availability, transportation and fate of metals in the aquatic environment. On another hand, variable physicochemical characteristics, adsorption and desorption reactions and mixing processes influence the particulate metal concentrations [12,13].

The goals of this paper were to (1) improve the description of the processes involved in the transportation and fate of trace metals in the river ecosystem, (2) estimate the partitioning coefficient of trace metals between the dissolved and particulate phases, and (3) provide useful data for water quality management [14], i.e., European Water Framework Directive 2000/60/EC [15], and the assessment of river sediments.

We decided to investigate two large basins in Western Europe: the Meuse District and the Mosel Basin (tributary of the Rhine). They are representative of the European aquatic ecosystem by their typology, hydromorphology and associated pressures.

2. Materials and Methods

2.1. Study Areas

The first investigated basin was the Meuse International District, located in Western Europe (Figure 1).

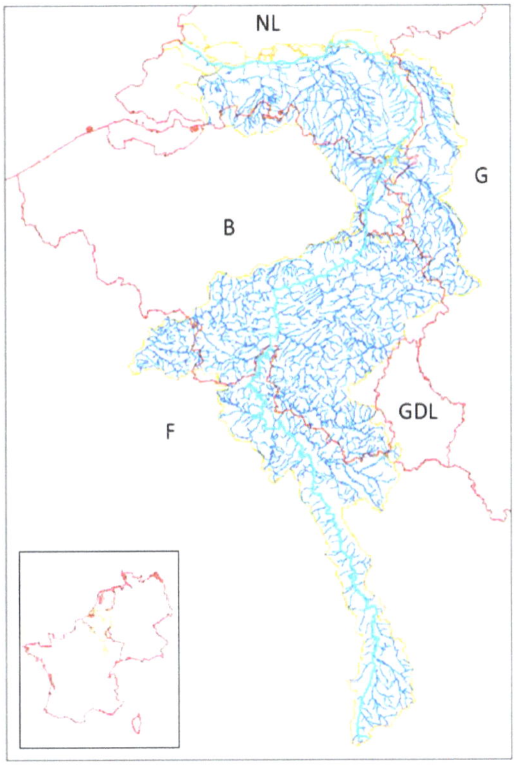

Figure 1. Map of the International River Basin District of the Meuse (France (F), Belgium (B), GD Luxembourg (GDL), Germany (G), the Netherlands (NL)). The Meuse River is marked in a light blueline, deep blue lines represent tributaries, red lines are the borders and yellow line is the International River Basin District boundary.

The Meuse River, which flows from the South to the North, has its source in France, near the Plateau de Langres. The Meuse length is 905 km, and its entire catchment has an area of 34,564 km^2.

The study site was the Eijsden measurement station, located at the Belgian–Dutch border, at the 617 km point. At this point, the Meuse has a watershed of 20,554 km^2 and a characteristic low flow (i.e. daily flow not reached more than 10 days per year, the 2.7 percentile, 10/365), estimated to be 54.7 m^3/s for the year 2021 [16]. The trace metals studied in this paper were copper and zinc. The natural background concentrations are generally about 0.06 μg/L [17] and 0.7–10 μL/L [18] in unpolluted river water, respectively, depending on the lithology of the bedrocks. The Cu and Zn average concentrations measured at Eijsden (~3 μg/L and ~20 μg/L, respectively, for the period 2002–2020) were higher than the lower thresholds provided from anthropogenic sources [19]. These values allowed us to estimate the flux, at this point, of Cu about 5175 kg/an and Zn about 34,500 kg/an.

The French part of the watershed is managed by the Rhine–Meuse Water Agency and the Artois–Picardie Water Agency. The Meuse crosses the Walloon region (in Belgium) and the Netherlands, to flow towards the North Sea, south of Rotterdam. Some tributaries of the Meuse have their sources in the Grand Duchy of Luxembourg, in France (Artois-Picardie Water Agency), in the Flemish region (Belgium) and in Germany.

The French part of the Meuse River basin is essentially agricultural and forest. The Belgian part of the basin becomes more urban and industrial, up to the Netherlands [20–24]. The Meuse River is then canalized and used for navigation and goods transportation.

The Mosel, the second investigated watershed, is a part of the Mosel–Saar Transnational Basin, located in Western Europe (Figure 2).

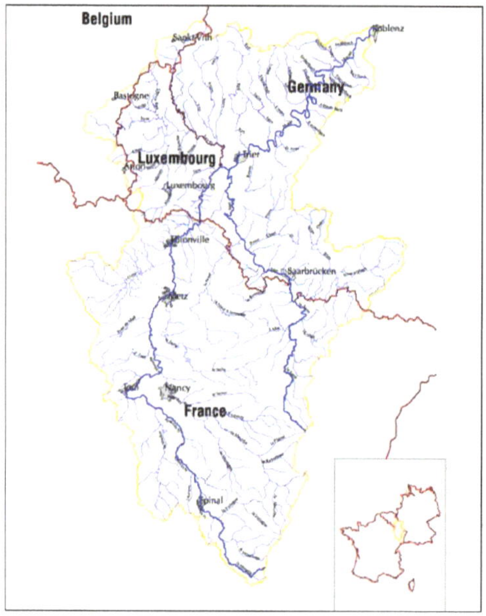

Figure 2. Map of the Mosel–Saar Transnational Basin (France, GD Luxembourg and Germany). The Mosel (at the West) and Saar (at the East) rivers are marked in dark bluelines, light blue lines represent tributaries, red lines are the borders and yellow line is the International Mosel-Saar Basin boundary.

The Mosel has its source in France where it is managed by the Rhine-Meuse Water Agency. Its length is 544 km long and its entire catchment has an area of 28,286 km^2. The study reach was 314 km upstream located in France. At the French outlet, the Mosel has a watershed of 11,280 km^2. Low flow (i.e. monthly minimum flow reached once every five years; statistical flow that gives information on the severity of the low flow) of the Mosel at this point is about 31.6 m^3/s [25]. The Cu and Zn annual fluxes are estimated at 13,000 kg/an and 30,000 kg/an, respectively [26].

The Mosel streams towards the GD Luxembourg and Germany, and flows into the Rhine at Koblenz (Germany). The Mosel is canalized from Nancy (150 km from its source). The Mosel watershed is mainly agricultural and forest. Urbanized and industrial areas are developed on the downstream part of the basin [27,28].

2.2. Data

The Meuse and the Mosel are well-instrumented watercourses. They provided the two separately datasets used in this study.

In the Meuse District, the Eijsden measurement station monitors daily water quality (Supplementary Materials, Figure S1). These measurements are freely downloadable on the website of the Rijkswaterstaat, Ministerie van Infrastructuur en Waterstaat [19].

Decades of weekly measurements of the dissolved and total concentrations of several substances are also downloadable. Total suspended sediment (SS) concentrations, mass concentrations of total SS, grain size (percentages of SS < 2 μm, < 16 μm and < 63 μm) are available.

The dissolved and total concentrations of copper and zinc decreased in the Meuse between the two periods illustrated in Figure 3. For the period 2002–2012, the values of total and dissolved copper concentrations ranged from 0.8 to 98 μg/L and 0.78 to 16 μg/L, respectively. The median value of the total and dissolved copper concentrations were 2.73 μg/L and 1.57 μg/L, respectively, and the average value of the total and dissolved copper concentrations were 3.70 μg/L and 1.74 μg/L, respectively. For the same period, the values of the total and dissolved zinc concentrations ranged from 4.7 to 533.0 μg/L and 1.0 to 110.0 μg/L, respectively. The median value of the total and dissolved zinc concentrations were 18 μg/L and 7 μg/L, respectively, and the average value of the total and dissolved zinc concentrations were 26.38 μg/L and 9.53 μg/L, respectively. For the period 2013–2020, the values of the total and dissolved copper concentrations ranged from 1.08 to 34.1 μg/L and 0.77 to 6.21 μg/L, respectively. The median value of the total and dissolved copper concentrations were 1.95 μg/L and 1.37 μg/L, respectively, and the average value of the total and dissolved copper concentrations were 2.52 μg/L and 1.44 μg/L, respectively. For the same period, the values of the total and dissolved zinc concentrations ranged from 2.09 to 143.0 μg/L and 1.42 to 18.1 μg/L, respectively. The median value of the total and dissolved zinc concentrations were 9.76 μg/L and 4.17 μg/L, respectively, and the average value of the total and dissolved zinc concentrations were 14.35 μg/L and 4.69 μg/L, respectively. This represented an average reduction in the total copper and zinc of 31.9% and 45.6%, respectively.

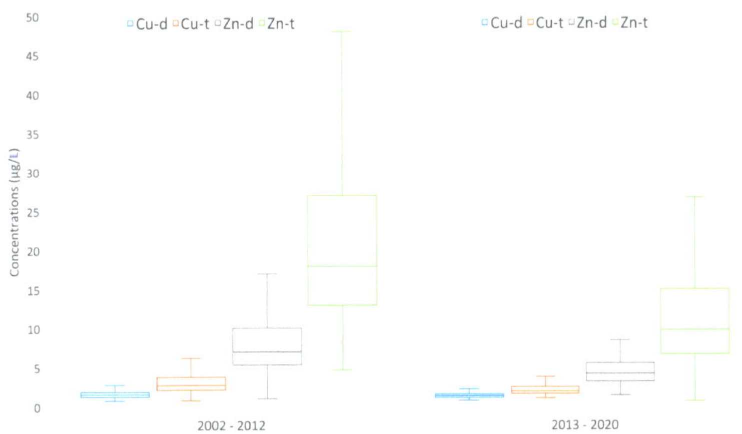

Figure 3. Copper and zinc dissolved and total concentrations at Eijsden station (Meuse) over the period 2002–2012 (left part of the figure) and over the period 2013–2020 (right part of the figure).

On the French Mosel watershed, the water quality is monitored by the Rhine–Meuse Water Agency and the measurements are freely downloadable on the website of the Agency [29]. The dissolved and total copper and zinc concentrations are available for the years 2012–2019 from about 20 stations of the monitoring network. Adsorbed copper and zinc mass concentrations on bed sediment (particles < 2 mm) and on suspended sediments (particles < 2 mm) are available for the years 2012–2017 (no data were available between 2017 and 2019) from six stations of the monitoring network.

For both investigated basins, in the "Inventories of emissions, losses and discharges in the Rhine–Meuse basin" [26] and "Etat des Lieux" [30], copper and zinc are inventoried as pollutant substances with a high toxic risk and their fluxes are among the ten largest emitted flows in the Rhine–Meuse basin. In this study, we decided to focus on copper and zinc for the Meuse and Mosel basins.

2.3. Methodology

Trace metals and sediments represent a major threat to the quality of river surface waters.

The chemical speciation of trace metals plays an important role in geochemistry and determines to a large extent their mobility, toxicity and bioavailability [31]. As it may vary continuously in space and time, measurements of the total metal concentrations are not sufficient to entirely describe the fate, biological effects and environmental impact of trace metals. The measurement of relevant specific metal species or groups of homologous metal species and their variation as a function of time is essential [8].

The present study focused on the complexation and adsorption processes of two metals (copper and zinc) to inorganic particles. The objective was to establish geochemical coefficients characterizing Cu and Zn trace metals in a river ecosystem. In order to be used in model working at the watershed scale, partitioning coefficients were calculated, considering trace metals as inert tracers and simulating them by applying mass–balance relationships in the rivers [32]. The trace metals behavior is strongly linked to the particles. Sedimentation and resuspension are two ubiquitous processes in a natural river. These processes may lead to the redistribution of trace metal concentrations between their particulate and dissolved phases in the water column and on the river bed [10,13,33], acting like a sink or a source of contamination [34].

The data acquirement comprised raw data extraction of targeted trace metals from recent monitored years. Data importation into spreadsheets enabled scientific consistency to check using algorithms developed on our own digital platforms: the consistency of the units, exclusion of periods without values, statistical processing, etc.

The first method was adapted to the large weekly dataset available at Eijsden. Copper and zinc mass concentrations of the suspended sediment (SS) were calculated by grain size, knowing the percentage of each grain size class and the total metal mass concentrations. These mass concentrations were used to calculate partitioning coefficient by grain size, divided by the dissolved metal concentrations. Based on these weekly results, statistics (median and average values) were then calculated to estimate the copper and zinc partitioning coefficients for the two selected periods (2002–2012 and 2013–2020). Note that the logarithmic values of the partitioning coefficients were compared to the literature data.

The second method was adapted to the dataset available for the Mosel River. Copper and zinc partitioning coefficients were calculated on one grain size class, less than 2 mm, for SS and BS, using monthly dissolved and mass concentrations. Statistics (median and average values) were calculated to estimate the copper and zinc partitioning coefficients of the SS (period 2012–2019) and bed sediment (BS) (period 2012–2017).

Four classes of fine mineral suspended sediments were considered (Table 1), according to Fournier et al. [35], who compiled data from Wenthworth [36], Friedman and Sanders [37], and Blott and Pye [38]:

Table 1. Classification of the mineral suspended sediments.

Classes	Diameters Range (μm)	Average Diameters (μm)
Clays	<2	1
Fine Silts	$2 \rightarrow 16$	9
Coarse Silts	$16 \rightarrow 63$	40
Sand	$63 \rightarrow 250$	157

Generally, the concentrations of trace metals in sediments (SS and BS) are mainly influenced by fine particles [10]. The smaller the particle size, the higher the surface area/volume ratio (and also the surface area/weight ratio): the concentration of metals in sediments (expressed in mg/kg) is therefore higher for small particles.

Metal partitioning coefficients, Kd, have been widely used to describe the distribution of metals between solid and solution phases in a number of fate and transportation models [32,39–41]. In this study, sorption processes were represented by a partitioning coefficient that is expressed as follows:

$$Kd = Css / Cw$$

where Css is the metal particulate concentration per mass (mg/kg) and Cw is the dissolved metal concentration per volume (mg/m^3).

The dissolved metals in the water column are transported by the advection and dispersion processes. The metals in the particulate phase are also governed by sediment dynamics (transport, settling processes, etc.). Metals behavior in water is strongly influenced by sorption kinetics. Regarding the spatio-temporal scales of the involved processes and parameters (temperature, pH, etc.), we can assume pseudo-steady-state conditions while calculating Kd coefficients [1,42].

3. Results and Discussion

3.1. Spatial Variation of Dissolved and Particulate Metal Concentrations in the Meuse and Mosel Watersheds

Beyond the classical parameters, trace metals are among the pollutants that should be monitored in order to obtain a coherent and comprehensive overview of the quality status of an aquatic system [13]. The dissolved fraction of metals is usually lower in natural water but is believed toxic to aquatic organisms and humans. They are also easily absorbed by SS, so attention has been paid to the solid–liquid partitioning coefficient of trace metals [10].

3.1.1. The Meuse Watershed

Fluxes of macro- and micropollutants were estimated at the outlet of the Meuse River at the French border during the inventory of environmental data [30]. The dissolved/total ratio, estimated from measurements at the monitoring stations (period 2017–2019), of Cu and Zn were 0.84 and 0.50, respectively (at Givet, 485 km on the Meuse River).

At Eijsden (Meuse River, 617 km), the period containing the most measurements covers years from 2002 to 2012. Then, this set of data was used in this study. There are measurements of Cu and Zn mass concentrations on the total SS, total SS concentrations, and percentages of SS < 2 μm, < 16 μm and < 63 μm.

Figures 4 and 5 show the Cu and Zn mass concentrations at Eijsden (Meuse River) on the SS for particles of average diameter equal to 1 μm (clays), 9 μm (fine silts) and 40 μm (coarse silts).

At Eijsden, for the period 2002–2012, the values of the copper mass concentrations on clays and fine silts ranged from 77.19 to 1898.92 mg/kg and 8.58 to 210.99 mg/kg, respectively. The median values of the copper mass concentrations in clays and fine silts were 361.60 mg/kg and 40.18 mg/kg, respectively, and the average values of the copper mass concentrations on clays and fine silts were 449.63 mg/kg and 49.96 mg/kg, respectively. The values of the zinc mass concentrations in clays and fine silts ranged from 847.78 to 14,588.48 mg/kg and 94.20 to 1620.94 mg/kg, respectively. The median values of the zinc mass concentrations on clays and fine silts were 2981.39 mg/kg and 331.27 mg/kg, respectively, and the average values of the zinc mass concentrations on clays and fine silts were, respectively 3376.15 mg/kg and 375.13 mg/kg.

The copper mass concentrations decreased after the end of 2010 (Supplementary Materials, Figure S2). The same observation was noted for zinc concentrations. These observations were potentially due to the reduction in metal release into the environment. Nevertheless, it is necessary to strengthen this study on the reduction in trace metal releases into surface water in occidental regions to assert this hypothesis.

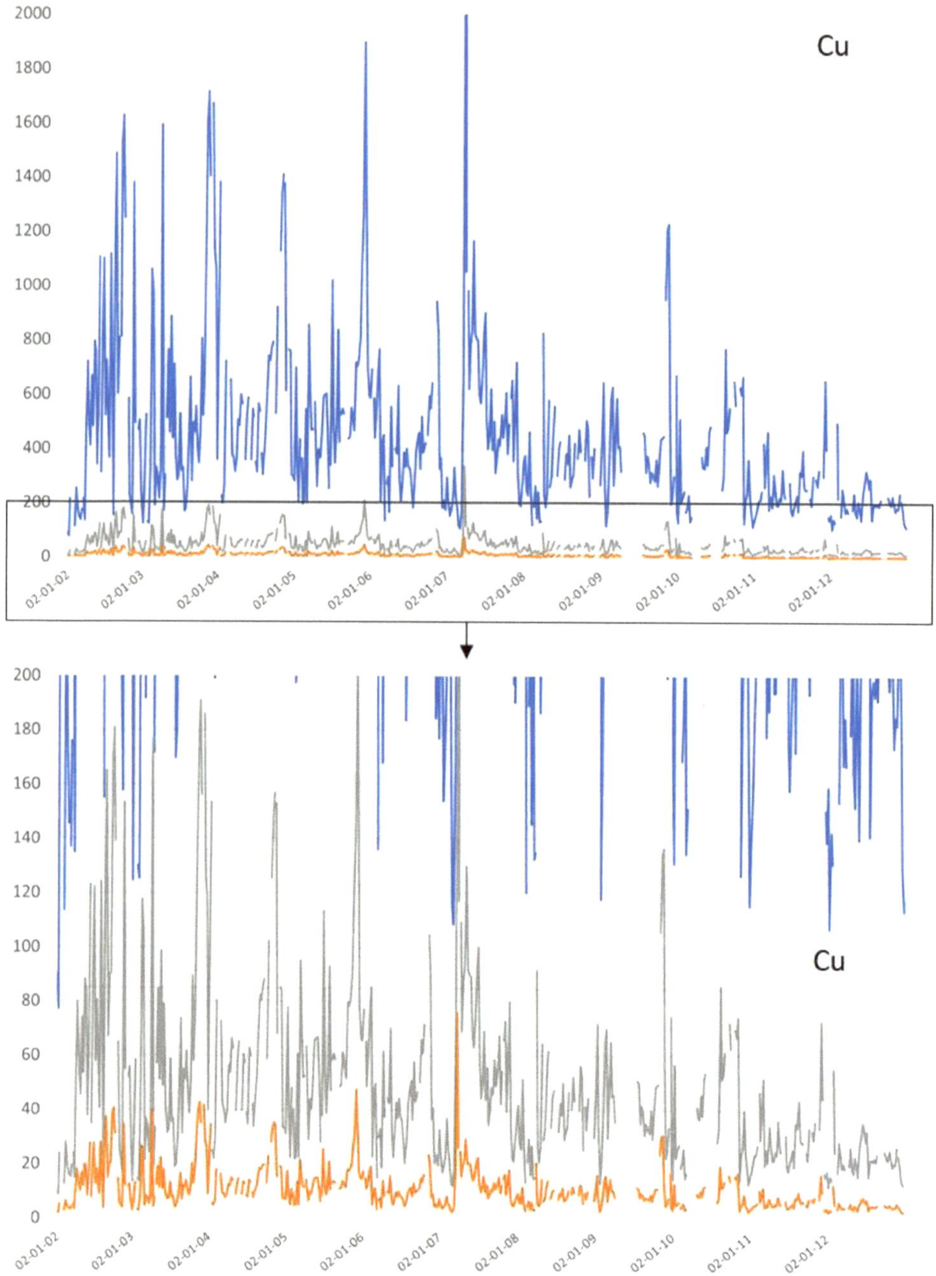

Figure 4. Copper mass concentrations (mg/kg) at Eijsden (Meuse river) of the suspended sediment:clays = particles < 2 µm (average diameter = 1 µm), fine silts = 2 µm < particles < 16 µm (average diameter = 9 µm) and coarse silts = 16 µm < particles < 63 µm (average diameter = 40 µm). The top figure's y-axis scale from 0 to 2000 mg/kg; the bottom figure's y-axis scale from 0 to 200 mg/kg.

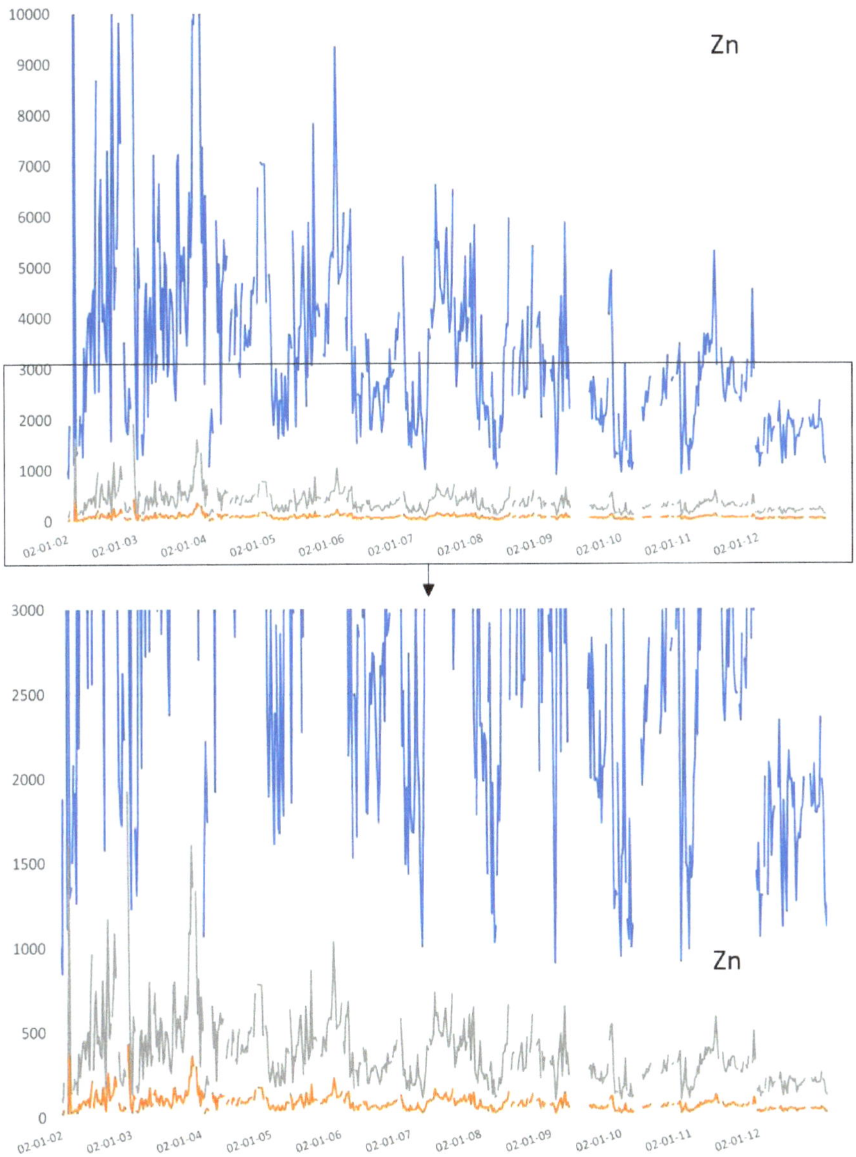

Figure 5. Zinc mass concentrations (mg/kg) at Eijsden (Meuse river) of the suspended sediment: clays = particles < 2 μm (average diameter = 1 μm), fine silts = 2 μm < particles < 16 μm (average diameter = 9 μm) and coarse silts = 16 μm < particles < 63 μm (average diameter = 40 μm). The top figure's y-axis scale from 0 to 10,000 mg/kg; the bottom figure's y-axis scale from 0 to 3000 mg/kg.

3.1.2. The Mosel Watershed

Fluxes of micropollutants were estimated at the outlet of the Mosel River in France (at Sierck, 309.4 km along the river) during the inventory of environmental data [26]. The dissolved/total ratios, estimated from measurements at monitoring stations (period 2017–2019), for Cu and Zn were 0.78 and 0.57, respectively.

Figure 6 shows the dissolved and total Cu concentrations (in µg/L) in the water column along the Mosel River. For the years 2012–2019, the values of the dissolved Cu concentrations ranged from 0.19 to 9.69 µg/L. The median value of the dissolved Cu concentrations was 1.0 µg/L and the average value of the dissolved Cu concentrations was 1.88 µg/L. Over the period 2017–2019, the values of the total Cu concentrations ranged from 0.42 to 4.48 µg/L. The median value of the total Cu concentrations was 1.08 µg/L and the average value of the total Cu concentrations was 1.71 µg/L. This figure shows the adsorbed Cu mass concentrations on the bed sediment (particles < 2 mm) and on the suspended sediment (particles < 2 mm) along the Mosel River, for the years 2012 to 2017. The values of the adsorbed Cu mass concentrations on the bed sediment and suspended sediment ranged from 7.27 to 256.80 mg/kg and 12.0 to 650.0 mg/kg, respectively. The median value of the adsorbed Cu mass concentrations on the bed sediment was 31.1 mg/kg. The median value of the adsorbed Cu mass concentrations on the suspended sediment was 69.0 mg/kg. The average value of the adsorbed Cu mass concentrations on the bed sediment was 47.63 mg/kg. The average value of the adsorbed Cu mass concentrations on the suspended sediment was 148.47 mg/kg.

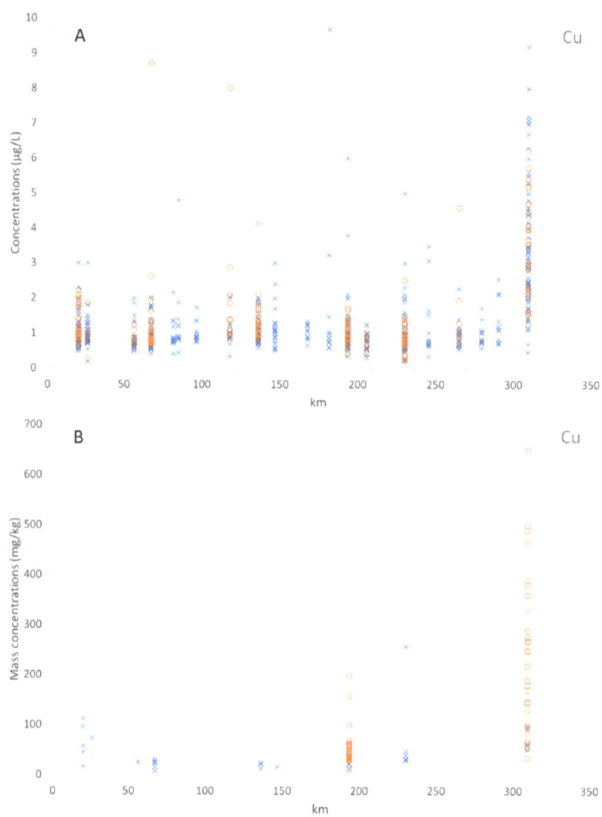

Figure 6. (**A**) Dissolved copper concentrations (blue crosses, µg/L, years 2012 to 2019) and total copper concentrations (orange circles, µg/L, years 2017 to 2019) in the water column along the Mosel river (from the source to the French border, in km), and (**B**) adsorbed copper mass concentrations on the bed sediment (blue crosses, mg/kg, particles < 2 mm) and adsorbed copper mass concentrations on the suspended sediment in the water column (orange circles, mg/kg, particles < 2 mm) along the Mosel river (from the source to the French border, in km), from the year 2012 to 2017.

Figure 7 shows the dissolved and total Zn concentrations (in µg/L) in the water column along the Mosel River. For the years 2012 to 2019, the values of the dissolved Zn concentrations ranged from 0.93 to 110.0 µg/L. The median value of the dissolved Zn concentrations was 2.83 µg/L and the average value of the dissolved Zn concentrations was 4.25 µg/L. Over the period 2017–2019, the values of the total Zn concentrations ranged from 1.0 to 17.7 µg/L. The median value of the total Zn concentrations was 2.97 µg/L and the average value of the total Zn concentrations as 3.92 µg/L. This figure shows also the adsorbed Zn mass concentrations on the bed sediment (particles < 2 mm) and on the suspended sediment (particles < 2 mm) along the Mosel River, for the years 2012 to 2017. The values of the adsorbed Zn mass concentrations on the bed sediment and on the suspended sediment ranged from 63.7 to 1183.26 mg/kg and 120.0 to 770.0 mg/kg, respectively. The median value of the adsorbed Zn mass concentrations on the bed sediment was 177.2 mg/kg. The median value of the adsorbed Zn mass concentrations on the suspended sediment was 273.35 mg/kg. The average value of the adsorbed Zn mass concentrations on bed sediment was 236.17 mg/kg. The average value of the adsorbed Zn mass concentrations on the suspended sediment as 331.66 mg/kg.

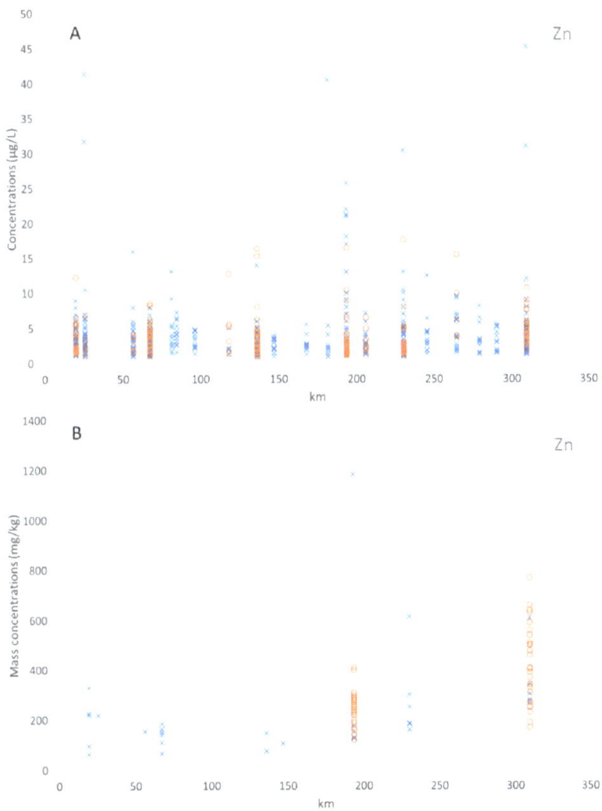

Figure 7. (**A**) Dissolved zinc concentrations (blue crosses, µg/L, years 2012 to 2019) and total zinc concentrations (orange circles, µg/L, years 2017 to 2019) in the water column along the Mosel river (from the source to the French border, in km), and (**B**) adsorbed zinc mass concentrations on the bed sediment (blue crosses, mg/kg, particles < 2 mm) and adsorbed zinc mass concentrations on the suspended sediment in the water column (orange circles, mg/kg, particles < 2 mm) along the Mosel river (from the source to the French border, in km), from the year 2012 to 2017.

The zinc concentrations were higher than the copper concentrations in the water column and on the sediments. The data analyses demonstrated that both Cu and Zn were more adsorbed by the suspended sediment than the bed sediment, due to the higher proportion of fine particles in the suspended sediment (coarser particles settling first).

3.2. Partitioning Coefficients of Metals between Water and Suspended Sediment

The partitioning coefficient provides empirical information regarding the combined effects of heterogeneous reactions on the solid/solution distribution of each trace metal [43]. A high Kd value indicates a strong affinity of the material for the particulate phase [10]. Suspended and bed particle size play a significant role in controlling particulate metals in the river. The smaller the particle size, the higher the content of trace metals [10].

At Eijsden (Meuse River), large copper and zinc datasets were analyzed from 2002 to 2012. The lgKd (logarithm of Kd) values of Cu and Zn (Figure 8), for total SS, ranged from 3.6 to 5.8 L/kg and 3.6 to 6.1 L/kg, respectively, with average values of 4.9 L/kg and 5.1 L/kg, respectively, and median values of 4.9 L/kg and 5.1 L/kg, respectively. All the lgKd values were higher than 3, indicating the strong adsorptive capacity of these trace metals for the SS in the Meuse River.

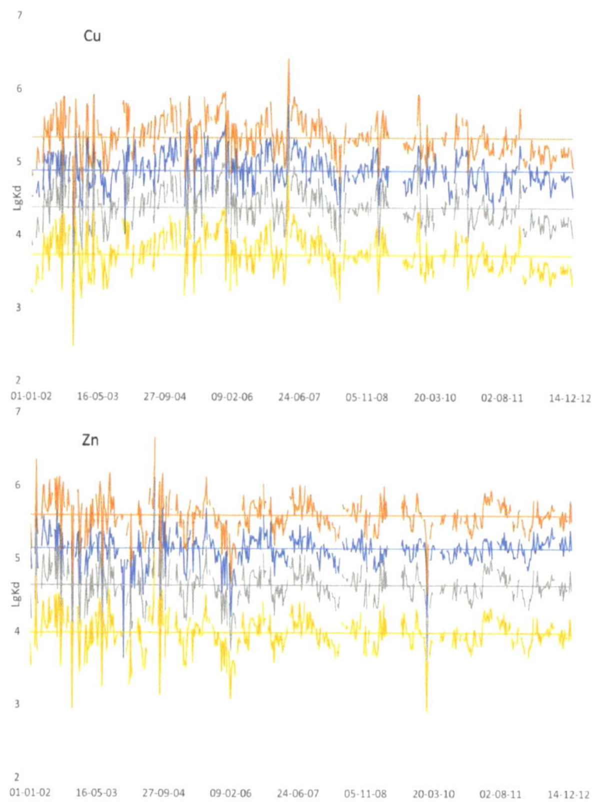

Figure 8. Partitioning coefficients (lgKd) of copper (upper graph) and zinc (lower graph) at Eijsden (Meuse). The blue line indicates the lgKd calculated for total SS, the orange line indicates the lgKd calculated for clays, the grey line indicates the lgKd calculated for fine silts, the yellow line indicates the lgKd calculated for coarse silts. The horizontal lines correspond to the median of lgKd for each class of particles.

Regarding particle size, the lgKd values of Cu and Zn were higher for clays and ranged from 4.1 to 6.4 L/kg and from 4.5 to 6.7 L/kg, respectively, with average values of 5.4 L/kg and 5.6 L/kg, and median values of 5.3 L/kg and 5.6 L/kg, respectively. For fine silts, values ranged from 3.1 to 5.5 L/kg and from 3.6 to 5.7 L/kg, respectively, with average values of 4.4 L/kg and 4.6 L/kg, and median values of 4.4 L/kg and 4.6 L/kg, respectively. For coarse silts, values ranged from 2.5 to 4.8 L/kg and from 2.9 to 5.1 L/kg, respectively, with average values of 3.7 L/kg and 4.0 L/kg, and median values of 3.7 L/kg and 4.0 L/kg, respectively.

Figure 8 shows the high variations of lgKd with time indicating that the instantaneous equilibrium is a theoretical assumption. The Kd depends on chemical processes affecting trace metal speciation. It is far from constant: it varies depending on the chemical composition of the water and the sediment. The Kd is a simple way to characterize particle–solution distributions of metals. The Kd values are also influenced by the effects of seasons and river flow changes [32]. The variation of the Kd should not be seen as resulting from a single geochemical process, but from a combination of simultaneous binding processes to ligands in the water and on the sediments.

Partitioning coefficient variations are also explained, at a second order of magnitude, by metals releases varying day by day.

In the Mosel watershed, the measurements of Cu and Zn mass concentrations are carried out on total SS < 2 mm. This granulometry does not correspond to (but includes) the fine grain size which is essentially concerned by the adsorption and desorption processes by sediments. The partitioning coefficients were calculated for Cu and Zn using these available data. The median values of the Cu and Zn lgKd were 4.7 L/kg (ranging from 3.2 to 5.3 L/kg) and 4.9 L/kg (ranging from 3.7 to 5.3 L/kg), respectively, and the average values were 4.6 L/kg and 4.8 L/kg, respectively. These values were lower regarding the ones calculated at Eijsden where a larger dataset was analyzed and a grain sizes description was carried out.

3.3. Partitioning Coefficients of Metals between Water and Bed Sediment

Measurements on the bed sediment < 2 mm (Figures 6B and 7B) are carried out once a year along the Mosel river at six stations of the monitoring network. The partitioning coefficients between the water column and the bed sediment were calculated for Cu and Zn using these available data. The median values of the Cu and Zn lgKd were 4.4 L/kg (ranging from 3.9 to 5.5 L/kg) and 4.8 L/kg (ranging from 4.0 to 5.3 L/kg), respectively, and the average values were 4.6 L/kg and 4.8 L/kg, respectively. The median values of copper and zinc partitioning coefficients were lower for BS than SS.

Generally, partitioning coefficients of trace metals between particulate and dissolved fractions in the water column appeared in the following order: Zn > Cu [10,44]. The present study confirmed that zinc is more easily adsorbed by sediments due to its stronger affinity for the particulate phase.

3.4. Discussion

Partitioning coefficient depends on the nature of the suspended solids or sediments, geochemical parameters of the water and specific characteristics of each element [11]. The influence of hydrometeorological conditions and metals releases will be further investigated through more appropriated modelling approaches. The Kd logarithmic values of Cu are shown in Figure 9. These values ranged from 2.5 to 6.4 L/kg. The Kd logarithmic values of Zn are shown in Figure 9. These values ranged from 2.9 to 6.7 L/kg. Zn exhibited higher values.

Regarding other studies in the world, these Kd values confirmed that our results in the Meuse (Belgian–Dutch border) and Mosel (France) were reliable. As an example, the Cu lgKd average value was 4.7 in the Garonne River (France) [41]. These new partitioning coefficients, calculated according to grain sizes (clay, fine silt and coarse silt), can be used in environmental modelling dedicated to the assessment of water quality. One remaining question is to estimate if the sediments constitute a sink or a source of trace metals.

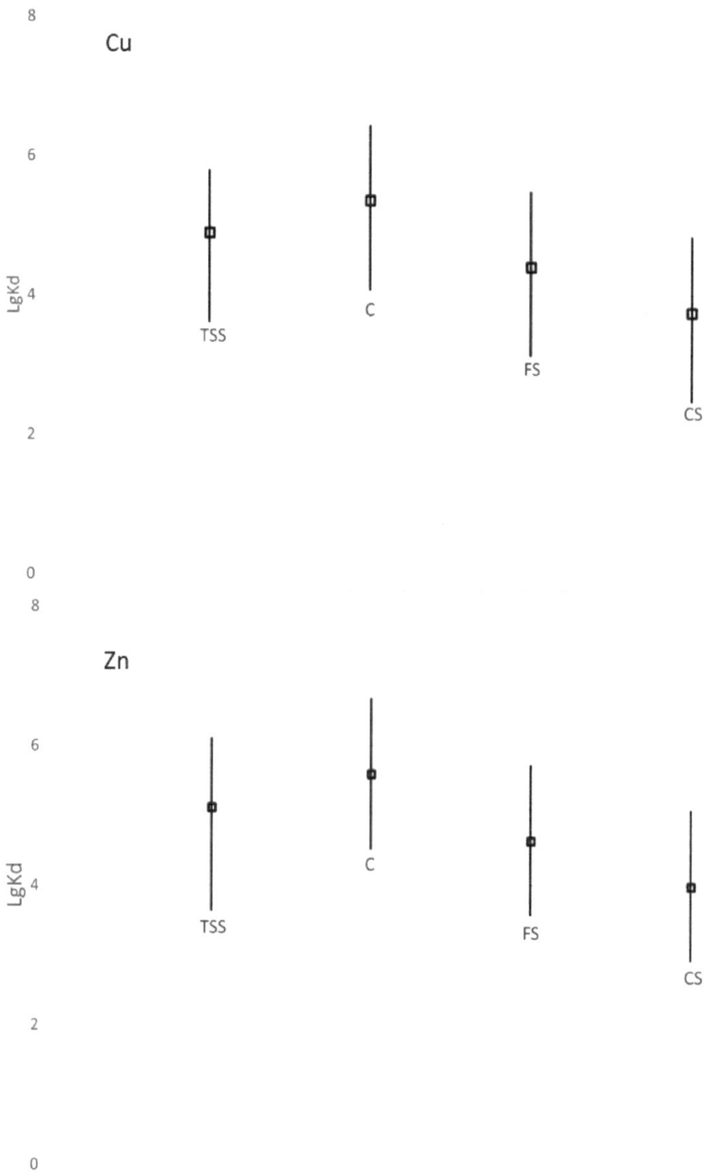

Figure 9. Partitioning coefficients (lgKd) of Cu (upper graph) and Zn (lower graph) for TSS (total suspended sediment), C (clays), FS (fine silts) and CS (coarse silts), with average values represented by squares.

The Cu and Zn lgKd values in some rivers of the world range from 3.9 to 6.3, 3.9 to 8.5, respectively (Table 2).

Table 2. The mean partitioning coefficients (lgKd) of trace metals in some rivers of the world.

River	Cu	Zn
EPA (Environmental Protection Agency) [11]	4.7	5.1
Tigris River, Iraq [45]	6.3	8.5
Day River, Vietnam [46]	5.4	5.1
Sava River, Yugoslavia [47]	3.9	3.9
Downstream of Yangtze River, China [48]	4.1	4.3
Jianglin River, China [48]	4.2	4.0
Beijiang River, China [10]	4.7	4.6
Meuse River, Belgium–Holland (this paper)	4.9	5.1
Mosel River, France (this paper)	4.6	4.8

4. Conclusions and Future Perspectives

In the present study, we determined new copper and zinc partitioning coefficients in two large representative rivers in western Europe, the Meuse and the Mosel. Large sets of data were analyzed. Both dissolved and particulate phases play key roles in the controlling of copper and zinc partitioning in fresh river waters.

In the Meuse river, the copper and zinc partitioning coefficients were calculated according to grain sizes in the water column. These coefficients are strongly affected by particle sizes, and it was demonstrated that clays and fine silts preferentially adsorb trace metals. As observed in other studies, it was confirmed that zinc is more easily adsorbed by sediments due to its stronger affinity for the particulate phase.

In the Mosel River, the copper and zinc partitioning coefficients were calculated in the water column and on the river bed. Compared to copper and zinc adsorption by suspended sediment, the partitioning coefficients were lower on the bed sediment. To strengthen this assessment, it is necessary to improve the granulometry study and focus on fine bed sediments (<63 μm).

This is a preliminary paper to further investigate the environmental modelling of trace metals. Prerequisite knowledge is necessary to tackle trace metal modelling. Among this knowledge, the calibration of the processes involved in trace metal fate are required: partitioning coefficients, sedimentation velocity, critical deposition, erosion velocities, etc. Some of these processes have been improved. Explicit characterization of these processes constitutes a guarantee to best assess water quality. This will allow us to establish consistent pressure–impact relationships between the loads discharging into the river from the watersheds and the reduction in copper and zinc concentrations over the years. Useful added information were also provided for water quality management. The predictive capacity of environmental modelling in the scope of management plans, priority plans, depollution concerning dangerous substances such as trace metals and associated sediments will be challenged.

Supplementary Materials: The following supporting information can be downloaded at: https://www.mdpi.com/article/10.3390/hydrology10020051/s1, Figure S1: Map of the International River Basin District of the Meuse (France, Belgium, GD Luxembourg, Germany, The Netherlands). Eijsden is illustrated by the thumbnail (on the left); Figure S2: Copper mass concentration (mg/kg) at Eijsden (Meuse river) on different classes of suspended sediment for the years 2002–2020.

Author Contributions: Investigation, A.G.; validation, A.G.; writing—original draft: A.G.; project administration, J.-F.D.; supervision, J.-F.D.; conceptualization, J.-F.D. and A.G.; methodology, J.-F.D. and A.G.; writing—review and editing, J.-F.D. and A.G. All authors have read and agreed to the published version of the manuscript.

Funding: This research received no external funding.

Institutional Review Board Statement: Not applicable.

Informed Consent Statement: Not applicable.

Data Availability Statement: The Rhine–Meuse Water Agency, France (https://www.eau-rhin-meuse.fr/, accessed on 24 September 2020); the Rijkswaterstaat, The Netherlands (https://www.rijkswaterstaat.nl/, accessed on 9 July 2021).

Acknowledgments: The authors wish to express their thanks to the administration that made this study possible: the Rhine–Meuse Water Agency, France.

Conflicts of Interest: The authors declare no conflict of interest.

Abbreviations

AERM	Agence de l'Eau Rhin-Meuse
B	Belgium
BS	Bed Sediment
C	Clays
CIM	Commission Internationale de la Meuse;
CS	Coarse Silts
F	France
FS	Fine Silts
G	Germany
GDL	Grand Duchy of Luxembourg
Kd	Partitioning Coefficient
LgKd	Logarithmic value of Kd
NL	The Netherlands
SS	Suspended Sediment
SPW	Service Public de Wallonie
TSS	Total Suspended Sediment

References

1. Trento, A.; Álvarez, A.M. A Numerical Model for the Transport of Chromium and Fine Sediments. *Environ. Model. Assess.* **2011**, *16*, 551–564. [CrossRef]
2. Barré, H.; Greaud, L.; Guarnieri, J.; Lepot, B.; Ughetto, E. Les Substances Dangereuses pour le Milieu Aquatique dans les Rejets des Stations de Traitement des Eaux Usées Urbaines. Rapport; INERIS (Institut National de l'Environnement industriel et des RISques): Verneuil-en-Halatte, France, 2016; 141p.
3. Cho, E.; Arhonditsis, G.B.; Khim, J.; Chung, S.; Heo, T.-Y. Modeling metal-sediment interaction processes: Parameter sensitivity assessment and uncertainty analysis. *Environ. Model. Softw.* **2016**, *80*, 159–174. [CrossRef]
4. Deliège, J.F.; Everbecq, E.; Magermans, P.; Grard, A.; Bourouag, T.; Blockx, C.; Smitz, J. PEGASE, an integrated river/basin model dedicated to surface water quality assessment: Application to cocaine. *Acta Clin. Belg.* **2010**, *65* (Suppl. S1), 42–48. [CrossRef]
5. Kim, S.; Yang, D.S.; Kim, Y.S. Distribution of metal contamination and grain size in the sediments of Nakdong River, Korea. *Environ. Monit. Assess.* **2020**, *192*, 502. [CrossRef] [PubMed]
6. Tipping, E. Humic ion-binding model VI: An improved description of the interactions of protons and metal ions with humic substances. *Aquat. Geochem.* **1998**, *4*, 3–48. [CrossRef]
7. Templeton, D.M.; Ariese, F.; Cornelis, R.; Danielsson, L.G.; Muntau, H.; Van Leeuwen, H.P.; Lobinski, R. IUPAC guidelines for terms related to speciation of trace elements. *Pure Appl. Chem.* **2000**, *72*, 1453–1470. [CrossRef]
8. Tercier-Waeber, M.L.; Stoll, S.; Slaveykova, V.I. Trace metal behavior in surface waters: Emphasis on dynamic speciation, sorption processes and bioavailability. *Arch. des Sci.* **2012**, *65*, 119–142.
9. Lu, Y.; Allen, H.E. Partitioning of copper onto suspended particulate matter in river waters. *Sci. Total Environ.* **2001**, *277*, 119–132. [CrossRef]
10. Li, R.; Tang, C.; Cao, Y.; Jiang, T.; Chen, J. The distribution and partitioning of trace metals (Pb, Cd, Cu, and Zn) and metalloid (As) in the Beijiang River. *Environ. Monit. Assess.* **2018**, *190*, 399. [CrossRef]
11. Allison, J.D.; Allison, T.L. *Partition Coefficients for Metals in Surface Water, Soil and Waste*; U.S. Environmental Protection Agency: Washington, DC, USA, 2005.
12. Nguyen, H.L.; Leermakers, M.; Osán, J.; Török, S.; Baeyens, W. Heavy metals in Lake Balaton: Water column, suspended matter, sediment and biota. *Sci. Total Environ.* **2005**, *340*, 213–230. [CrossRef]
13. Pertsemli, E.; Voutsa, D. Distribution of heavy metals in Lakes Doirani and Kerkini, Northern Greece. *J. Hazard. Mater.* **2007**, *148*, 529–537. [CrossRef] [PubMed]

14. Deliège, J.F.; Everbecq, E.; Magermans, P.; Grard, A.; Bourouag, M.; Blockx, C. PEGASE, a software dedicated to Surface Water Quality Assessment and to European database reporting. In Proceedings of the European Conference of the Czech Presidency of the Council of the EU Towards Environment, Opportunities of SEIS and SIZE: Integrating Environmental Knowledge in Europe, Masaryk University, Brno, Czech Republic, 26 March 2009; pp. 24–32.
15. Directive 2000/60/EC of the European Parliament and of the Council of 23 October 2000 Establishing a Framework for Community Action in the Field of Water Policy. Official Journal of the European Communities L 327. p. 72. Available online: https://eur-lex.europa.eu/legal-content/EN/TXT/PDF/?uri=OJ:L:2000:327:FULL&from=EN (accessed on 1 July 2022).
16. Etat de l'Environnement Wallon. Direction de l'Etat Environnemental, SPW. 2021. Available online: http://etat.environnement.wallonie.be/contents/indicatorsheets/EAU%202.html (accessed on 12 October 2022).
17. *Rapport sur l'évolution de la Qualité des eaux de la Meuse 1998–2007 sur la base des Résultats issus du Réseau de Mesures Homogène*; CIM: Liège, Belgium, 2011.
18. Li, J.; Li, F.; Liu, Q.; Zhang, Y. Trace metal in surface water and groundwater and its transfer in a Yellow River alluvial fan: Evidence from isotopes and hydrochemistry. *Sci. Total Environ.* **2014**, *472*, 979–988. [CrossRef] [PubMed]
19. Waterinfo, Rijkswaterstaat, Ministerie van Infrastructuur en Waterstaat. Available online: https://waterinfo.rws.nl/#!/nav/index/ (accessed on 7 July 2021).
20. Institut National de la Statistique et des Etudes Economiques (INSEE, Montrouge, France). Available online: https://www.insee.fr/fr/statistiques/1405599?geo=DEP-55 (accessed on 23 November 2022).
21. Contrat de Rivière Haute Meuse (CRHM, Namur, Belgium). Available online: https://www.crhm.be/index.php/52-articles/le-crhm/52-le-territoire-de-la-haute-meuse (accessed on 23 November 2022).
22. EPAMA - EPTB Meuse (Etablissement Public pour l'Aménagement de la Meuse et de ses Affluents - Etablissement Public Territorial de Bassin, Charleville-Mézières, France). Available online: https://www.epama.fr/comprendre-la-meuse-et-son-territoire (accessed on 23 November 2022).
23. District Hydrographique International de la Meuse—Analyse, Rapport faîtier, Commission Internationale de la Meuse (CIM). 2005. Available online: http://www.meuse-maas.be (accessed on 15 July 2022).
24. District Hydrographique International de la Meuse: Fiche de Caractérisation de la Masse d'eau MV35R, Meuse II. Plan de gestion 2016–2021. SPW, Direction Générale opérationnelle "Agriculture, Ressources Naturelles & Environnement". Document réalisé par la Direction des Eaux de Surface. 2016. Available online: http://eau.wallonie.be (accessed on 1 October 2022).
25. INRAE (Institut National de Recherche pour l'Agriculture, l'Alimentation et l'Environnement, Antony, France): The Sunshine Web Apps. Available online: https://sunshine.inrae.fr/app/profilsHydro (accessed on 12 October 2022).
26. Inventaires des Émissions, pertes et rejets sur le bassin Rhin-Meuse (Inventories of Emissions, Losses and Discharges in the Rhine-Meuse Basin). Méthodologie et Résultats. November 2019. Available online: http://cdi.eau-rhin-meuse.fr (accessed on 15 June 2022).
27. Agence de l'Eau Rhin-Meuse. Caractéristiques et Priorités dans la Territoire d'interventions Moselle Amont. 2003. Available online: https://www.documentation.eauetbiodiversite.fr/notice/00000000017db0a477f07192919dce8b (accessed on 10 October 2022).
28. Syndicat Mixte Moselle Aval. Dossier de Candidature à la Labellisation «Programme d'Actions de Prévention des Inondations (PAPI)». 2019. Available online: http://www.moselleaval.fr (accessed on 15 October 2022).
29. Agence de l'Eau Rhin-Meuse, SIERM (Système d'information sur l'Eau Rhin-Meuse), Metz, France. Available online: https://rhin-meuse.eaufrance.fr/ (accessed on 13 July 2021).
30. Etat des Lieux. Éléments de Diagnostic de la partie Française du District du Rhin et du District de la Meuse. Adopté au Comité de bassin de l'Agence de l'Eau Rhin-Meuse. 6 December 2019. Available online: https://www.eau-rhin-meuse.fr/etat-des-lieux-des-districts-du-rhin-et-de-la-meuse (accessed on 15 June 2022).
31. Campbell, P.G.C. Interactions between trace metals and aquatic organisms: A critique of the free-ion activity model. *IUPAC Ser. Anal. Phys. Chem. Environ. Syst.* **1995**, *2*, 45–102.
32. Garneau, G.; Sauvage, S.; Sánchez-Pérez, J.M.; Lofts, S.; Brito, D.; Neves, R. Modelling trace metal transfer in large rivers under dynamic hydrology: A coupled hydrodynamic and chemical equilibrium model. *Environ. Model. Softw.* **2017**, *89*, 77–96. [CrossRef]
33. Sun, Z.; Xu, G.; Hao, T.; Huang, Z.; Fang, H.; Wang, G. Release of heavy metals from sediment bed under wave induced liquefaction. *Mar. Pollut. Bull.* **2015**, *97*, 209–216. [CrossRef]
34. Westrich, B.; Förstner, U. *Sediment Dynamics and Pollutant Mobility in Rivers, an Interdisciplinary Approach*; Springer: Berlin, Germany, 2007.
35. Fournier, J.; Bonnot-Courtois, C.; Paris, R.; Le Vot, M. *Analyses Granulométriques: Principes et Méthodes*; Centre National de la Recherche Scientifique-Ecole Pratique des Hautes Etudes: Paris, France, 2012.
36. Wenthworth, C.K. A scale of grade and class terms for clastic sediments. *J. Geol.* **1922**, *30*, 377–392. [CrossRef]
37. Friedman, G.M.; Sanders, J.E. *Principles of Sedimentology*; Wiley: New York, NY, USA, 1978.
38. Blott, S.J.; Pye, K. Gradistat: A Grain Size Distribution and Statistics Package for the Analysis of Unconsolidated Sediments. *Earth Surf. Process. Landf.* **2001**, *26*, 1237–1248. [CrossRef]
39. Sauvé, S.; Hendershot, W.; Allen, H.E. Solid-Solution Partitioning of Metals in Contaminated Soils: Dependence on pH, Total Metal Burden, and Organic Matter. *Environ. Sci. Technol.* **2000**, *34*, 1125–1131. [CrossRef]
40. Lu, Y.; Allen, H.E. A predictive model for copper partitioning to suspended particulate matter in river waters. *Environ. Pollut.* **2006**, *143*, 60–72. [CrossRef]

41. Garneau, C.; Sauvage, S.; Probst, A.; Sánchez-Pérez, J.M. Modelling of trace metal transfer in a large river under different hydrological conditions (the Garonne River in southwest France). *Ecol. Model.* **2015**, *306*, 195–204. [CrossRef]
42. Schnoor, J.L. Environmental modeling. In *Fate and Transport of Pollutants in Water Air and Soil*; Schnoor, I.J., Zehnder, A., Eds.; Modelling trace metals; Wiley: New York, NY, USA, 1996; pp. 381–483.
43. Nguyen, H.L.; Leermakers, M.; Elskens, M. Correlations, partitioning and bioaccumulation of heavy metals between different compartments of Lake Balaton. *Sci. Total Environ.* **2005**, *341*, 211–266. [CrossRef] [PubMed]
44. Zheng, S.; Wang, P.; Wang, C.; Hou, J.; Qian, J. Distribution of metals in water and suspended particulate matter during the resuspension processes in Taihu Lake sediment, China. *Quat. Int.* **2013**, *286*, 94–102. [CrossRef]
45. Hamad, S.H.; Schauer, J.J.; Shafer, M.M.; Al-Raheem, E.A.; Satar, H. The distribution between the dissolved and the particulate forms of 49 metals across the Tigris River, Baghdad, Iraq. *Sci. World J.* **2012**, *2012*, 246059. [CrossRef] [PubMed]
46. Duc, T.A.; Loi, V.D.; Thao, T.T. Partition of heavy metals in a tropical river system impacted by municipal waste. *Environ. Monit. Assess.* **2013**, *185*, 1907–1925. [CrossRef]
47. Drndarski, N.; Stojic, D.; Zupancic, M.; Cupic, S. Determination of partition coefficients of metals in the Sava River environment. *J. Radioanal. Nucl. Chem.* **1990**, *140*, 341–348. [CrossRef]
48. Huo, W.; Chen, J. Water-particulate distribution coefficient of heavy metal and application in sediment quality criteria in China river. *Environ. Sci.* **1997**, *18*, 11–14.

Disclaimer/Publisher's Note: The statements, opinions and data contained in all publications are solely those of the individual author(s) and contributor(s) and not of MDPI and/or the editor(s). MDPI and/or the editor(s) disclaim responsibility for any injury to people or property resulting from any ideas, methods, instructions or products referred to in the content.

hydrology MDPI

Article

Thermally Enhanced Spreading of Miscible Plumes in Porous Media

Ryan G. Tigera, Whitney L. Benson and David C. Mays *

Department of Civil Engineering, University of Colorado Denver, Denver, CO 80217-3364, USA
* Correspondence: david.mays@ucdenver.edu; Tel.: +1-303-315-7570

Abstract: In situ groundwater remediation often calls for a chemical or biological amendment to be injected as an aqueous solution into a contaminated groundwater aquifer. Accordingly, remediation depends on mixing the amendment into the contaminated groundwater, which, in turn, depends on spreading the plume of the injected amendment effectively. Here, we present proof-of-principle results from a laboratory study showing that amendment plume spreading can be enhanced by heating the injected water, which is consistent with the mechanism of miscible viscous fingering. The heated water has a lower viscosity, rendering a mobility ratio (i.e., log viscosity ratio) of 1.2 that generates elongated plume perimeters for essentially consistent plume areas. Using a quasi-two-dimensional apparatus and recording photographs after each increment of the injection volume, two image analysis techniques were employed to measure the area and perimeter of the injected plume, and the results are compared to isothermal controls, showing that the plume perimeter increased by 47% when determined by binary image analysis or 56% when determined by morphological image analysis. Accordingly, this study offers evidence that heating the injected water enhances miscible plume spreading in porous media.

Keywords: groundwater remediation; reactive transport; plume spreading; viscous fingering

Citation: Tigera, R.G.; Benson, W.L.; Mays, D.C. Thermally Enhanced Spreading of Miscible Plumes in Porous Media. *Hydrology* **2023**, *10*, 98. https://doi.org/10.3390/hydrology10040098

Academic Editor: Meysam Vadiati

Received: 26 March 2023
Revised: 13 April 2023
Accepted: 17 April 2023
Published: 21 April 2023

Copyright: © 2023 by the authors. Licensee MDPI, Basel, Switzerland. This article is an open access article distributed under the terms and conditions of the Creative Commons Attribution (CC BY) license (https://creativecommons.org/licenses/by/4.0/).

1. Introduction

Reactive transport in porous media is important for a number of natural and engineering processes, including geochemical cycling, in situ mining, and groundwater remediation. In any of these applications, a plume of the reagent is introduced—either naturally or deliberately—into the resident groundwater. The reaction depends on mixing the reagent with the groundwater, which fundamentally depends on molecular diffusion, but practically depends on a process called plume spreading. Plume spreading transforms the reagent plume into a fractal-like network of lamella that is thin enough for molecular diffusion to bring reagents together. Because flows in porous media are typically laminar, which precludes the turbulence that provides mixing in other engineered reactors, reactions in porous media are transport-limited. Accordingly, the transport of reagents in porous media is governed by the process of plume spreading.

Plume spreading can be classified as passive or active. Passive spreading results from the heterogeneity that is inherent in essentially any natural porous media. Finding the paths of least resistance, the fluid establishes channels of preferential flow, and the resulting velocity contrasts enhanced plume spreading compared to a hypothetical baseline of homogeneous media. In this context, mass transport by transverse dispersion is known to be an important process [1]. By contrast, active spreading results from the deliberate manipulation of the velocity field through an approach called engineered injection and extraction, for example, through vertically separated segments of the well screen [2] through a manifold of wells [3,4] confirmed by laboratory testing [5,6], or through a rotated dipole mixer [7] confirmed by field testing [8]. The present study proposes a new approach to active plume spreading by heating the injected water. Rather than imposing an engineered velocity field, this approach seeks to enhance plume spreading through the fundamental physics of fluid displacement.

Fluid displacement is the process by which a certain fluid, called the defending fluid, is replaced by a different fluid, called the invading fluid. This process can be classified as stable or unstable. Stable displacement causes the complete replacement of the defending fluid by the invading fluid, for example, when a more viscous fluid displaces a less viscous fluid of equal density or when a dense fluid displaces a light fluid from below. Neglecting hydrodynamic dispersion, stable displacement manifests itself as the plug flow, which is the default conceptual model for many environmental treatment unit operations and for groundwater remediation hydraulics, including pump-and-treat and engineered injection and extraction. By contrast, unstable displacement causes an unstable interface between the defending and invading fluids, causing incomplete replacement (Figure 1). In the context of engineered reactive transport, this unstable interface generates the additional plume spreading of the invading fluid into the defending fluid, called *fingering*, which provides an opportunity for enhanced mixing by molecular diffusion, and, consequently, a more complete reaction.

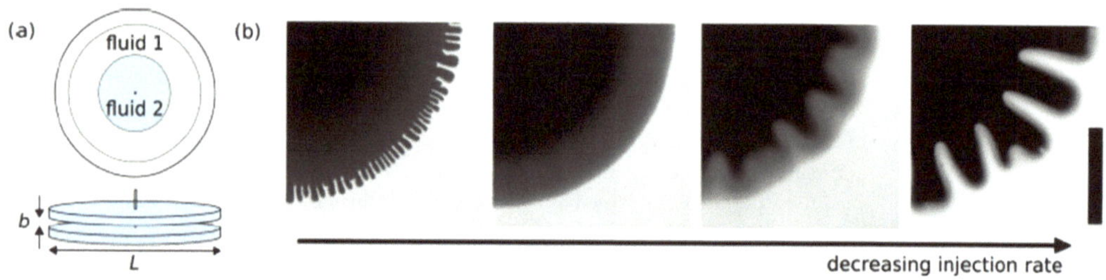

Figure 1. Viscous fingering for miscible fluids with mobility ratio $R = 1.6$ in a Hele–Shaw cell (**a**) with an aperture width of $b = 205$ μm. As the injection rate decreases from 10 mL/min in the left-most panel to 1 μL/min in the right-most panel (**b**), the viscous fingers become more pronounced. Reproduced from Videbæk and Nagel [9] with permission from the American Physical Society.

Unstable displacement results from various combinations of interfacial tension, density difference, or viscosity difference. When the interfacial tension is nonzero, the displacement is called immiscible, examples of which include enhanced oil recovery and the removal of non-aqueous phase liquids (NAPLs). When the interfacial tension is zero, the displacement is called miscible, for example, when an aqueous chemical or biological amendment is injected into a contaminated groundwater aquifer, which is the focus of the present study. When miscible fluids have an unequal density, a lighter invading fluid fingers into a denser defending fluid during the upward flow, and a denser invading fluid fingers into a lighter defending fluid during the downward flow through a process called gravity fingering [10]. In contrast, for constant-density fluids, gravity fingering is prevented. While it is certainly possible to imagine a groundwater remediation application where the injected aqueous amendment has a significant density difference from the defending groundwater, the focus of the present study is on the viscous fingering of miscible fluids with a constant density.

Viscous fingering results when a less viscous fluid breaks through the miscible interface and creates a new pathway into the more viscous fluid [11,12]; viscous fingering always occurs when a less viscous fluid displaces a more viscous one, regardless of miscibility [13]. Viscous fingering can result from the native viscosity difference of the fluids [14,15] or from varying the injection rate over several orders of magnitude to create new flow regimes [16]. Here, we considered the viscous fingering caused by an imposed temperature that renders a viscosity difference between otherwise identical defending and invading fluids.

Viscous fingering between miscible fluids (Figure 1) depends on two dimensionless numbers: the mobility ratio R, and the Péclet number, Pe [15,17]. The mobility ratio, also called the log-viscosity ratio [17], quantified the viscosity difference between two miscible fluids, where $R > 0$ was required for viscous fingering, and larger values of R indicated that viscous fingering was more likely. The mobility ratio can be defined as:

$$R = ln\left(\frac{\mu_1}{\mu_2}\right), \tag{1}$$

where μ_1 is the dynamic viscosity of the defending fluid and μ_2 is the dynamic viscosity of the invading fluid. The Péclet number is a dimensionless ratio of advection to diffusion, which quantifies the general pattern of advection, imposing fine structure on plumes, while diffusion smooths out fine structure. The Péclet number is defined as:

$$Pe = \frac{vL}{D}, \tag{2}$$

where v is the fluid velocity, L is a characteristic length, and D is the fluid's self-diffusion constant.

There have been several studies involving temperature as a factor in groundwater remediation. Kaslusky and Udell [18] injected steam to enhance the removal of volatile organic compounds (VOCs), especially dense non-aqueous phase liquids (DNAPLs), from groundwater but did not specifically address mixing, spreading, or fingering. Kosegi et al. [19] found that increasing the temperature across the entire system in an aquifer remediation simulation resulted in faster cleanup times, but their study only noted changes in viscosity without considering changes in plume morphology. Similarly, Payne et al. [20] identified thermal effects in groundwater remediation hydraulics but did not consider plume morphology. Jackson et al. [21] modeled temperature difference between two immiscible fluids that already had a difference in viscosity, and they found that increasing the temperature difference increased the interfacial area, echoing similar results achieved by a model by Islam and Azaiez [22] that assumed the fluids were miscible. Both of these studies graphically analyzed the interfacial length between the invading plume and the defending fluid. However, a review of the literature has yet to identify research in terms of which enhanced plume spreading can be achieved by heating the invading fluid, thus lowering its viscosity. Accordingly, the novel aspect of the present study was to explore plume spreading by injecting a hot invading fluid into a cold defending fluid.

2. Materials and Methods

Thermally enhanced plume spreading was investigated by injecting a yellow-dyed hot invading fluid over a range of injection rates and into a quasi-2D apparatus packed with porous media and saturated with a blue-dyed cold-defending fluid. Photographs were recorded after each 10 mL of injection, and images were analyzed using k-means clustering with both binary and morphological analysis. The results were reported as an increased plume perimeter compared to an isothermal control. Details on each of these points are presented below, and additional information is provided elsewhere [23].

2.1. Apparatus

A quasi-2D porous media apparatus was constructed by etching a 305 mm × 305 mm square into a clear acrylic sheet to a depth of 12 mm. The resulting chamber was filled with glass beads of a diameter of 1 mm, rendering a porosity of $n = 0.36$ and assuming close random packing [24]. To provide a watertight seal for the acrylic lid, two parallel channels were carved outside the perimeter of the etched square for inner and outer rubber gaskets with circular cross-sections. The lid was secured by 24 round washer head screws (six on each side). Three ports were drilled along a diagonal between opposite corners of the square chamber (one in each corner and one in the center) and were tapped with threads to accommodate them, nominally as a $\frac{1}{4}$, in PVC plastic barbs.

The chamber was filled with saturated media by temporarily mounting the apparatus on a vertical jig, sealing the lower barb with Parafilm, removing the middle and upper barbs, and adding a slurry of beads and blue-dyed water through a tube inserted first into the middle port, and then into the upper port. A piece of stainless-steel mesh was added to each port before reinstalling the middle and top barbs to prevent the beads from being extracted during experiments. Once the barbed tube connection was tight, the cell was slowly lowered to a horizontal position for testing.

Once the cell was placed horizontally, a tripod was used to mount a 20.4-megapixel camera (Sony DSC-HX300, Tokyo, Japan) set to a high-resolution automatic portrait mode, with pictures were taken using a remote shutter to avoid movement. A consistent photograph orientation was provided by the middle barb and a mark placed on the apparatus lid that appeared in the upper-right of each image.

2.2. Experiments

Four experiments were performed over a range of decreasing injection rates (Table 1). In each experiment, the defending fluid was tap water dyed with 10 drops of blue dye (Standard Blue 106002, Kingscote Chemicals, Miamisburg, OH, USA) per 100 mL, and the invading fluid was tap water dyed with 10 drops of yellow dye (Yellow Color 53–140, Honeyville Grain, Ogden, UT, USA) per 100 mL. Identical dye concentrations were chosen to avoid density-driven flows.

Table 1. Overview of plume spreading experiments.

Experiment	Discharge [mL/min]	Péclet Number	Figure
1	60	320	S1
2	30	160	S2
3	15	80	S3
4	7.5	40	S4

In the control experiments, both fluids were maintained at room temperature at approximately 22 °C. The invading fluid was injected through the center port (Figure 2), while the defending fluid was simultaneously removed from the right port, as required by continuity, to prevent over-pressurizing the apparatus. Close to the center port, the flow was approximately radial; farther from the center port, the flow approximated a dipole (disturbed by the boundary of the square chamber). In each experiment, a 60 mL syringe was filled with invading fluid and placed in the syringe pump, while an empty syringe was placed at the other side of the syringe pump to receive the defending fluid. Injection rates (Table 1) corresponded to 60 mL injected over 1, 2, 4, and 8 min. Photographs were recorded after each 10 mL of injection up to 60 mL. Once the invading fluid was injected and the defending fluid was extracted, to prepare for the next experiment, the pump was reversed to re-inject the blue defending fluid and extract the yellow invading fluid (some of which will have mixed to green) to waste. The receiving vessel was elevated to maintain a positive gauge pressure in the apparatus in order to avoid introducing air bubbles.

In thermally enhanced experiments, the steps above were followed with several modifications. The defending fluid was cooled to a temperature of 11 °C, somewhat below the average groundwater temperature of approximately 15 °C in Colorado, USA [25]. A total of 60 mL of cold blue fluid from this chilled reservoir was then injected by the syringe pump through the left port, while room-temperature blue fluid was extracted through the right port into a vessel placed above the flow chamber. This process of injecting 60 mL batches of cold blue fluid was repeated six times to ensure the entire cell was isothermal. Meanwhile, the invading fluid was boiled on a hot plate and then, after the fluid handling described below, was injected at 73 °C. The injection fluid was assumed to have the same density as the defending fluid because the slight density decrease (2.4% from 11 °C to 72 °C) from heating was assumed to compensate for the slight density increase from the evaporative concentration of the yellow dye. The syringe was filled with hot water first, then the tube from the syringe to the four-way valve (Figure 2), and then the tube from the four-way valve to the flow chamber. The tube from the syringe to the four-way valve was fitted with insulation, but the tube from the four-way valve to the chamber was not insulated to minimize blocking the plume from the camera.

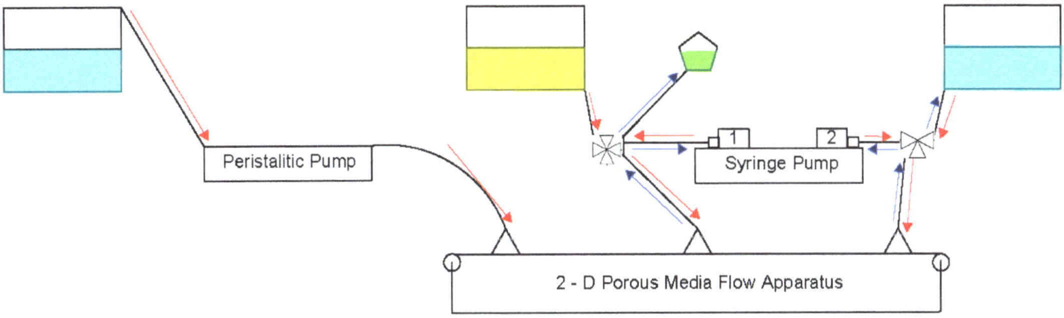

Figure 2. Schematic of experimental apparatus, where the cold defending fluid is blue, the hot invading fluid is yellow, and the mixed fluid is green. Red arrows are injections; blue arrows are extraction. Syringe pump 1 connects via a four-way valve to the reservoir of hot invading fluid, the center injection port, and the waste line; syringe pump 2 connects via a three-way valve to the reservoir of cold defending fluid and the right extraction port.

2.3. Analysis

The viscosity of water as a function of temperature was estimated using the correlation of Sharqawy et al. [26]:

$$\mu = 4.2844 \times 10^{-5} + \frac{1}{0.157(T + 64.993)^2 - 91.296} \tag{3}$$

where μ is the dynamic viscosity of water [kg m^{-1} s^{-1}], and T is the temperature [°C] in the range of $0 \leq T \leq 100$ °C at sea level with an atmospheric pressure of 0.1 MPa. Equation (3) is assumed to be valid at Denver's atmospheric pressure of approximately 84 kPa. For the defending fluid at 11 °C and the invading fluid at 73 °C, this equation gives $\mu_1 = 1.3 \times 10^{-3}$ kg m^{-1} s^{-1} and $\mu_2 = 3.9 \times 10^{-4}$ kg m^{-1} s^{-1}, respectively, rendering the mobility ratio $R = 1.2$.

Because the injection flow is approximately radial, velocity declines with the distance from the center port, so it is necessary to define a characteristic radius at which to evaluate the velocity. This characteristic radius was chosen to be 54 mm, corresponding to the radius of a theoretical circular cylinder 12 mm tall with a porosity of 0.36 after an injection volume of 40 mL, after which thermally enhanced plume spreading was observed, as presented below. The characteristic length $L = 0.1$ cm was taken as the diameter of the glass beads, and the self-diffusion coefficient $D = 2.14 \times 10^{-5}$ cm^2/s was taken for the water at 22 °C [27]. These assumptions define a characteristic Péclet number for each of the experiments (Table 1).

Photographs were analyzed to quantify the plume geometry as the area and perimeter of the invading plume, where thermally enhanced experiments were compared to isothermal controls. These geometric results were determined using the Image Region Analyzer in Matlab R2019b [28] for images generated by each of the two methods. Both methods began with color segmentation using k-means clustering and were implemented with the L*a*b* color space and cluster command in Matlab R2019b. This command separated the $k = 3$ colors of blue (defending), yellow (invading), and green (mixed) in the raw photograph into clusters. In the first image analysis method, the cluster representing the yellow invading plume was converted to a binary image and used to quantify plume geometry. The second image analysis method was morphological structuring, which was implemented with the command strel in Matlab R2019b. This command is specific to shapes and assigns a value to each pixel in relation to the other pixels in its vicinity.

Once the image was flattened with the strel command, the imfill command was used to fill the holes within the image to create a continuous shape, accounting for the space occupied by the invading fluid supply tube and the barb fitting.

3. Results

The invading plume after the final injection volume of 60 mL is shown for each experiment 1–4 in Figure 3, and complete results are provided in the Supplementary Information (Figures S1–S4). By construction, the volume of the invading fluid was constant across experiments; the quasi-2D nature of the flow was confirmed by noting that areas, determined by both the image analysis methods, are consistent at approximately 6300 ± 200 mm^2 (plus or minus one standard error) for all thermally enhanced experiments and isothermal controls (Table 2). No significant differences were observed in the plume areas between the image analysis methods or between the tests and controls ($p < 0.05$). The consistently measured plume area was approximately half that of a theoretical circular cylinder 12 mm tall with a porosity of 0.36 after an injection volume of 60 mL, reflecting certain imperfections that render the flow quasi-2D rather than strictly 2D. In contrast, the differences in the invading plume perimeter reflect increases with the decreasing Péclet number. Relative to the control, the plume perimeters were lower, comparable, and lower for Runs 1, 2, and 3, respectively, which could be attributed to experimental variation. For Run 4, at the lowest Péclet number, the plume perimeter increased up to 47% for the binary image analysis and up to 56% for the morphological image analysis.

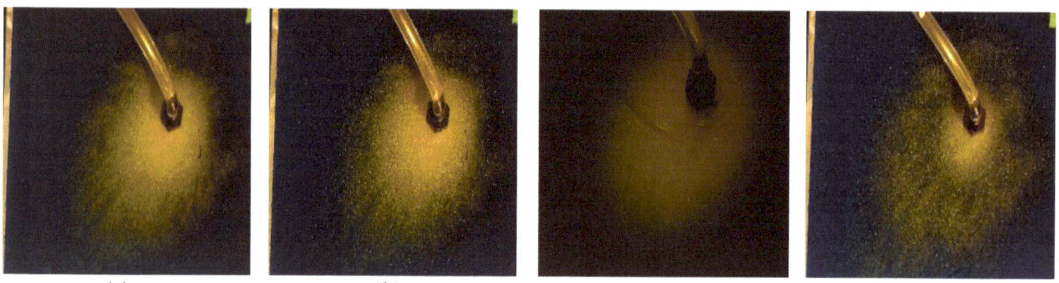

(a) (b) (c) (d)

Figure 3. Thermally enhanced plumes after injection of 60 mL for (**a**) Experiment 1 at Pe = 320, (**b**) Experiment 2 at Pe = 160, (**c**) Experiment 3 at Pe = 80, and (**d**) Experiment 4 at Pe = 40.

Table 2. Mean areas of invading plumes [mm^2] after 60 mL of injection.

	Image Analysis Method *	
	Binary	**Morphological**
control	6610 ± 150	6540 ± 202
test	6108 ± 654	6025 ± 728

* Results shown plus or minus one standard error.

3.1. Binary Image Analysis

To illustrate the first image analysis method, Figure 4 shows the binary image of plumes for experiment 4 after the injection of 60 mL with the isothermal control plume (in blue) superimposed on the thermally enhanced plume (in white). The elongation of the plume toward the lower-left corner of the image reflects the approximate dipole flow to the extraction port, which is not shown in order to discern more detail in the vicinity of the injection port. Figure 5 shows how the perimeter of the invading plume in the thermally enhanced experiment evolved with time in comparison to the isothermal control. We speculate that the slight reduction in perimeter between 50 and 60 mL of the injection results from the dispersive blurring of the interface with time. The perimeter measured by binary image analysis was approximately ten times that measured by the morphological image analysis, as discussed below. This difference reflects the pixel-by-pixel nature of binary image analysis, which renders a much rougher perimeter. Nevertheless, in experiment 4, the perimeters were similar for injection volumes up to 30 mL, after which the thermally enhanced perimeter was up to 47% greater than the isothermal control.

Figure 4. Binary image of the control plume (blue) superimposed on the thermally enhanced plume (white) for experiment 4 with $Pe = 40$ after injection of 60 mL.

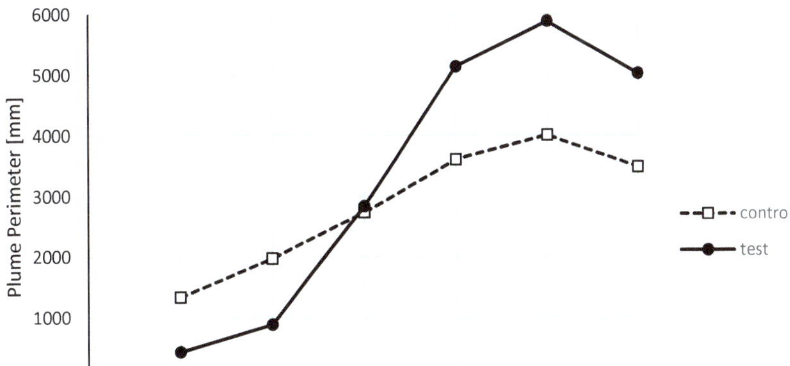

Figure 5. Plume perimeter determined by binary image analysis versus injection volume for experiment 4 with $Pe = 40$. The thermally enhanced test plume had a maximum perimeter of 5883 mm, which is 47% more than the corresponding control perimeter plume length of 4005 mm.

3.2. Morphological Image Analysis

To illustrate the second image analysis method, Figure 6 shows the isothermal control plume (in blue) superimposed on the thermally enhanced plume (in white), and Figure 7 shows how the perimeter of the invading plume in the thermally enhanced experiment evolved with time in comparison to the isothermal control. Again, we speculate that the slight reduction in perimeter between 50 and 60 mL of the injection resulted from dispersive blurring. The perimeter of the isothermal control was slightly larger than that of a theoretical circular cylinder, appearing 12 mm tall with a porosity of 0.36, which is consistent with the flow approximating a dipole, which has a larger perimeter per area than a circular cylinder. The much smaller perimeter compared to binary image analysis reflects the feature of morphological image analysis that seeks to create contiguous regions, for example, by filling holes. However, the qualitative results matched the binary image analysis, with similar perimeters for injection volumes up to 30 mL, after which the thermally enhanced perimeter was up to 56% greater than the isothermal control.

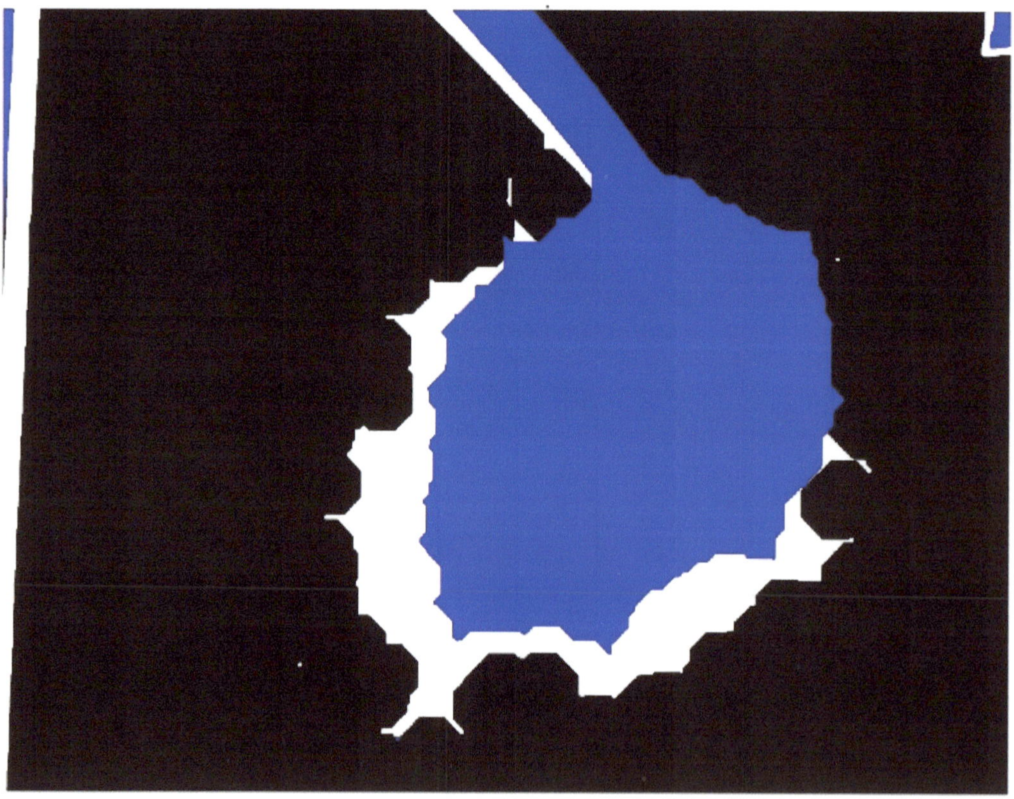

Figure 6. Morphological image of the control plume (blue) superimposed on the thermally enhanced plume (white) for experiment 4 with *Pe* = 40 after injection of 60 mL.

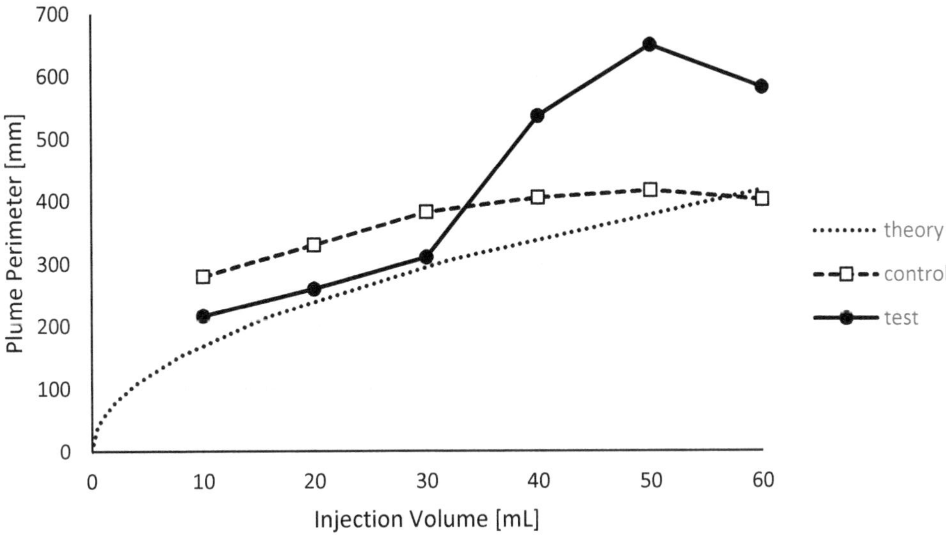

Figure 7. Plume perimeter determined by morphological image analysis versus injection volume for experiment 4 with $Pe = 40$. The thermally enhanced test plume had a maximum perimeter of 649 mm, which is 56% more than the corresponding control perimeter plume length of 416 mm. For comparison, the dotted line is the perimeter of a theoretical circular cylinder that is 12 mm tall with a porosity of 0.36.

4. Discussion

The proof-of-principle experiments presented here show the potential for the thermally enhanced spreading of injected plumes of miscible fluids in porous media. Elongated plume perimeters occur with a mobility ratio of $R = 1.2$ and a Péclet number (as defined in Table 1) of $Pe = 40$, and this observation is independent of the image analysis method chosen, since both binary and morphological analysis result in similar results. The binary image analysis (Figures 4 and 5) renders a fractal-like plume geometry with increases in perimeter compared to the isothermal control. The complementary morphological image analysis (Figures 6 and 7) renders a solid-like plume geometry also with increases in the perimeter. The combination of these methods provides a more in-depth understanding of the thermally enhanced plume spreading of miscible plumes in porous media.

It is notable that elongated plume perimeters can be generated even within the limited temperature range of liquid water that constrains the maximum possible mobility ratio. Using Equations (1) and (3), the maximum temperature range of 0 °C to 100 °C corresponds to a maximum viscosity range of $\mu_1 = 1.8 \times 10^{-3}$ kg m^{-1} s^{-1} to $\mu_2 = 2.8 \times 10^{-4}$ kg m^{-1} s^{-1}, which corresponds to a maximum theoretical mobility ratio of $R = 1.9$. Practically, the lower temperature is more or less fixed, perhaps reflecting some seasonal variation, but seldom comes close to freezing. The temperature $T_1 = 11$ °C used here is probably a reasonable figure for temperate climates. Similarly, although the injection fluid can be heated to boiling, its temperature upon injection is limited by heat loss during fluid handling. If the tubes used in a field application are larger than the 6.4 mm (1/4 in) diameter tubes used here, the smaller area-to-volume ratio limits heat loss; if the delivery time $t = V/Q$ in a field application is smaller, this could also limit heat loss. Practically, the higher temperature may be higher than the $T_2 = 73$ °C used here but seldom comes close to boiling. This limitation contrasts the present study with prior work by others, where higher temperatures generated steam and, consequently, introduced the immiscible displacement of water by steam. Such higher temperatures have been used in the remediation of NAPLs [18]. In contrast, the present study demonstrates the ability to elongate plume interfaces within a temperature range that one might expect in real aquifers and allows strictly miscible displacement.

The experiments reported here show more plume spreading with a decreasing Péclet number opposite the expectation for miscible plume spreading by viscous fingering [29,30], but is at least qualitatively consistent with the results of Videbæk and Nagel [9], as shown in Figure 1, where the left panels show the suppression of 3D fingers and the right panels show the development of 2D fingers with a decreasing Péclet number. The present study differs from these three in at least two respects. First, the present study measures plume spreading in porous media rather than Hele–Shaw cells. Second, the present study generated plume spreading thermally, so the viscosity difference, and therefore, mobility ratio, depends on both fluid mixing and thermodynamics. That is, given enough time, the two fluids would reach thermal equilibrium with an equal viscosity and mobility ratio $R = 0$. Accordingly, the results presented here may be somewhat counterintuitive because lower Péclet numbers imply lower velocities and correspondingly more time for the two fluids to reach thermal equilibrium, which drives the mobility ratio back toward zero. The observation of increased plume perimeter suggests that the time scale for elongating plume interfaces is shorter than the time scale for the thermal equilibrium, at least in the experiments reported here.

Another manifestation of the Péclet number effects could be observed in both the binary image analysis (Figure 5) and the morphological image analysis (Figure 7). In both figures, the perimeter of the thermally enhanced invading plume began to exceed that of the isothermal control after 40 mL of the cumulative injection volume. This transition was observed only in experiment 4 with the smallest Péclet number $Pe = 40$; it was not observed in other experiments with larger Péclet numbers (Table 1). This observation suggests that there is a critical Péclet number above which little thermally enhanced plume spreading occurs. In experiment 4, when the cumulative injection volume was 30 mL or less, the Péclet number was too high; when the cumulative injection volume was 40 mL or more, the Péclet number was low enough. In experiment 3, when the cumulative injection volume was 60 mL or less, the Péclet number was too high, which was similar to experiments 2 and 1, which had even higher injection rates. Accordingly, experiment 4 suggests a critical Péclet number in the range of 40–46, while experiment 3 suggests that the critical Péclet number is less than 65. Taken together, these results suggest that thermally enhanced plume spreading might have been expected in experiment 3 at a cumulative injection volume of 160 mL (although this larger volume would correspond to a longer injection time which could allow thermodynamics to eliminate the mobility ratio as discussed above).

Further experiments are required to address the limitations of this proof-of-principle study. First, the assumption of the equal density of invading and defending fluids should be tested. Second, a modified apparatus could prevent the inlet fitting and supply tubing from appearing in the plume images and using a deeper chamber could determine whether a fully 3D apparatus might reveal experimental artifacts in our quasi-2D apparatus. Third, additional experiments are required to further constrain the critical Péclet number and to determine whether enhanced plume spreading at lower injection rates (i.e., lower Péclet number) could be suppressed by the thermal equilibrium resulting from the additional injection time. Fourth, additional experiments are required to extend these results to 3D flows and reactive transport. For example, delivering hot amendments could accelerate reactions not only by improving plume spreading but also by hastening reaction kinetics. On the other hand, boiling (or nearly boiling) the injection fluid could preclude injecting amendments that are volatile, thermally unstable, or biologically active. Having stated these limitations, the observation of elongated plume interfaces in this experiment suggests that heating the injection fluid may increase the size and extent of the reactive interface between the injected plume and the native groundwater, which, in turn, may result in a larger volume of remediated groundwater. The results of the present study are the first steps toward quantifying the effectiveness of thermally enhanced plume spreading as a tool for in situ groundwater remediation.

Supplementary Materials: The following supporting information can be downloaded at https://www.mdpi.com/article/10.3390/hydrology10040098/s1: A portable document format (PDF) file containing Figures S1–S4. A Microsoft Excel workbook containing experimental data.

Author Contributions: All authors contributed to the study conception and design. Material preparation, data collection and analysis were performed by W.L.B. in 2017–2018 and by R.G.T. in 2018–2020, both supervised by D.C.M. The first draft of the manuscript was written by R.G.T. All authors have read and agreed to the published version of the manuscript.

Funding: The authors declare that no funds, grants, or other support were received during the preparation of this manuscript.

Data Availability Statement: The Supplementary Information provides (1) photographs of control and test plumes after each 10 mL of injection in experiments 1–4 and (2) a spreadsheet with details on geometry, mobility ratio, Reynolds number, Péclet number, and image analysis.

Acknowledgments: The authors thank Jac Corless for fabricating the apparatus and three anonymous referees for their constructive feedback.

Conflicts of Interest: The authors have no relevant financial or non-financial interests to disclose.

References

1. Cirpka, O.A. Effects of sorption on transverse mixing in transient flows. *J. Contam. Hydrol.* **2005**, *78*, 207–229. [CrossRef]
2. Sposito, G. Chaotic solute advection by unsteady groundwater flow. *Water Resour. Res.* **2006**, *42*, W06D03. [CrossRef]
3. Bagtzoglou, A.C.; Oates, P.M. Chaotic advection and enhanced groundwater remediation. *J. Mater. Civ. Eng.* **2007**, *19*, 75–82. [CrossRef]
4. Mays, D.C.; Neupauer, R.M. Plume spreading in groundwater by stretching and folding. *Water Resour. Res.* **2012**, *48*, W07501. [CrossRef]
5. Zhang, P.F.; DeVries, S.L.; Dathe, A.; Bagtzoglou, A.C. Enhanced mixing and plume containment in porous media under time-dependent oscillatory flow. *Environ. Sci. Technol.* **2009**, *43*, 6283–6288. [CrossRef]
6. Sather, L.J.; Roth, E.J.; Neupauer, R.M.; Crimaldi, J.P.; Mays, D.C. Experiments and simulations on plume spreading by engineered injection and extraction in refractive index matched porous media. *Water Resour. Res.* **2023**, *59*, e2022WR032943. [CrossRef]
7. Lester, D.; Metcalfe, G.; Trefry, M.G.; Ord, A.; Hobbs, B.; Rudman, M. Lagrangian topology of a periodically reoriented potential flows Symmetry, optimization, and mixing. *Phys. Rev. E* **2009**, *80*, 036298. [CrossRef]
8. Cho, M.S.; Solano, F.; Thomson, N.R.; Trefry, M.G.; Lester, D.R.; Metcalfe, G. Field trials of chaotic advection to enhance reagent delivery. *Ground Water Monit. Remediat.* **2019**, *39*, 23–39. [CrossRef]
9. Videbæk, T.E.; Nagel, S.R. Diffusion-driven transition between two regimes of viscous fingering. *Phys. Rev. Fluids* **2019**, *4*, 033902. [CrossRef]
10. Flowers, T.C.; Hunt, J.R. Viscous and gravitational contributions to mixing during vertical brine transport in water-saturated porous media. *Water Resour. Res.* **2007**, *43*, W01407. [CrossRef]
11. Saffman, P.C. Viscous fingering in Hele-Shaw cells. *J. Fluid Mech.* **1986**, *173*, 73–94. [CrossRef]
12. Chen, J.-D. Radial viscous fingering patterns in Hele-Shaw cells. *Exp. Fluids* **1987**, *5*, 363–371. [CrossRef]
13. Homsy, G.M. Viscous fingering in porous media. *Annu. Rev. Fluid Mech.* **1987**, *19*, 271–311. [CrossRef]
14. Saghir, M.Z.; Chaalal, O.; Islam, M.R. Numerical and experimental modeling of viscous fingering during liquid–liquid miscible displacement. *J. Pet. Sci. Eng.* **2000**, *26*, 253–262. [CrossRef]
15. Jha, B.; Cueto-Felgueroso, L.; Juanes, R. Fluid mixing from viscous fingering. *Phys. Rev. Lett.* **2011**, *106*, 194502. [CrossRef]
16. Sandnes, B.; Flekkøy, E.G.; Knudsen, H.A.; Måløy, K.J.; See, H. Patterns and flow in frictional fluid dynamics. *Nat. Commun.* **2011**, *2*, 288. [CrossRef]
17. Afshari, S.; Hejazi, S.H.; Kantzas, A. Role of medium heterogeneity and viscosity contrast in miscible flow regimes and mixing zone growth: A computational pore-scale approach. *Phys. Rev. Fluids* **2018**, *3*, 054501. [CrossRef]
18. Kaslusky, S.F.; Udell, K.S. Co-Injection of air and steam for the prevention of the downward migration of DNAPLs during steam enhanced extraction: An experimental evaluation of optimum injection ratio predictions. *J. Contam. Hydrol.* **2005**, *77*, 325–347. [CrossRef]
19. Kosegi, J.M.; Minsker, B.S.; Dougherty, D.E. Feasibility study of thermal in situ bioremediation. *J. Environ. Eng.* **2000**, *126*, 601–610. [CrossRef]
20. Payne, F.C.; Quinnan, J.A.; Potter, S.T. *Remediation Hydraulics*; CRC Press: New York, NY, USA, 2008.
21. Jackson, S.J.; Power, H.; Giddings, D. Immiscible thermo-viscous fingering in Hele-Shaw cells. *Comput. Fluids* **2017**, *156*, 621–641. [CrossRef]
22. Islam, M.N.; Azaiez, J. Thermo-viscous fingering in quarter five-spot miscible displacements. *Eur. J. Mech. B/Fluids* **2011**, *30*, 107–119. [CrossRef]
23. Tigera, R.G. Temperature Dependent Viscosity Driven Mixing in a Quasi-2D Porous Media Apparatus. Master's Thesis, Department of Civil Engineering, University of Colorado Denver, Denver, CO, USA, 2020.

24. Dullien, F.A.L. *Porous Media: Fluid Transport and Pore Structure*, 2nd ed.; Academic Press: Cambridge, MA, USA, 1992.
25. Repplier, F.N.; Rell, M.M.; Columbia, R.K. MS-21 Groundwater Temperatures Map of Colorado. In *Colorado Geological Survey*; Department of Natural Resources: Denver, CO, USA, 1981; Available online: https://coloradogeologicalsurvey.org/publications/groundwater-temperatures-map-colorado (accessed on 30 September 2022).
26. Sharqawy, M.H.; Lienhard, V.J.H.; Zubair, S.M. The thermophysical properties of seawater: A review of existing correlations and data. *Desalination Water Treat.* **2010**, *16*, 354–380. [CrossRef]
27. Dahal, U.; Adhikari, N.P. Molecular dynamics study of diffusion of heavy water in normal water at different temperatures. *J. Mol. Liq.* **2012**, *167*, 34–39. [CrossRef]
28. *MathWorks*; Matlab 9.7.0.1319299 (R2019b) Update 5; MathWorks: Natick, MA, USA, 2020.
29. Sharma, V.; Nand, S.; Pramanik, S.; Chen, C.-Y.; Mishra, M. Control of radial miscible viscous fingering. *J. Fluid Mech.* **2020**, *884*, A16. [CrossRef]
30. An, B.; Solorzano, D.; Yuan, Q. Viscous fingering dynamics and flow regimes of miscible displacements in a sealed Hele-Shaw cell. *Energies* **2022**, *15*, 5798. [CrossRef]

Disclaimer/Publisher's Note: The statements, opinions and data contained in all publications are solely those of the individual author(s) and contributor(s) and not of MDPI and/or the editor(s). MDPI and/or the editor(s) disclaim responsibility for any injury to people or property resulting from any ideas, methods, instructions or products referred to in the content.

hydrology

MDPI

Article

Impact of Gravel Pits on Water Quality in Alluvial Aquifers

Igor Karlović [1,*], Tamara Marković [1], Andrew C. Smith [2] and Krešimir Maldini [3]

1 Croatian Geological Survey, 10000 Zagreb, Croatia
2 British Geological Survey, Nottingham NG12 5GG, UK
3 Main Water Laboratory, Department of Monitoring, Josip Juraj Strossmayer Water Institute, 10000 Zagreb, Croatia
* Correspondence: ikarlovic@hgi-cgs.hr

Abstract: Gravel pits are considered potentially hazardous in terms of groundwater quality protection as they represent an open part of the aquifer system, increasing the aquifer's vulnerability to contamination from the surface. The aim of this research was to determine the biogeochemical processes in gravel pits that have a positive effect on the groundwater quality in the alluvial aquifer in NW Croatia. The aquifer is situated below developed agricultural land, with high groundwater nitrate concentrations having been recorded over the last decades. The differences between two gravel pits and the surrounding groundwater were studied using in situ, hydrochemical, and isotopic parameters ($\delta^{15}N$-NO_3 and $\delta^{18}O$-NO_3), together with existing microbial data. The analyses of nitrogen species indicated that nitrate attenuation processes take place in gravel pits. Bacterial denitrification and nitrate uptake by algae were responsible for significant decreases in nitrate concentration. These processes were more effective in the inactive gravel pit, which has a longer water residence time and during warm periods, when microbial biomass, abundance, and activity were high. The seasonally variable microbial activity also affected trace metals, removing them from groundwater, possibly through the biosorption of metal ions. The presented research shows that the observed biogeochemical processes are associated with seasonal changes that affect the types and number of microbial communities and the chemical composition of water, resulting in gravel pits being groundwater remediation points.

Keywords: gravel pit; surface and groundwater quality; nitrogen species; denitrification; biosorption

Citation: Karlović, I.; Marković, T.; Smith, A.C.; Maldini, K. Impact of Gravel Pits on Water Quality in Alluvial Aquifers. *Hydrology* **2023**, *10*, 99. https://doi.org/10.3390/hydrology10040099

Academic Editor: Meysam Vadiati

Received: 28 March 2023
Revised: 19 April 2023
Accepted: 20 April 2023
Published: 21 April 2023

Copyright: © 2023 by the authors. Licensee MDPI, Basel, Switzerland. This article is an open access article distributed under the terms and conditions of the Creative Commons Attribution (CC BY) license (https://creativecommons.org/licenses/by/4.0/).

1. Introduction

Gravel and sand are non-renewable natural resources which are used in the construction industry for various infrastructure projects (e.g., buildings, roads, and other concrete-based structures). With population growth, the demand for these mineral resources is increasing. The U.S. Geological Survey estimates that the world produced about 265 million metric tons of sand and gravel in 2020 [1]. The results of gravel and sand excavation include gravel pits, which change the morphology and drainage pattern of catchments [2]. The locations of gravel pits are conditioned by the position of natural deposits of sand and gravel, such as alluvial river deposits, streambeds, glaciofluvial deposits, etc. When sand and gravel extraction occur below the water table, groundwater naturally fills the gravel pit, forming a lake. The creation of gravel pits can therefore affect groundwater quality, which is especially important in areas that use groundwater as a source of drinking water. As a result of excavation, the protective soil cover is removed, which exposes the aquifer to the atmosphere and increases its vulnerability to contamination [3]. Another threat to groundwater quality is related to the illegal waste disposal in inactive gravel pits [4,5], which has an impact on both the gravel pit water and the downstream groundwater. Thus, it is very important to investigate the interaction between gravel pits and groundwater, particularly when gravel pits are formed in areas close to groundwater abstraction sites [6]. In general, limited attention has been paid to the positive effects of gravel pits on water quality compared to the negative ones. However, several studies have recognized that gravel pits can mitigate groundwater nitrate contamination originating from agricultural activity through assimilation by algae and/or bacterial denitrification [7–9]. Additionally, the inflow and accumulation of nutrients into the gravel pits may cause eutrophication and interfere with metal cycling [2]. There is potential to use gravel pits as nutrient filters and to regulate groundwater nitrogen pollution, but it is important to understand the factors responsible for nitrogen removal [10].

Groundwater reserves in Croatia are related either to alluvial aquifers in river valleys with primary porosity or to karst aquifers with secondary porosity, the former of which are suitable for gravel and sand extraction. In 2013, approximately 2.6 million m^3 of sand and gravel were excavated in Croatia, one third of which was extracted from an alluvial aquifer in the Drava River valley in the Varaždin region [11]. The groundwater of this area belongs to strategic groundwater resources but also has a nitrate contamination problem, originating mainly from agriculture. The origin, fate, and transport of nitrate within the Varaždin aquifer have been studied extensively using hydraulic, geochemical, isotope, microbiological, statistical, and modelling techniques [12], but a natural mechanism to significantly reduce high nitrate concentrations in groundwater has not been identified. Given the agricultural production and groundwater residence time [13], nitrate contamination in the Varaždin aquifer will likely continue for decades, so it is important to further investigate possible processes that can contribute to the improvement of groundwater quality. Therefore, the focus of this research is to investigate gravel pits as potential groundwater remediation sites. In this context, we aimed to improve the understanding of the biochemical processes affecting dissolved nitrate and metal concentrations in gravel pits. Nitrate attenuation was evaluated by comparing our results with existing isotope and microbial data, with emphasis on the influence of in situ parameters, seasonal differences, and gravel pit activity. A sampling methodology that included collecting both filtered and unfiltered water samples from the gravel pit allowed us to observe changes in metal concentrations and discuss their relation to seasonal nitrate variations.

2. Materials and Methods

2.1. Study Area

The study was carried out within the Varaždin aquifer in the northwestern part of Croatia (Figure 1). This area has a long-standing problem with nitrate groundwater contamination, resulting mainly from sources related to agricultural production and the sewage system [13,14]. The groundwater abstracted at well fields is used in the public water supply network, excluding the Varaždin well field due to its high nitrate concentrations. The aquifer was formed in the Quaternary as a result of the accumulation processes of the Drava River [15], so its lithology is dominated by alluvial gravel-sand deposits. Mean annual precipitation over the basin is 832 mm/a, with typical seasonal variations in air temperature [16]. The groundwater flows in the SE direction and is recharged by surface water and by the infiltration of precipitation [17]. Several gravel pits have been excavated along the Drava River for sand and gravel mining (Figure 1). Among them, two small gravel pits located in the central part of the aquifer and its adjacent observation wells were selected as representative case studies. The water level in gravel pits represents a surface of equipotential heads. Land use in the vicinity of the studied gravel pits is dominated by agriculture. The agricultural fields are intensively cultivated (cabbage, maize, pumpkin, potato), fertilized and are in contact with groundwater by rainfall infiltration through the unsaturated zone. The gravel pits are recharged by groundwater and precipitation, without inflow from surface waters. The origin of the nitrate in gravel pits is mainly related to groundwater inflow, with concentrations in the central part of the aquifer exceeding the threshold value of 50 mg/L [13]. However, flushing of the surface during rain events occurs too.

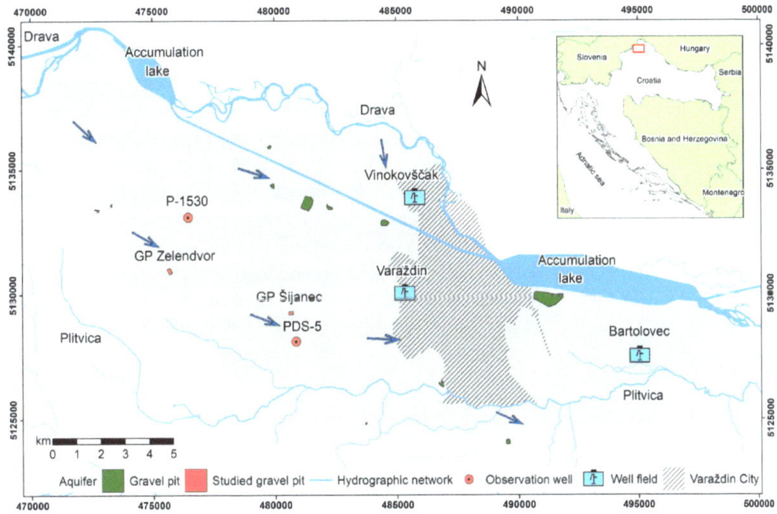

Figure 1. Geographical position of the study area with indicated locations of the investigated gravel pits and observation wells. Blue arrows indicate groundwater flow direction according to Karlović et al. [13].

The gravel pit in the village of Zelendvor is active, i.e., the excavation of gravel and sand is ongoing. The current surface area of the gravel pit is ca. 25,000 m^2. The thickness of the Quaternary sediments is around 12 m in this part of the study area, which limits the maximum depth of the gravel pit to between 10 and 12 m. The closest observation well, P-1530, is located about 2 km away in NE direction. The well is screened at its bottom at a depth of 7.5 m. The measured depth to groundwater, i.e., the thickness of the unsaturated zone, was between 5.80 and 6.33 m within the study period.

The gravel pit in the village of Šijanec is inactive and is used for recreational fishing. It is generally shallow, with water depths of up to 4 m, covering a surface area of approximately 12,000 m^2. The aquifer thickness at this site is around 35 m. There are high nutrient loads, and diverse communities of algae and bacteria that inhabit this gravel pit, most noticeable in the summer period when algal blooms of Microcystis sp. occur [18]. The closest observation well, PDS-5, is situated about 1 km from the gravel pit in the SE direction. The well is 31.0 m deep, with a 6 m long screen at a depth from 13.7 to 19.7 m. The measured depth to groundwater was between 2.63 and 5.52 m within the study period.

2.2. Water Sampling and Laboratory Analyses

Gravel pit water and groundwater samples were collected once a month for chemical analyses of nitrogen species and metal concentrations in water. Gravel pits were sampled from a boat in the central part of each gravel pit by submerging a bailer sampler below the water level, taking composite samples in the period from June 2017 to February 2020 at gravel pit Šijanec, and between June 2017 and December 2017 at gravel pit Zelendvor. The shorter period of sampling in the active pit was due to our inability to access the site from 2018 onwards. Neighboring observation wells were sampled on the same days as the gravel pits. Groundwater samples were collected after pumping at least three times the well volume, i.e., until the stabilization of in situ parameters. In situ parameters (temperature—T, pH, electrical conductivity—EC, dissolved oxygen—DO) were measured using a multiparameter WTW probe. Water samples for nitrate (NO_3^-), nitrite (NO_2^-), and ammonia (NH_4^+) analyses were filtered in the field through a 0.45 μm cellulose membrane filter into 200 mL HDPE bottles. Separate water samples for metal analyses were collected into 100 mL HDPE bottles, both unfiltered and filtered through a 0.45 μm filter, and acidified with 6 M ultra-pure HNO_3. During the sampling campaigns, duplicate samples (unfiltered and filtered) were taken eight times from the inactive gravel pit and two times from the active gravel pit. Samples were kept cool in a refrigerator, transported to the laboratory, and measured on the same day. Gravel pit water samples for the isotopic analyses of $\delta^{15}N$ and $\delta^{18}O$ isotopes in nitrates were filtered through 0.2 μm membrane filters into HPDE 1000 mL bottles in the field, and upon arrival at the laboratory, they were frozen. The NO_3^- concentrations were measured using ion chromatography on Dionex ICS 6000, while NO_2^- and NH_4^+ concentrations were analyzed using a spectrophotometer, the HACH DR 9000. The concentrations of metals in the water were measured using an inductively coupled plasma-mass spectrometry on the Agilent 8900 ICP-MS Triple Quad, following the procedure described in Karlović et al. [19]. $\delta^{15}N$ and $\delta^{18}O$ isotope analyses were conducted at the Stable Isotope Facility of the British Geological Survey, following the analytical method described in Marković et al. [16].

2.3. Methodological Approach

For this work, water sampling was conducted to compare the hydrochemical parameters of samples from the gravel pits with those of their closest observation wells, selected to represent the surrounding groundwater. In particular, in situ parameters and nitrogen species were observed over time to examine their interrelation and the changes in the water chemistry, and to explain possible mechanisms that reduce nitrate concentrations in gravel pits. Numerous authors have studied bacterial denitrification as an effective nitrate reduction process (e.g., [20–22]). The process is mediated by denitrifying bacteria and is mostly efficient under anaerobic conditions with available electron donors, such as dissolved organic carbon [23,24]. During denitrification, the decrease in nitrate concentration is accompanied by the enrichment of $\delta^{15}N$ and $\delta^{18}O$ [25]. Moreover, $\delta^{15}N/\delta^{18}O$ ratios between 1.3 and 2.1 suggest the occurrence of denitrification [26–28]. Nitrate reduction in gravel pits has also been associated with nutrient uptake by primary producers such as algae [3,9,29]. In order to identify the biochemical processes that govern nitrate dynamics in gravel pit waters, the obtained results were compared to existing microbial [18,30] and groundwater nitrogen isotope data [16]. Besides nitrate reduction processes, the potential bioaccumulation of metals in gravel pits was studied by comparing metal concentrations between unfiltered and filtered water samples.

3. Results and Discussion

3.1. In Situ Parameters and Nitrogen Species

The measured in situ parameters and analyzed nitrogen species are presented in Table 1. The in situ parameters within individual gravel pits did not show significant changes with depth, which points to homogenized water in the gravel pit lakes. The high EC values indicate dissolved solids at the sampled locations, which is expected given that they are located in an agricultural area with an intensive input of nutrients into the system. The highest EC was measured in the groundwater, followed by a lower EC in the active pit, and the lowest EC in the inactive gravel pit. Higher variations in water temperature were observed in the gravel pits due to the influence of seasonal changes in air temperature, whereas the mean groundwater temperature indicates the mean annual temperature of the aquifer recharge area. The gravel pits had higher pH values than the surrounding groundwater, with a significant shift to more alkaline in the inactive gravel pit. All measurements of DO suggest that groundwater and gravel pit waters are oxygen saturated.

Table 1. In situ parameters and nitrogen species recorded in investigated gravel pits and in groundwater (observation wells). Data are shown as minimum and maximum values with mean values in parentheses.

Parameter	Active Gravel Pit	P-1530	Inactive Gravel Pit	PDS-5
EC (μS/cm)	447 (572) 649	689 (693) 697	217 (334) 501	661 (685) 694
T (°C)	5.5 (16.8) 24.1	9.2 (14.5) 22.9	1.2 (15.5) 32.1	11.8 (12.6) 13.9
pH	7.52 (7.80) 8.14	7.33 (7.45) 7.53	7.04 (8.64) 10.67	6.91 (7.28) 7.45
DO (mg/L)	9.9 (12.8) 18.5	8.5 (9.7) 11.6	7.1 (12.9) 19.5	3.5 (8.2) 9.9
$NO3^{-}$ (mg/L)	33.7 (54.2) 64.5	48.1 (88.7) 109	0.6 (15.8) 50.1	42.1 (76.5) 210
NO_2^{-} (mg/L)	0.06 (0.16) 0.22	<0.01 (<0.01) 0.01	<0.01 (0.09) 0.38	<0.01 (0.01) 0.02
NH_4^{+} (mg/L)	0.02 (0.15) 0.34	<0.01 (0.02) 0.05	<0.01 (0.27) 0.70	<0.01 (0.04) 0.10

Although gravel pits are mainly recharged by groundwater, and the observation wells are not far away, significant differences in nitrate concentrations were observed (Table 1; Figure 2). Both observation wells had high nitrate levels, with mean concentrations exceeding the threshold value of 50 mg/L. This is influenced mainly by agricultural activity in this part of the aquifer [13,16]. Seasonal peaks of groundwater nitrate concentrations are associated with the rainy season, during which nutrient leaching from cultivated areas is increased, but also with the dry season, during which the absence of rain events is replaced by intensive irrigation. As a result, the highest nitrate concentration of around 210 mg/L was measured in a groundwater sample collected in summer (June 2017). Conversely, the nitrate concentrations of the gravel pits were lower than those of groundwater throughout the whole study period. Mean nitrate concentrations reduced more than 30 mg/L within the active gravel pit. An even greater decrease in nitrate concentrations was observed in the inactive gravel pit, with maximum values around the 50 mg/L threshold value. The nitrate concentrations of the gravel pits varied among hydrological periods, which is likely controlled by the inflow of groundwater nitrate concentrations, the dilution of lake water by rainwater, and nitrogen transformation processes. Low NO_2^{-} and NH_4^{+} concentrations have been recorded in groundwater samples, often below the detection limit. However, low NO_2^{-} and NH_4^{+} concentrations in gravel pit waters are generally observed during the colder parts of the year, when higher nitrate concentrations are recorded (Figure 3); conversely, peak NO_2^{-} and NH_4^{+} values are measured during the warmer parts of the year, when nitrate concentrations are seen to decrease. This reversal indicates seasonal changes in the dynamics among nitrogen species, possibly due to nitrogen transformation processes.

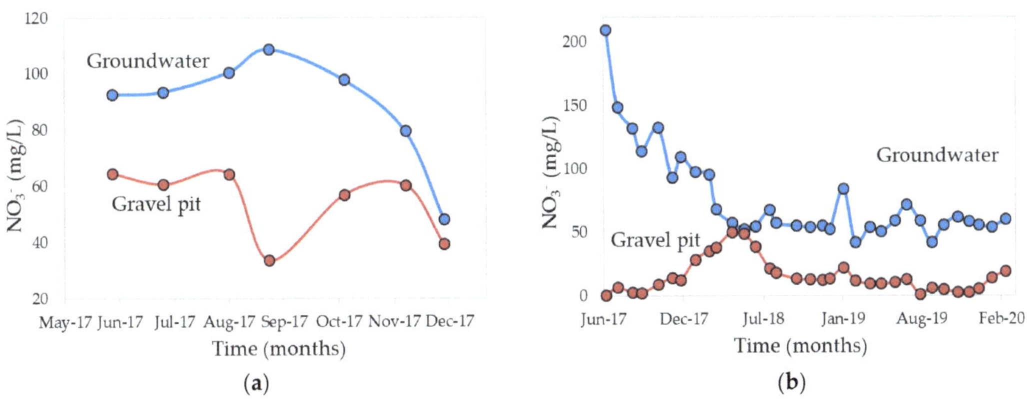

Figure 2. Time series showing the decrease in nitrate concentrations of the groundwater in the (**a**) active gravel pit and (**b**) inactive gravel pit.

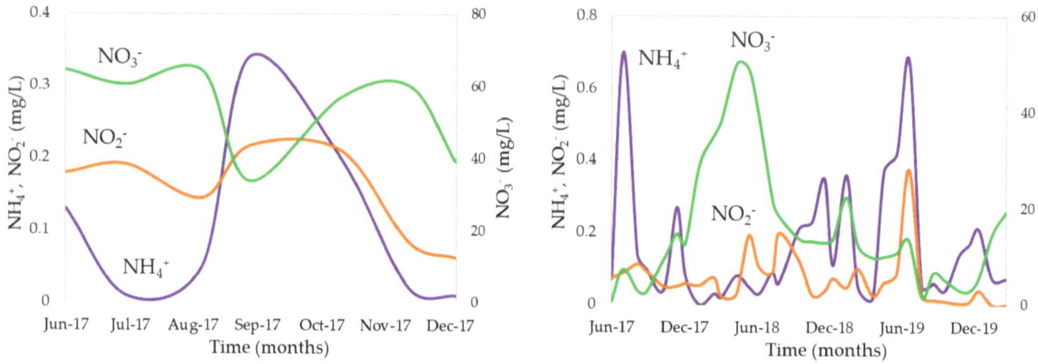

(a) (b)

Figure 3. Relationship between the nitrate, ammonia, and nitrite concentrations of the (**a**) active gravel pit and (**b**) inactive gravel pit.

3.2. Nitrate Attenuation in Gravel Pits

The two main nitrate attenuation processes studied in this piece of research are denitrification and nutrient uptake by algae in gravel pits. Since both processes are microbial-catalyzed, the inclusion of previous microbial research in the studied gravel pits [18,30] in conjunction with nitrogen isotope data contributed to the interpretation of nitrate reduction in gravel pits. Nitrate entering the gravel pits is readily available for microorganisms to utilize, and high dissolved organic carbon concentrations [31] suggest that electron donors are also available. However, high levels of dissolved oxygen may be a limiting factor for denitrification at these sites, as the occurrence of denitrification is more favorable at dissolved oxygen concentration levels below 1–2 mg/L O_2 [20]. Nevertheless, denitrification was assessed using nitrate isotope data from groundwater and inactive gravel pit (Figure 4). The pronounced separation of gravel pit samples from groundwater samples clearly shows changes in stable isotope composition. The gravel pit samples were enriched with $\delta^{15}N$ while having low nitrate concentrations, an indication of denitrification. Moreover, the $\delta^{15}N/\delta^{18}O$ ratio from a sample collected in winter (December 2019) was 1.1, having shifted to a value very close to the denitrification range. The $\delta^{15}N/\delta^{18}O$ ratio of another gravel pit sample, collected in summer (June 2019), suggests the absence of denitrification.

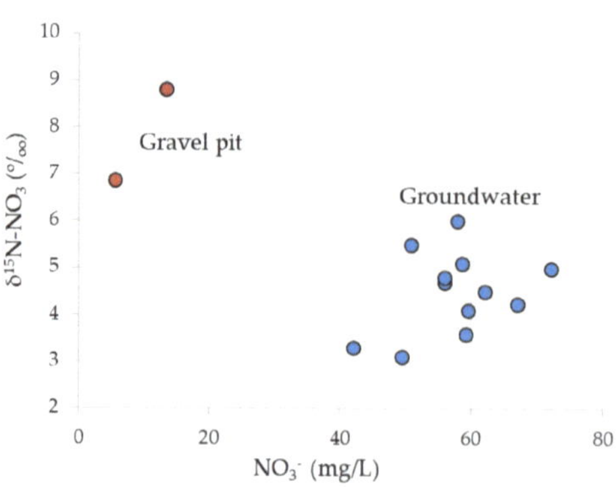

Figure 4. Relationship between the nitrate concentration and nitrate isotopic composition ($\delta^{15}N$-NO_3) of the inactive gravel pit and nearby groundwater (values for groundwater are interpreted in Marković et al. [16]).

Although the potential for denitrification is higher in anaerobic conditions, the process has also been observed in aerobic conditions [32]. Furthermore, denitrification in gravel pits has been previously documented [7,8,10,33], implying that our high dissolved oxygen measurements do not completely exclude the possibility of denitrification. High DO measurements are due to the process of photosynthesis, where algal and bacterial communities generate oxygen as a byproduct, causing the supersaturation of oxygen in the surrounding water.

Seasonal variation in denitrification activity suggests that another process is responsible for nitrate attenuation in gravel pits throughout the year. In the inactive gravel pit, algal and bacterial communities are in competition for nutrients, with Cyanobacteria dominating in the summer period, followed by their breakdown and replacement with diatoms, dinoflagelatte, Bacillariophyceae, and Actinobacteriota in the winter period [18]. This competition is crucial for understanding how microbial dynamics govern nitrate reduction in gravel pits. Water temperature affects the rate of biochemical processes [34], but it also influences the survival time of microorganisms [35]. Both studied gravel pits showed decreases in phytoplankton biomass and abundance in the winter season [30], which coincided with the higher nitrate concentration. In the summer, high water temperatures, exposure to sunlight, low water levels, and increased nutrient load present favorable conditions for algal development in gravel pits and the consequent nutrient uptake by algae, which is responsible for the observed nitrate reduction. Additionally, higher summer pH in the inactive gravel pit, combined with lower EC and nitrate concentration, suggest that the activity of algae is more efficient in the inactive gravel pit. In the winter period, the effective nitrate utilizers diatoms [36] and dinoflagelatte [37] may assist in bacterial denitrification in nitrate reduction.

The rate of nitrate decrease is also related to the activity of gravel pits. Nitrate reduction is affected by the lake water residence time, i.e., a longer mean residence time likely results in increased nitrate uptake and a decrease in nitrate concentrations [3]. The residence time of water in the gravel pit may increase with time, as the permeability of its banks changes due to clogging [2]. Additionally, the post-excavation age of the gravel pit influences the ecosystem metabolism of the gravel pits [9], and Cyanobacteria favor lakes with a long residence time [38]. It is fair to assume that the active gravel pit has shorter water residence times due to the constant excavation of gravel and sand, which enables a continuous supply of nitrate through the inflow of fresh groundwater and affects the productivity of algae, resulting in a smaller decrease in nitrate concentration.

Overall, different mechanisms dominate nitrate attenuation in gravel pits depending on the season. Nitrate reduction in summer is due to nutrient uptake by algae, while the combined effect of bacteria and algae is present in winter. Although the research emphasis was on the comparison of the summer and winter seasons, the nitrate fate in gravel pits in other seasons is likely transitional between these two processes. The presented results are in accordance with [7,10], who identified denitrification and assimilation by algae as the main nitrogen removal mechanisms in gravel pits. Based on our observations, gravel pits act as a sink for nitrate within the studied aquifer system, therefore having a positive effect on groundwater quality.

3.3. Metal Bio-Removal in Gravel Pits

The seasonally variable activity of algae was also seen in the analysis of selected trace metals (Al, As, Cd, Cr, Cu, Fe, Mn, and Pb) in the gravel pit waters. The groundwater contained very low concentrations of such trace metals, measuring from below the detection limit of the instrument to the highest concentration, which was for iron around 34 μg/L (Table 2). Filtered gravel pit water samples were characterized by lower trace metal concentrations compared to unfiltered samples, close to the concentrations found in groundwater. During filtration, colloidal particles, algae, and bacteria are removed from samples, together with trace metals, which are bound to them. Gravel pit lakes are generally clear with low turbidity (low colloidal particle content), but during algal blooms, the turbidity is higher. It is observed that during these blooms in the warm period of the year, trace metals were higher in unfiltered samples than in filtered ones. During the colder period of the year, the concentrations of trace metals in unfiltered and filtered samples were closer. This suggests the sequestration of metals from gravel pit waters by algae and Cyanobacteria, which is especially pronounced in the warm period when their activity is high.

Table 2. Comparison of metal concentrations in gravel pits (unfiltered and filtered pairs) and groundwater samples. Data are shown as minimum and maximum values with mean values in parentheses.

Parameter	Inactive Gravel Pit		Active Gravel Pit		Groundwater
	Unfiltered	Filtered	Unfiltered	Filtered	Filtered
Al (μg/L)	11.1 (144) 516	4.72 (9.98) 29.5	28.8 (84.3) 197	1.77 (1.96) 2.15	0.40 (3.62) 18.8
As (μg/L)	0.62 (1.32) 2.28	0.60 (0.83) 1.71	0.61 (0.81) 1.07	0.84 (0.88) 0.92	0.11 (0.17) 0.39
Cd (μg/L)	<0.01 (0.02) 0.08	<0.01 (0.02) 0.03	0.01 (0.02) 0.02	<0.01 (<0.01) <0.01	<0.01 (0.03) 0.12
Cr (μg/L)	0.09 (0.38) 0.68	0.04 (0.12) 0.28	0.18 (0.30) 0.40	0.23 (0.24) 0.24	0.37 (0.52) 0.70
Cu(μg/L)	1.07 (2.77) 12.8	0.52 (1.63) 2.26	0.94 (1.50) 2.05	0.39 (0.58) 0.87	0.11 (0.75) 6.65
Fe (μg/L)	67.9 (366) 836	11.9 (23.2) 33.8	36.1 (112) 258	2.50 (3.05) 3.59	1.50 (9.60) 34.0
Mn (μg/L)	1.94 (20.7) 47.3	0.10 (0.47) 1.06	11.3 (25.8) 62.4	1.06 (1.81) 2.56	0.29 (0.64) 1.58
Pb (μg/L)	0.25 (1.19) 2.68	0.06 (0.57) 3.14	0.23 (0.36) 0.57	0.06 (0.13) 0.20	0.06 (0.17) 0.60

Determining the actual microbial processes of metal removal was outside the scope of this study, but biosorption may be one of them. According to Al-Amin et al. [39], different cyanobacterial species are reported to sequester metal ions by biosorption (occurring on the cell surface) and/or bioaccumulation (occurring inside the cell). Among them, Microcystis, the most dominant species in the inactive gravel pit during the summer period [18], is reported to have removal efficiencies by biosorption of Cd (II), Cu (II), and Cr (VI) between 24–76% [40].

Using the presented results and identified biochemical processes, natural groundwater remediation mechanisms within the studied gravel pits are depicted in a conceptual model (Figure 5).

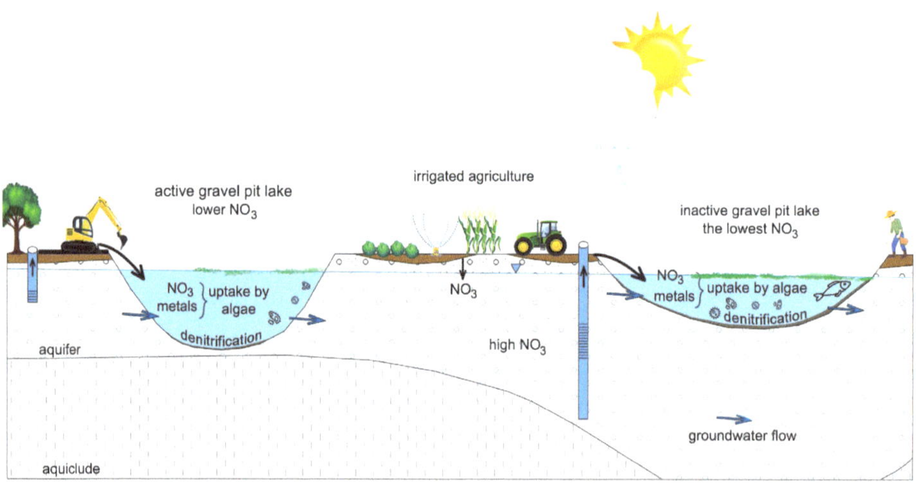

Figure 5. A conceptual model illustrating natural groundwater remediation processes observed in gravel pits surrounded by agricultural land. Nitrate-contaminated groundwater and surface wash-out

recharges the gravel pit, where nitrate is reduced through uptake by algae in the summer period, combined with denitrification in the winter period. The nitrate decrease is more pronounced in the inactive gravel pit and in the warm period, when microbial activity is high. The metal ions are transported into gravel pits via groundwater, surface washout during rain events, and from machines for gravel extraction. The bio-removal of metals from gravel pit water is closely related to algal activity in summer, i.e., the presence of Cyanobacteria and their uptake capacity.

4. Conclusions

The main objective of this research was to explore the biochemical processes which take place in gravel pits and have a positive effect on groundwater quality. The conducted research provided the following conclusions:

- Highly active microbial systems are present in gravel pit lakes, where bacterial denitrification and nitrate uptake by algae are responsible for significant decreases in nitrate concentration, thus serving as a sink for nitrate within the studied aquifer system.
- These processes were more efficient in the inactive gravel pit that has a longer water residence time, resulting in increased nitrate uptake and decreases in nitrate concentrations.
- The bio-removal of dissolved metals from gravel pit water is mediated by cyanobacteria, probably by the biosorption of metal ions.
- All observed processes are more pronounced in the warm period when microbial biomass, abundance, and activity are high, which confirms that when favorable conditions are met, microorganisms are the key factor that governs the fate of nitrate and metals in the studied gravel pits.

Although this study has demonstrated the positive effects of gravel pits, there are some negative aspects that pose a potential risk to groundwater quality. The excavation of gravel pits removes the protective soil cover, thus increasing the vulnerability of aquifers to contamination from the surface. In our case, nutrients from agricultural land are easily transferred to groundwater by rainfall or irrigation. Opening new gravel pits could lead to evaporation losses from the lake water surface and to the emission of N_2O from the potential denitrification process. However, it has been considered that the denitrification activity of gravel pits does not modify the world stock of N_2O [33]. Additionally, human activities such as fish farming can affect groundwater quality by adding extra nutrients into gravel pits, leading to eutrophication. Of particular concern are cyanobacteria, as some species are toxic [38,39].

Gravel pits have the potential to be significant nitrate sinks in aquifers below agricultural land and represent a unique approach to groundwater remediation. Considering both the positive and negative aspects of gravel pits, their overall impact on groundwater quality remains unclear. At this study site, future research efforts should focus on quantifying the nitrate removal capacity of gravel pits. The ability to use a series of gravel pits to provide groundwater remediation services at a regional level can be evaluated by numerical groundwater modelling. Currently, it is questionable whether the small number of gravel pits in the study area can significantly affect nitrate concentrations in the aquifer, given the large volume of groundwater. From an economic perspective, the increase in the prices of gravel and sand as construction materials opens up the possibility of excavating new gravel pits, which is a likely scenario in the future as gravel and sand mining represents important industrial activity in Croatia. Our research showed that the monitoring of water quality in gravel pits is a prerequisite for understanding the processes within gravel pits and establishing appropriate protection measures, which can ultimately contribute to improving the quality of groundwater.

Author Contributions: I.K. conceptualization, methodology, data collection, writing—original draft preparation; T.M. methodology, data collection, writing—original draft preparation; A.C.S. analyses, writing—review and editing; K.M. analyses, writing—review and editing. All authors have read and agreed to the published version of the manuscript.

Funding: This research was financially supported by the Croatian Scientific Foundation (HRZZ) under grant number HRZZ-IP-2016-06-5365 and supported by the Young Researchers Career Development Project—Training of New PhDs—HRZZ and ESF.

Data Availability Statement: The data presented in this study are available in this article. Additional data are available on request from the corresponding author.

Conflicts of Interest: The authors declare no conflict of interest.

References

1. U.S. Geological Survey. *Mineral Commodity Summaries 2021*; U.S. Geological Survey: Reston, VA, USA, 2021; 200p. [CrossRef]
2. Mollema, P.N.; Antonellini, M. Water and (bio)chemical cycling in gravel pit lakes: A review and outlook. *Earth Sci. Rev.* **2016**, *159*, 247–270. [CrossRef]
3. Muellegger, C.; Weilhartner, A.; Battin, T.J.; Hofmann, T. Positive and negative impacts of five Austrian gravel pit lakes on groundwater quality. *Sci. Total Environ.* **2013**, *443*, 14–23. [CrossRef] [PubMed]
4. Navarro, A.; Carbonell, M. Assessment of groundwater contamination caused by uncontrolled dumping in old gravel quarries in the Besòs aquifers (Barcelona, Spain). *Environ. Geochem. Health* **2007**, *30*, 273–289. [CrossRef] [PubMed]
5. Urbanc, M.; Breg, M. Gravel plains in urban areas: Gravel pits as an element of degraded landscapes. *Acta Geogr. Slov.* **2005**, *45*, 35–61. [CrossRef]
6. Søndergaard, M.; Lauridsen, T.L.; Johansson, L.S.; Jeppesen, E. Gravel pit lakes in Denmark: Chemical and biological state. *Sci. Total Environ.* **2018**, *612*, 9–17. [CrossRef] [PubMed]
7. Helmer, C.; Labroue, L. Denitrification in gravel-pit lakes. *Hydrobiologia* **1993**, *251*, 35–44. [CrossRef]
8. Nizzoli, D.; Carraro, E.; Nigro, V.; Viaroli, P. Effect of organic enrichment and thermal regime on denitrification and dissimilatory nitrate reduction to ammonium (DNRA) in hypolimnetic sediments of two lowland lakes. *Water Res.* **2010**, *44*, 2715–2724. [CrossRef]
9. Weilhartner, A.; Muellegger, C.; Kainz, M.; Mathieu, F.; Hofmann, T.; Battin, T.J. Gravel pit lake ecosystems reduce nitrate and phosphate concentrations in the outflowing groundwater. *Sci. Total Environ.* **2012**, *420*, 222–228. [CrossRef]
10. Nizzoli, D.; Welsh, D.T.; Viaroli, P. Denitrification and benthic metabolism in lowland pit lakes: The role of trophic conditions. *Sci. Total Environ.* **2020**, *703*, 134804. [CrossRef]
11. Dedić, Ž.; Kruk, B.; Kruk, L.; Kovačević Galović, E. *Rudarsko-Geološka Studija Varaždinske Županije*; Croatian Geological Survey: Zagreb, Croatia, 2016; 372p. (In Croatian)
12. Karlović, I. Origin, Fate and Transport Modelling of Nitrate in the Varaždin Aquifer. Doctoral Dissertation, Faculty of Mining, Geology and Petroleum Engineering, University of Zagreb, Zagreb, Croatia, 14 July 2022.
13. Karlović, I.; Posavec, K.; Larva, O.; Marković, T. Numerical groundwater flow and nitrate transport assessment in alluvial aquifer of Varaždin region, NW Croatia. *J. Hydrol. Reg. Stud.* **2022**, *41*, 101084. [CrossRef]
14. Marković, T.; Karlović, I.; Orlić, S.; Kajan, K.; Smith, A. Tracking the nitrogen cycle in a vulnerable alluvial system using a multi proxy approach: Case study Varaždin alluvial aquifer, Croatia. *Sci. Total Environ.* **2022**, *853*, 158632. [CrossRef] [PubMed]
15. Prelogović, E.; Velić, J. Quaternary tectonic activity in western part of Drava basin. *Geol. Vjesn.* **1988**, *41*, 237–253.
16. Marković, T.; Karlović, I.; Perčec Tadić, M.; Larva, O. Application of Stable Water Isotopes to Improve Conceptual Model of Alluvial Aquifer in the Varaždin Area. *Water* **2020**, *12*, 379. [CrossRef]
17. Karlović, I.; Marković, T.; Vujnović, T.; Larva, O. Development of a Hydrogeological Conceptual Model of the Varaždin Alluvial Aquifer. *Hydrology* **2021**, *8*, 19. [CrossRef]
18. Kulaš, A.; Marković, T.; Žutinić, P.; Kajan, K.; Karlović, I.; Orlić, S.; Keskin, E.; Filipović, V.; Gligora Udovič, M. Succession of Microbial Community in a Small Water Body within the Alluvial Aquifer of a Large River. *Water* **2021**, *13*, 115. [CrossRef]
19. Karlović, I.; Marković, T.; Šparica Miko, M.; Maldini, K. Geochemical Characteristics of Alluvial Aquifer in the Varaždin Region. *Water* **2021**, *13*, 1508. [CrossRef]
20. Rivett, M.O.; Buss, S.R.; Morgan, P.; Smith, J.W.N.; Bemment, C.D. Nitrate attenuation in groundwater: A review of biogeochemical controlling processes. *Water Res.* **2008**, *42*, 4215–4232. [CrossRef]
21. Otero, N.; Torrentò, C.; Soler, A.; Menciò, A.; Mas-Pla, J. Monitoring groundwater nitrate attenuation in a regional system coupling hydrogeology with multi-isotopic methods: The case of Plana de Vic (Osona, Spain). *Agric. Ecosyst. Environ.* **2009**, *133*, 103–113. [CrossRef]
22. Pastén-Zapata, E.; Ledesma-Ruiz, R.; Harter, T.; Rampata, A.I.; Mahlknecht, J. Assessment of sources and fate of nitrate in shallow groundwater of an agricultural area by using a multi-tracer approach. *Sci. Total Environ.* **2014**, *470*, 855–864. [CrossRef]
23. Mariotti, A.; Landreau, A.; Simon, B. ^{15}N isotope biogeochemistry and natural denitrification processes in groundwater: Application to the chalk aquifer of northern France. *Geochim. Cosmochim. Acta.* **1988**, *52*, 1869–1878. [CrossRef]
24. Zhang, Y.; Zhou, A.; Zhou, J.; Liu, C.; Cai, H.; Liu, Y.; Xu, W. Evaluating the Sources and Fate of Nitrate in the Alluvial Aquifers in the Shijiazhuang Rural and Suburban Area, China: Hydrochemical and Multi-Isotopic Approaches. *Water* **2015**, *7*, 1515–1537. [CrossRef]
25. Chen, D.J.Z.; MacQuarrie, K.T.B. Correlation between δ^{15}N and δ^{18}O in NO_3 during denitrification in groundwater. *J. Environ. Eng. Sci.* **2005**, *4*, 221–226. [CrossRef]

26. Liu, C.Q.; Li, S.L.; Lang, Y.C.; Xiao, H.Y. Using δ^{15}N- and δ^{18}O- values to identify nitrate sources in karst ground water, Guiyang, Southwest China. *Environ. Sci. Technol.* **2006**, *40*, 6928–6933. [CrossRef]
27. Fukada, T.; Kevin, M.; Dennis, P.F.; Grischek, T. A dual isotope approach to identify denitrification in ground water at a riverbank infiltration site. *Water Res.* **2003**, *37*, 3070–3078. [CrossRef] [PubMed]
28. Aravena, R.; Robertson, W.D. Use of multiple isotope tracers to evaluate denitrification in groundwater: Study of nitrate from a large—Flux septic system plume. *Ground Water* **1998**, *36*, 975–982. [CrossRef]
29. Harrison, J.A.; Maranger, R.J.; Alexander, R.B.; Giblin, A.E.; Jacinthe, P.A.; Mayorga, E.; Seitzinger, S.P.; Sobota, D.J.; Wollheim, W.M. The regional and global significance of nitrogen removal in lakes and reservoirs. *Biogeochemistry* **2009**, *93*, 143–157. [CrossRef]
30. Kulaš, A. *Phytoplankton Community Structure: Report of the TRANITAL Project*; Croatian Geological Survey: Zagreb, Croatia, 2021. (In Croatian)
31. Karlović, I.; Marković, T.; Kanduč, T.; Vreča, P. Assessment of Seasonal Changes on the Carbon Cycle in the Critical Zone of a Surface Water (SW)–Groundwater (GW) System. *Water* **2022**, *14*, 3372. [CrossRef]
32. Robertson, L.A.; Van Niel, E.W.J.; Torremans, R.A.M.; Kuenen, G.J. Simultaneous nitrification and denitrification in aerobic chemostat culture of *Thiosphaera pantotropha*. *Appl. Environ. Microbiol.* **1988**, *54*, 2812–2818. [CrossRef]
33. Labroue, L.; Delmas, R.; Serca, D.; Dagnac, J. Nitrate contamination of groundwater as a factor of atmospheric pollution. *C. R. Acad. Sci. 3* **1991**, *313*, 119–124.
34. Sprenger, C.; Lorenzen, G.; Hülshoff, I.; Grützmacher, G.; Ronghang, M.; Pekdeger, A. Vulnerability of bank filtration systems to climate change. *Sci. Total Environ.* **2011**, *409*, 655–663. [CrossRef]
35. Brugger, A.; Reitner, B.; Kolar, I.; Quéric, N.; Herndl, G.J. Seasonal and spatial distribution of dissolved and particulate organic carbon and bacteria in the bank of an impounding reservoir on the Enns River, Austria. *Freshwat. Biol.* **2001**, *46*, 997–1016. [CrossRef]
36. Glibert, P.M.; Wilkerson, F.P.; Dugdale, R.C.; Parker, A.E.; Alexander, J.A.; Blaser, S.; Murasko, S. Phytoplankton communities from San Francisco Bay Delta respond differently to oxidized and reduced nitrogen substrates-even under conditions that would otherwise suggest nitrogen sufficiency. *Front. Mar. Sci.* **2014**, *1*, 17. [CrossRef]
37. Dortch, Q. The interaction between ammonium and nitrate uptake in phytoplankton. *Mar. Ecol. Prog. Ser.* **1990**, *61*, 183–201. [CrossRef]
38. Cross, I.D.; McGowan, S.; Needham, T.; Pointer, C.M. The effects of hydrological extremes on former gravel pit lake ecology: Management implications. *Fundam. Appl. Limnol.* **2014**, *185*, 71–90. [CrossRef]
39. Al-Amin, A.; Parvin, F.; Chakraborty, J.; Kim, Y. Cyanobacteria mediated heavy metal removal: A review on mechanism, biosynthesis, and removal capability. *Environ. Technol. Rev.* **2021**, *10*, 44–57. [CrossRef]
40. Rai, P.K.; Tripathi, B.D. Removal of heavy metals by the nuisance cyanobacteria Microcystis in continuous cultures: An eco-sustainable technology. *Environ. Sci.* **2007**, *4*, 53–59. [CrossRef]

Disclaimer/Publisher's Note: The statements, opinions and data contained in all publications are solely those of the individual author(s) and contributor(s) and not of MDPI and/or the editor(s). MDPI and/or the editor(s) disclaim responsibility for any injury to people or property resulting from any ideas, methods, instructions or products referred to in the content.

hydrology

MDPI

Project Report

Application of a Novel Amendment for the Remediation of Mercury Mine Sites with Hydrologic Controls

Stephen McCord [1,*], Gregory Reller [2], Jon Miller [3] and Kim Pingree [3]

1 McCord Environmental, Inc., Davis, CA 95616, USA
2 Terracon Consultants, Inc., Sacramento, CA 95864, USA; gr@burlesonconsulting.com
3 Albemarle Corporation, Charlotte, NC 28209, USA; jon.miller@albemarle.com (J.M.); kim.pingree@albemarle.com (K.P.)
* Correspondence: sam@mccenv.com; Tel.: +1-530-220-3165

Abstract: MercLok^{TM} P-640 (MercLok) is a proprietary product developed by Albemarle as a mercury (Hg) treatment technology. MercLok captures mercury and sequesters it for a long period under ambient environmental conditions. For this project, MercLok was applied to Hg-contaminated calcines at two abandoned Hg mine sites in northern California to evaluate its efficacy in rendering such contaminated materials less hazardous and thereby reducing remediation project costs. The first application (Site 1) consisted of two calcines amended with MercLok in isolated reactor buckets under two hydrologic remediation approaches ("repository cap" and "reactive barrier") while exposed to ambient environmental conditions. Non-amended and amended calcines and their leachates were analyzed for Hg content and related conditions during a five-month study period, demonstrating >95% reduction in leachable Hg. The second application (Site 2) involved full-scale site remediation with the application of both approaches and additional hydrologic controls to minimize run-on, erosion, and runoff. Confirmation sampling and subsequent observations indicate that the amendments and hydrologic controls effectively stabilized the site and minimized Hg releases. These application projects demonstrate the efficacy of MercLok as a component of hydrologic controls for treating Hg-contaminated material to achieve long-term mine site remediation objectives.

Keywords: mercury; leachate; mine; remediation; repository; reactive barrier; hydrologic controls

Citation: McCord, S.; Reller, G.; Miller, J.; Pingree, K. Application of a Novel Amendment for the Remediation of Mercury Mine Sites with Hydrologic Controls. *Hydrology* **2023**, *10*, 155. https://doi.org/10.3390/hydrology10070155

Academic Editor: Meysam Vadiati

Received: 24 June 2023
Revised: 12 July 2023
Accepted: 18 July 2023
Published: 22 July 2023

Copyright: © 2023 by the authors. Licensee MDPI, Basel, Switzerland. This article is an open access article distributed under the terms and conditions of the Creative Commons Attribution (CC BY) license (https://creativecommons.org/licenses/by/4.0/).

1. Introduction

California experienced significant mining activity during the 19th and 20th centuries, including hard rock and hydraulic gold mining (primarily in the Klamath Mountains and Sierra Nevada), mercury (Hg) mining in the Coast Range, and hard rock mining for copper, silver, and other metals in portions of the Sierras and northern Coast Range. California's Coast Range represents one of the world's five major Hg-mining areas [1].

The basic Hg processing systems crush Hg-rich ore, roast it in a retort or furnace to volatilize the Hg, and recover the Hg vapor in condenser tubes [2]. The roasted material, called "calcine", is typically dumped in a pile on site. These Hg recovery processes produce calcines that are exposed to the environment at the site. Mercury recovery also requires cooling water to facilitate the condensation of the Hg vapor. For this reason, most processing areas and calcines are located adjacent to water sources. While the most abundant Hg ore mineral is cinnabar (HgS), the calcines after roasting contain a variety of other Hg species, including readily soluble Hg sulfate and Hg chloride species [3]. Thus, calcines at abandoned Hg mine sites provide a source of environmentally reactive Hg that is readily mobilized via erosion and leaches into nearby surface water.

One pervasive legacy of these mining activities is elevated concentrations of Hg in soils, streams, and reservoirs covering vast areas of the state. Most waterbodies downstream of these mines are now listed by the state as impaired [4], and fish consumption advisories are posted [5].

Remediating mine sites can provide environmental improvements on site and downstream [6]. The successful remediation of Gambonini Mercury Mine in 2000 by state and federal agencies [7] provided an example in the same region, wherein consolidating mine waste and deploying subsequent hydrologic and erosion controls reduced the amount of particulate-bound Hg transported off-site via surface water. The remediation of the Gambonini Mine did not address the leachability of Hg from calcines.

Albemarle researched and developed the remediation amendment product MercLok^{TM} P-640 (hereafter "MercLok") to capture and sequester Hg. The purposes of the projects reported here were to quantify the effectiveness of MercLok in minimizing the leachability of Hg, and to gain experience in applying MercLok to remediate Hg mine sites. The successes in both cases are significant because they demonstrate the effectiveness of a new remediation tool that efficiently enhances the effectiveness of hydrologic controls at mine sites.

2. Project Approach

This project applied MercLok at two abandoned Hg mine sites in California's Coast Range. This region is characterized generally as having a Mediterranean climate with arid, warm summers, relatively wet winters, and moderate temperatures. Both sites were mined at multiple times with a variety of devices [8], leaving multiple piles of calcines. Both sites are subject to regulatory pressure to address impaired downstream waters. One site owner wishes to remain anonymous. Thus, for consistency, they are referred to herein as Site 1 and Site 2.

The first project, at Site 1, conducted field trials in containers. The study goal was to evaluate the absolute and relative effectiveness of multiple treatment options (material amended, amendment product, amendment dose and configuration) for controlling the hydrology-driven leaching of Hg, and the erosion of three generations of Hg-rich calcines. The configurations are meant to represent the performance of a "repository cap" and a permeable "reactive barrier". Site features were identified based on a literature review and observations made during site assessment visits. In particular, two piles of calcines (generated during different mining eras) were identified for treatment trials as Areas of Concern 6 and X (AOC-6 and AOC-X).

The second project, at Site 2, implemented full-scale treatment (using multiple hydrologic controls, including MercLok in both repository cap and reactive barrier configurations) amending a pile of calcines. Background soil samples (serpentinite, silica carbonate rock, and sandstone-siltstone) reflected unmineralized rock with Hg concentrations below 1 milligram per kilogram (mg/kg) and mineralized rock with Hg concentrations of 120 to 350 mg/kg. The remediation goals for this site were to reduce the risks of (1) water quality degradation posed by calcines and (2) human and wildlife exposure to Hg in the on-site calcines. These goals were achieved by implementing three primary hydrologic controls: (1) a repository cap—amending calcines with MercLok to reduce Hg leachability from direct precipitation and throughflow; (2) a reactive barrier—amending the material in permeable, subsurface, perimeter trenches; and (3) surface flow controls—installing surface hydrologic controls to minimize run-on, runoff and associated erosion of the amended calcines.

The project proponents addressed the following environmental regulations:

- The Comprehensive Environmental Response, Compensation, and Liability Act (Superfund) regulating releases of hazardous substances that pose unacceptable risks to human health and the environment;
- The Clean Water Act regulating discharges of pollutants into waters of the USA;
- California's Water Code regulating mine waste and drainage, as well as discharges of pollutants into waters of the state.

3. Activities

At both sites, samples were analyzed for:

- Total metals—Required for a general understanding of the metal concentrations and comparison with California Title 22 total threshold limit concentrations (TTLC, a hazardous waste criterion). The threshold for total mercury is 20 mg/kg.
- Toxicity characteristic leaching potential (TCLP)—Required for federal hazardous waste characterization. This procedure is used to simulate a landfill environment and to determine how much of a regulated compound will leach into the environment. None of the Site 1 or Site 2 samples were classified as hazardous based on TCLP analysis; therefore, TCLP results are not discussed further herein.
- Waste extraction test (WET) metals—Used to evaluate the change in hazard classification of calcines due to treatment with MercLok. Compounds are analyzed by extracting Hg from a sample with a 10:1 ratio of citric acid solution to solid. The California hazardous soluble threshold limit concentration (STLC) for the leachability of Hg is 0.2 mg/L.
- De-ionized waste extraction test (DI WET) metals—A WET test in which DI water is used as the extracting solution, which characterizes the amounts of metals that would leach from material under the conditions most likely to be encountered at the site, was used to assess potential threats to water quality. The California Toxics Rule limitation is 0.05 μg/L.
- Synthetic precipitation leaching procedure (SPLP)—Developed by the federal Environmental Protection Agency (EPA) to evaluate the impact of contaminated soils on groundwater.

Based on these analyses and field observations of surface and groundwater hydrology, calcines at Site 1 were the focus of amendment testing to compare the two hydrologic controls, namely a repository cap and reactive barrier. Calcines at Site 2 were treated with all three hydrologic controls (that is, both treatment configurations plus surface flow controls).

1. Site 1 Activities

Analyses showed that AOC-X and AOC-6 materials were calcines from different generations of ore processing. Two composite surface samples of soil were collected uphill of the mining site and were used to represent the background soil Hg concentrations. Table 1 below shows the Hg-specific results of lab testing.

Table 1. Key analytical results for the testing of Site 1 materials during setup.

Material	Treatment	Location	Total Hg (mg/kg)	WET (mg/L)	DI-WET (μg/L)
Calcines	Non-amended	AOC-6	360	0.025	0.3
		AOC-X	2700	1.1	51
	MercLok	AOC-6	78–190	0.0055–0.012	<0.18–0.20
		AOC-X	2600–5000	0.18–0.27	0.3–0.85
Background Soil		B1	11	0.0024	0.26
Regulatory Limit			<20	<0.2	<0.05

Based on DI WET extract Hg concentrations, both non-amended calcines pose a threat to water quality at the site and are considered Group B mine waste, while non-amended AOC-X calcines are considered as California hazardous waste. MercLok-amended calcines are classified as Group C mine waste based on DI WET extract Hg concentrations (and amended AOC X calcines were no longer characterized as hazardous). California mining regulations require much less stringent containment for Group C waste compared to Group B waste.

Field test material was obtained using hand tools (pick and shovel). MercLok amendment doses (as weight percent amendment) decided based on laboratory bench tests were 0 (non-amended "controls"), 1 wt%, 5 wt%, and 10 wt%. The materials and amendments were weighed and placed into a cement mixer. The amended and non-amended materials were placed directly into reactor buckets for testing in two configurations.

The repository cap configuration consisted of a layer of amended calcines on top of clean sand. The reactive barrier configuration consisted of a layer of amended calcines on top of clean sand, covered with a layer of non-amended calcines. Non-amended buckets were configured in the same way.

These two configurations were mimicked on site, as shown in Figure 1. Site photos are provided separately in Figure S1. Each reactor was a plastic 49 L bucket with a 30 cm layer of coarse sand in the bottom. A PVC pipe with a 1.3 cm inner diameter that was open at the bottom and perforated along the bottom 25 cm was placed in the sand/gravel layer. The PVC pipe extended to the top of the bucket to serve as a measuring and sampling port. All reactor buckets were contained in stock tanks to prevent the discharge of study-related materials to the site.

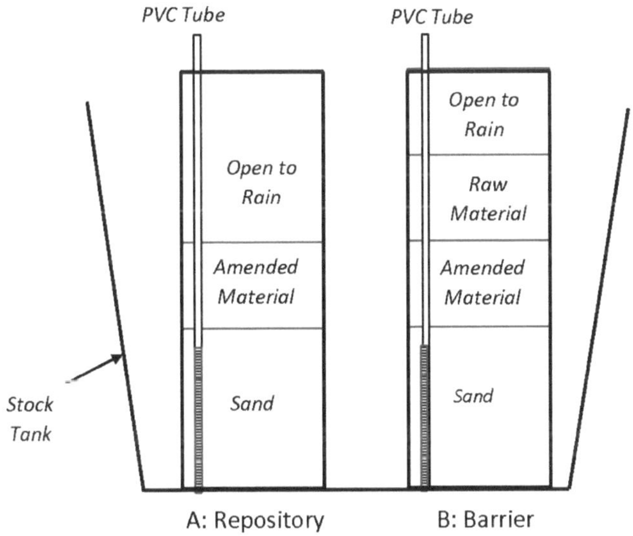

Figure 1. Schematic diagram of Site 1 amended material in buckets configured to mimic a repository cap (**A**) and reactive barrier. (**B**) Not to scale.

Each repository cap, reactive barrier and non-amended sample (14 L each) was placed into its respective bucket and manually distributed as needed to cover the sand and touch the sides of the bucket. The resulting thickness was about 16.5 cm. Reactive barrier layers (including non-amended controls) were then covered with another 16.5 cm of non-amended material.

Monitoring consisted of approximately monthly visits during the wet season. Solids were sampled at the beginning and end of the five-month study period. Samples were analyzed at California Laboratory Services (California Water Boards Environmental Laboratory Accreditation Program (ELAP) certificate # 1233) for the TCLP of metals, WET metals, DI WET metals, and total metals. Samples were analyzed at Eurofins Environment Testing in West Sacramento, CA, using the synthetic precipitation leaching procedure (SPLP) (ELAP # 2897).

The leachate was sampled, approximately monthly after the first month, by placing tubing through the PVC pipe and using a peristaltic pump to withdraw water from the sand layer. Because the study occurred during a significant drought, the lack of rainfall was supplemented with DI water to represent normal seasonal precipitation levels. Samples were analyzed for (among other metals) total and filtered Hg, total suspended solids, cations and anions, and field parameters (pH, temperature and specific conductance).

The volume of water needed to fill the sample containers for these analytes was about 2.5 L. Field parameters (pH, specific conductance, and temperature) were measured with calibrated field meters.

2. Site 2 Activities

At Site 2, calcines and background soil samples were sampled, and a subset were amended with MercLok. Several analyses were used to characterize the material, but key for this paper is that based on WET extract concentrations, the calcines were characterized as a California hazardous waste, and based on DI WET extract Hg concentrations, the calcines posed a threat to water quality at the site. Site 2 non-amended calcines were considered Group B mine waste. MercLok-amended calcines are classified as Group C mine waste based on WET and DI WET extract concentrations. In accordance with Title 27 §22480(d), another factor considered in the classification of the amended calcines as Group C mine waste is that amending the calcines with MercLok constitutes treatment to minimize the threat to water quality and minimizes the need to install waste containment structures.

Hydrologic controls implemented for Site 2 are shown in Figure 2. Site photos are provided separately in Figure S2. Hydrologic controls included, from upstream to downstream:

- An access road low water crossing to reconnect perennial spring discharge with a natural channel away from edge of calcines.
- Grading, mulching, seeding native grasses, and straw wattles on all bare soils to retain moisture, slow overland flow and promote stabilizing vegetation.
- A berm and ditch above the site to divert runoff around the pile.
- An upstream reactive barrier and an overlying berm to direct surface run-on and interflow through the reactive barrier for treatment.
- A repository cap to treat the pile surface.
- A rock retaining wall on downstream pile slope.
- A downstream reactive barrier and an overlying berm to direct surface runoff and interflow through the reactive barrier for treatment.

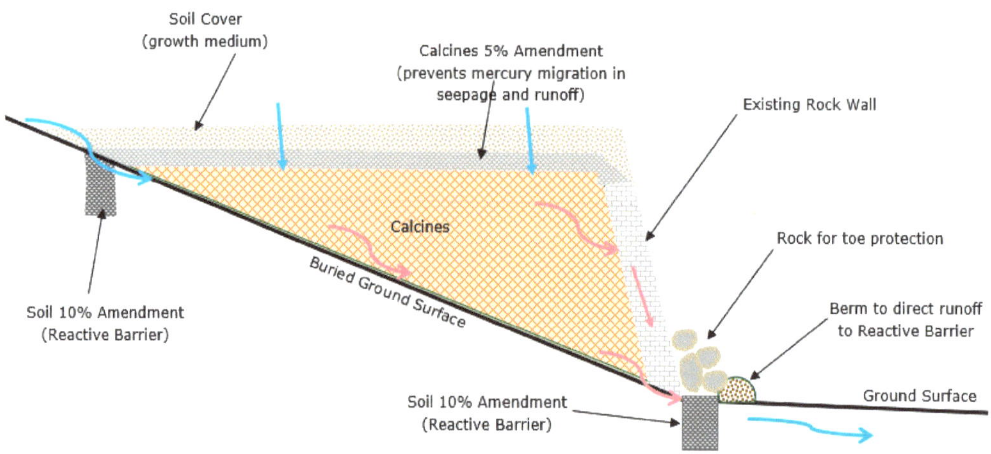

Figure 2. Site 2 amended calcine pile's vertical cross-sectional design. Blue arrows show the flow paths of infiltrating water that does not contain leachable Hg. Pink arrows show the flow paths of seepage that could contain Hg. The reactive barrier at the downslope edge of the calcines is intended to prevent the migration of Hg through the amended seepage face.

4. Technologies

Albemarle's amendment product, MercLok^{TM} P-640 (hereafter "MercLok"; www.albemarle.com), is a brominated porous material with a unique ability to capture and sequester mercury. Multiple forms of Hg found in a range of soils and industrial wastes can be transformed and interact chemically with MercLok, resulting in the long-term sequestration (i.e., remediation) of mercury by the product. With high efficacy and low loading requirements, the powder-based material can be delivered in 454 kg "supersacks" and applied using basic techniques for moistening and mixing into contaminated material. We believe that in most cases, MercLok's cost-in-use value is lower than that of other amendments due to MercLok's higher efficacy in capturing and sequestering Hg in the long term.

Using industry-accepted analyses [9–11], studies predict long-term stability across a range of soils and conditions. MercLok is not classified as a skin or respiratory sensitizer or as toxic through inhalation, oral or dermal exposure. Environmental toxicology tests on indicator species in water and soil (daphnia, fathead minnow, redworms) suggest that the current product has a favorable safety profile for its intended use at contaminated sites. No acute toxicity with up to 10 wt% MercLok was observed from soil biota using OECD Method 207 [12]. Ecotoxicity studies with MercLok using indicator species (water fleas and fatheaded minnows) confirmed that MercLok is not categorized as an aquatic toxin [13,14]. See Supplementary Technology Information for additional reading.

5. Project Details

Field test results (Site 1) and remediation performance results (Site 2) are presented in this section.

1. Site 1 Details

For Site 1, samples of the amended material quantify Hg mobility prior to and after amendment and during and after several months of exposure to weather at the site. The dates of such samples are set forth in Figure 3. The effectiveness of the amendments is assessed via the comparison of Hg concentrations detected in leachates created by the exposure of the materials to weather at the site, with background soil leachates and water quality protection-based thresholds, and through the comparison of the pre- and post-trial solid results with relevant thresholds.

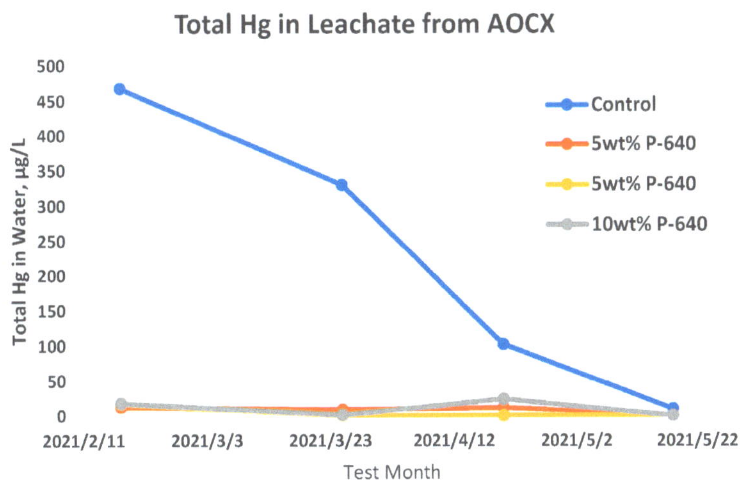

Figure 3. Site 1 total Hg in AOC-X leachate shows that MercLok significantly reduced leachability.

Samples of the leachate that passed through the buckets were collected four times during the field study. The total Hg leached by the rainwater through the AOC-X buckets with MercLok-amended calcines (two duplicates with 5 wt% and a third with 10 wt% doses) was compared with that in the control buckets (Figure 3). These results demonstrated a 97% reduction in leachable Hg as compared to the control.

Total and leachable Hg concentrations from amended AOC-6 were much lower than non-amended AOC-X. Thus, the differentiation between the AOC-6 control and the AOC-6 amended calcines was commensurately lower as well. AOC-6 calcines were amended with 1 wt% and 5 wt% MercLok and arranged in both configurations. Both configurations using MercLok demonstrated significant reductions in the leachable Hg to levels even below the leachability of background soil (Figure 4). Even by assuming non-detected values at the detection limit and including the outlier in the amended samples, there is a significant reduction from the control to amended AOC-6 DI WET Hg (p = 0.0479; alpha 0.05%). Furthermore, even though the reactive barrier configuration nominally held twice as much total Hg as the repository cap configuration, there was no difference in the leachable Hg (except for one outlier).

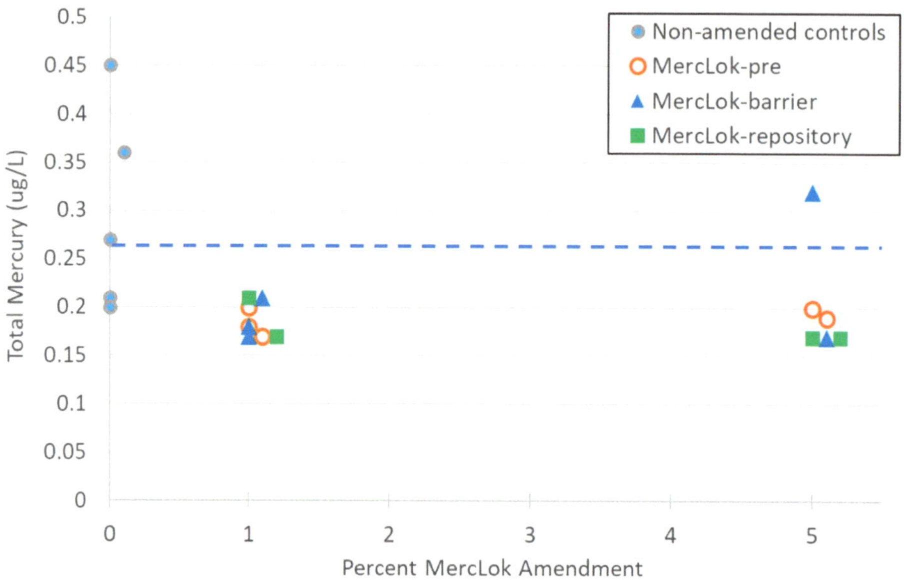

Figure 4. Site 1 Hg in DI-WET leachate from each reactor bucket shows that MercLok reduced the leaching of Hg from AOC-6 to concentrations below that of background soil.

Samples for methylmercury (methyl Hg) in leachate were taken during the second and final sampling periods and only from buckets with 5 wt% of MercLok dosed into AOC-6 and AOC-X. Methyl Hg concentrations in leachates were reduced by 77% and 78% from AOC-6 (Figure 5) and 76% from AOC-X (Figure 6). Since other testing has shown that methyl Hg reduction has a similar slope to total Hg, it is expected that the analysis of methyl Hg from the 10 wt% dosages of MercLok in AOC-X would have resulted in even greater reductions. The analytical results for other total metals in leachate are available in Supplementary Table S1.

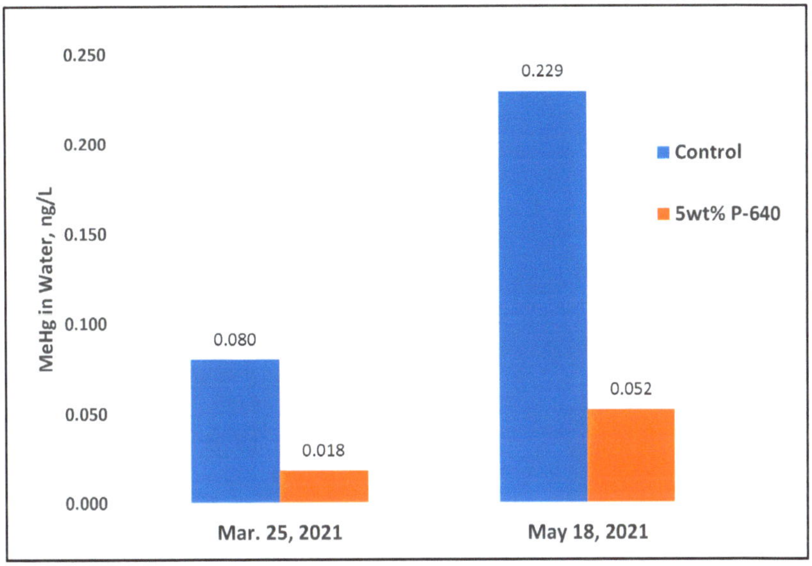

Figure 5. Site 1 MeHg was significantly reduced in AOC-6 leachate due to MercLok.

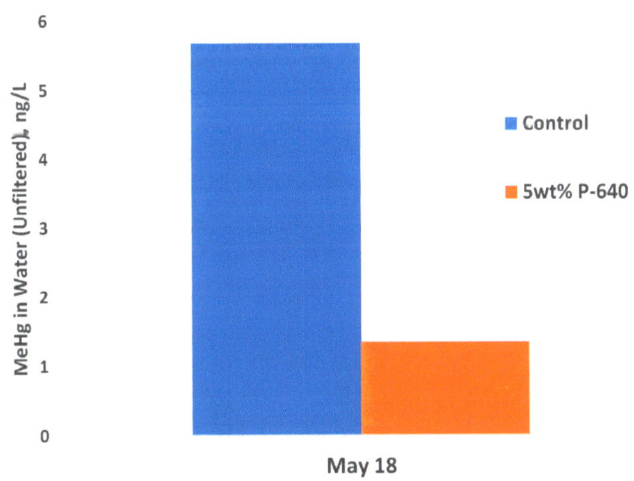

Figure 6. Site 1 MeHg was significantly reduced in AOC-X leachate due to MercLok.

Sampling and analysis of AOC-X calcines immediately after treatment with MercLok at 10 wt% showed that leachable Hg reduced below the CCR Title 22 limit of 0.2 mg/L (Figure 7).

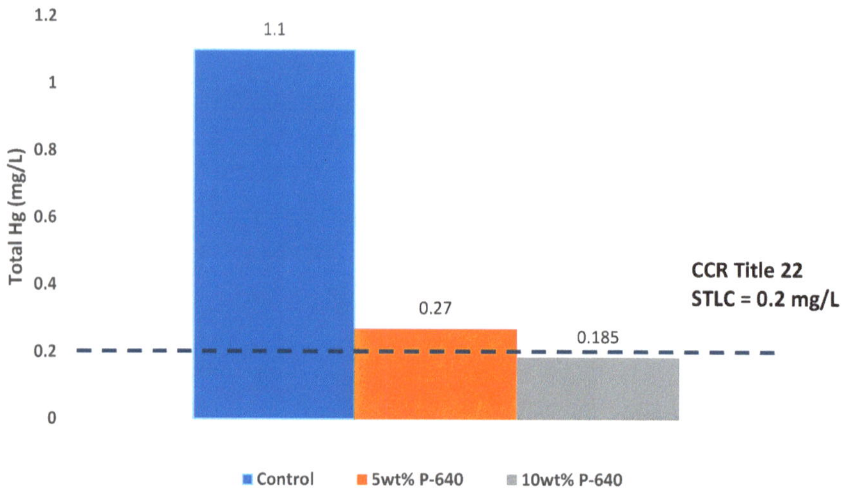

Figure 7. Site 1 AOC-X solids results for DI-WET extracts during trial setup (5 January 2021).

2. Site 2 Details

For Site 2, water was added to each MercLok supersack to attain a pourable slurry of 40% solids. The reactive barrier was excavated into the calcine pile's perimeter soil. Calcines (upgradient) or local sandy soil (downgradient) was mixed with MercLok at approximately 10:1 (base:MercLok) dry weight ratio. The moistened mixture was placed into the trench. About 1800 kg of MercLok was applied in a 30 m trench. The predominately clayey soil excavated from the downgradient trench was compacted into its overlying berm.

The MercLok slurry was placed onto the calcine pile using a backhoe and hand shovels, and scored into the upper one foot of calcines until uniform in appearance (uniform dark color) to achieve about 5% MercLok by weight. Water was also sprayed onto the amended surface, which served to mobilize MercLok from concentrated chunks to less concentrated pockets. About 4500 kg was applied to the calcine pile surface. The amended surface area is about 279 m^2.

The amended calcines were track-walked for compaction and graded to ensure positive drainage so that ponding did not occur within the amended calcines and to prepare the amended calcines for the installation of surface flow/erosion control measures including straw mulch, straw wattles and the broadcast seeding of native grasses and forbs. Erosion was monitored through the first rainy season and control features (berms, ditches, and retaining wall) were maintained. Vegetation grew readily on all amended areas.

Upon construction completion, confirmation samples of amended calcines were collected from four areas (one each from the upslope and downslope reactive barriers, and two from the amended calcines), each as a four-point composite sample. The subsamples were then sieved using a 2 mm sieve to remove coarse particles, homogenized in the field and submitted as composite samples for analyses. For quality control, a field duplicate sample and a non-amended sample were also obtained. Standard lab quality control and quality assurance measures were also performed.

The confirmation sample collected from the downgradient trench was analyzed for CAM 17 metals using EPA Methods 6020 and 7471, and for total solids using EPA Method 2540. Confirmation samples collected from both the calcine pile and the upgradient trench were analyzed for Hg using EPA Method 7471 (for comparison with the TTLC) and for total solids using EPA Method 2540, while the waste extraction test (WET) (for comparison with the STLC), and DI WET were employed for comparison with the background and criteria protective of water quality (Table 2).

Table 2. Comparison of Site 2 Hg in non-amended and amended calcines in confirmation samples.

Material	Total Hg (mg/kg Wet Weight)	Hg in TCLP Extract (mg/L)	Hg in WET Extract (mg/L)	Hg in DI WET Extract (mg/L)
Non-amended Retort Calcines	379	<0.010	0.441	0.2
Amended Retort Calcines	239 to 318	NA	0.119 to 0.184	<0.0010
N Pile Calcines	865	0.0496	1.13	3.46
Upgradient Trench	18.1	NA	NA	NA
Downgradient Trench	25.3	NA	0.00548	<0.0010
North Area	66.9	NA	0.00710	<0.0010
North Area Duplicate	27.7	NA	0.00621	<0.0010
South Area	48.5	NA	0.00493	<0.0010
Background Soil	259	<0.010	0.0224	0.0445
Regulatory Threshold	20	0.2	0.2	0.00005

Notes: Bold values exceed background and regulatory thresholds. TCLP = Toxicity Characteristic Leaching Procedure. STLC = Soluble threshold limit concentration. WET = California waste extraction test. DI WET = Deionized waste extraction Test. mg/kg = milligrams per kilogram. mg/L = milligrams per liter. < = less than the given value. NA = not analyzed.

The mercury concentration in WET extracts from non-amended calcines was 0.44–1.13 mg/L, above the 0.2 mg/L soluble threshold limit concentration (STLC) criterion for California hazardous waste classification (as per 22 CCR Section 66261.24(a)(2)). The mercury concentration in WET extracts from amended calcines was 0.005–0.007 mg/L, which is below the STLC criterion (and less than that in the WET extract from background soil). The mercury concentration in DI WET extracts from non-amended calcines was 0.2–3.46 mg/L, while it was not detected (<0.0010 mg/L) for amended calcines. The mercury concentration in this amended sample extract was about 40 times less than that observed for background soil (0.044 mg/L).

6. Conclusions

For Site 1, the proximity of erosive calcine tailings to surface water and the shallow depths of groundwater at the site indicate that basic site improvements such as erosion control and revegetation are not likely sufficiently protective of water quality. Without amendment, acceptable remedies for these calcines would include excavation, transport and disposal at a suitably permitted off-site disposal facility or an on-site liner and cap system followed by long-term maintenance and monitoring. Both remedies are costly and environmentally disruptive. The ability to amend the calcines with 5 wt% MercLok in either repository cap or reactive barrier hydrologic control configurations in order to remove hazardous leaching characteristics and stabilize Hg at levels protective of water quality under the site conditions makes it possible to construct a Group C on-site repository efficiently and at a lower cost.

The project at Site 2 successfully addressed its goals and objectives. The project team adequately prepared the site, followed its work plan and worker safety protocols, applied MercLok in the intended areas and concentrations using both treatment configurations, and controlled erosion with surface flow hydrologic controls. Confirmation samples document that MercLok treatment effectively controlled Hg leachability in calcines. The amendment of the calcines with MercLok reduced Hg in the leachate to concentrations below those detected in the leachate from background soils and below California hazardous waste classification criteria. With long-term monitoring and maintenance, the site is expected to serve for years to come as a positive example of effective Hg mine site remediation.

The excavation of mine waste and transport for off-site disposal at a permitted facility is often required to remediate contaminated mine sites, and this can be prohibitively expensive, dangerous, and energy-intensive. Similarly, the construction of an engineered, on-site, multi-layered containment system with integrated leachate control and removal system for hazardous wastes is also costly. The amendment of calcines consolidated on-site, treated with MercLok and protected from erosion using routine best management practices, renders the remediation of remote sites feasible.

7. Recommendations for Future Direction

Site 1 warrants remediation based on the site assessments and field testing performed to date. Hydrologic controls recommended to stabilize the site and minimize the downstream transport of mining material and its leachate include (1) recontouring steep waste material piles and revegetating with native, drought-tolerant plants, (2) maintaining the earthen berm separating the adjacent road runoff, (3) partially filling a ditch to prevent further headcut, (4) maintaining the site's culvert entrance and outlet to ensure that runoff from the upper watershed does not flow through the mining material, and (5) isolating groundwater seepage from the mine waste and repairing rills and gullies in the mine area to prevent continued erosion. On-site calcines should be consolidated and then amended with MercLok as a repository cap with a reactive barrier to control any leachate, and the best management practices should be applied to minimize erosion.

There are many abandoned Hg mines in the Coast Range stemming from its legacy of Hg mining, and a broad set of stakeholders are interested in addressing them as a source of contamination. At prospective Hg mine remediation sites, MercLok may be tested on mine wastes such as waste rock, ore, and overburdens that leach Hg above the threshold for hazardous waste or at concentrations that pose a threat to water quality, as well as calcines to determine if hazardous characteristics can be removed through amendment, and/or if leachable Hg can be reduced to concentrations protective of water quality under site conditions.

Supplementary Materials: The following supporting information can be downloaded at: https://www.mdpi.com/article/10.3390/hydrology10070155/s1. Figure S1: Site 1 photos. (A) Calcine pile hardened and fractured above the stream channel. (B) Sampling leachate from test buckets at Site 1; Figure S2: Site 2 photos. (A) Handling MercLok supersacks for the calcine cap. (B) Digging the lower permeable reactive barrier (left) and filling it with MercLok-amended sandy soil (right). (C) Views of the remediated calcine pile from above (upper) and below (lower) with key components identified. Supplementary Technology Information with additional MercLok stability testing results. Supplementary Table S1 with the total metals in the leachate for Site 1.

Author Contributions: Conceptualization, G.R.; methodology, all; formal analysis, G.R.; investigation, all; resources, J.M. and K.P.; data curation, G.R. and S.M.; writing—original draft preparation, S.M.; writing—review and editing, G.R., J.M. and K.P.; visualization, G.R.; supervision, S.M.; project administration, K.P.; funding acquisition, J.M. and K.P. All authors have read and agreed to the published version of the manuscript.

Funding: This research was partially funded under contract by Albemarle Amendments, LLC, with in-kind contributions from the private landowners of the mine sites. The APC was funded by Albemarle Amendments, LLC.

Data Availability Statement: The data presented in this study are available on request from the corresponding author. The data are not publicly available due to on-going intellectual property work.

Acknowledgments: Lindsay Whalin, Sarah Acker and Keith Roberson at the San Francisco Bay Regional Water Quality Control Board (Board) and Jeff Huggins and Kyle Johnson at the Central Valley Board provided helpful technical review and regulatory oversight. Landowners (anonymous) provided significant in-kind support for accessing the sites, obtaining historical records, protecting equipment, and implementing site improvements.

Conflicts of Interest: Authors S.M. and G.R. were contracted to Albemarle to undertake the projects described. Authors J.M. and K.P. were employed by Albemarle while providing technical input to design and implement the projects, writing the manuscript; and deciding to publish the results. Chemical analyses were performed primarily at certified commercial labs as well as some feasibility and supplemental testing at Albemarle's research lab.

References

1. Jasinksi, S.M. The material flow of mercury in the United States. *Resour. Conserv. Recycl.* **1995**, *15*, 145–179.
2. California State Mining Bureau. In *The Quicksilver Resources of California*, Bulletin No. 27, 2nd ed.; California State Mining Bureau: San Francisco, CA, USA, 1908.
3. Kim, C.S.; Rytuba, J.J.; Brown, G.E. Geological and Geological and Anthropogenic Factors Influencing Mercury Speciation in Mine Wastes: An EXAFS Spectroscopy Study. *Appl. Geochem.* **2004**, *19*, 379. [CrossRef]
4. California Environmental Protection Agency. Available online: https://www.waterboards.ca.gov/water_issues/programs/water_quality_assessment/2020_2022_integrated_report.html (accessed on 25 April 2023).
5. California Office of Environmental Health Hazard Assessment. Available online: https://oehha.ca.gov/fish/advisories (accessed on 25 April 2023).
6. Eckley, C.S.; Gilmour, C.C.; Janssen, S.; Luxton, T.P.; Randall, P.M.; Whalin, L.; Austin, C. The assessment and remediation of mercury contaminated sites: A review of current approaches. *Sci. Total Environ.* **2020**, *707*, 136031. [CrossRef] [PubMed]
7. Kirchner, J.W.; Austin, C.; Myers, A.; Whyte, D.C. Quantifying Remediation Effectiveness under Variable External Forcing Using Contaminant Rating Curves. *Environ. Sci. Technol.* **2011**, *45*, 7874–7881. [CrossRef] [PubMed]
8. U.S. Bureau of Mines. *Mercury Potential of the United States*; Information Circular 8252; Bureau of Mines, U.S. Department of the Interior: Washington, DC, USA, 1965.
9. U.S. Environmental Protection Agency. SW-846 Test Method 1311: Toxicity Characteristic Leaching Procedure, Part of Test Methods for Evaluating Solid Waste, Physical/Chemical Methods; Washington, DC, USA, September 2022. Available online: https://www.epa.gov/hw-sw846/sw-846-test-method-1311-toxicity-characteristic-leaching-procedure (accessed on 24 June 2023).
10. U.S. Environmental Protection Agency. SW-846 Test Method 1313: Liquid-Solid Partitioning as a Function of Extract pH Using a Parallel Batch Extraction Procedure; Washington, DC, USA, May 2023. Available online: https://www.epa.gov/hw-sw846/sw-846-test-method-1313-liquid-solid-partitioning-function-extract-ph-using-parallel-batch (accessed on 24 June 2023).
11. SW-846 Test Method 1314: Liquid-Solid Partitioning as a Function of Liquid-Solid Ratio for Constituents in Solid Materials Using an Up-Flow Percolation Column Procedure; Washington, DC, USA, May 2022. Available online: https://www.epa.gov/hw-sw846/sw-846-test-method-1314-liquid-solid-partitioning-function-liquid-solid-ratio-constituents (accessed on 24 June 2023).
12. Organization for Economic Cooperation and Development. *Guideline for Testing of Chemicals. Earthworm Acute Toxicity Tests*; OECD: Paris, France, 1984.
13. U.S. Environmental Protection Agency. Method Guidance and Recommendations for Whole Effluent Toxicity Testing; 40 CFR Part 136, EPA 821-B-00-004; Washington, DC, USA, 2000. Available online: https://www.epa.gov/sites/default/files/2016-02/documents/method-guidance-recommendations-wet-testing_2000.pdf (accessed on 24 June 2023).
14. U.S. Environmental Protection Agency. Methods for Measuring the Acute Toxicity of Effluents and Receiving Waters to Freshwater and Marine Organisms, 5th ed.; EPA-821-R-02-012; Washington, DC, USA, 2002. 2002. Available online: https://www.epa.gov/sites/default/files/2015-08/documents/acute-freshwater-and-marine-wet-manual_2002.pdf (accessed on 24 June 2023).

Disclaimer/Publisher's Note: The statements, opinions and data contained in all publications are solely those of the individual author(s) and contributor(s) and not of MDPI and/or the editor(s). MDPI and/or the editor(s) disclaim responsibility for any injury to people or property resulting from any ideas, methods, instructions or products referred to in the content.

hydrology MDPI

Article

Simple and Complex Substrates (Sugar, Acetate and Milk Whey) for In Situ Bioremediation of Groundwater with Nitrate and Actinide Contamination

Ivan Myasnikov [1], Grigory Artemiev [2], Elena Lavrinovich [1], Irina Kazinskaya [1], Alexander Novikov [1] and Alexey Safonov [2,*]

[1] V.I. Vernadsky Institute of Geochemistry and Analytical Chemistry, Russian Academy of Sciences, Moscow 119991, Russia; myasnikov@geokhi.ru (I.M.); lavrelen@mail.ru (E.L.); novikov@geokhi.ru (I.K.); novikovap@mail.ru (A.N.)

[2] A.N. Frumkin Institute of Physical Chemistry and Electrochemistry, Russian Academy of Sciences, Moscow 119991, Russia; artemyev56@gmail.com

* Correspondence: alexeysafonof@gmail.com

Abstract: The complex contamination of groundwater near radioactive waste repositories by nitrates and actinides is a common problem for many nuclear fuel cycle facilities. One of the effective methods to remove nitrates and reduce actinide migration activity is bioremediation through the activation of native microbial communities by soluble electron donors and carbon sources. This work evaluated the effectiveness of using simple and complex electron donors to remove nitrate in the microbial community in an aquifer near the B2 storage of the Siberian Chemical Combine (Seversk, Siberia). The addition of sugar and milk whey led to the maximum efficiency of nitrate-ion removal and a decrease in the redox potential of the system, creating optimal conditions for the immobilization of actinide. Special attention was paid to the behavior of uranium, plutonium, neptunium, and americium under conditions simulating groundwater when sugar, acetate, and milk whey were added and when microbial metabolic products were formed. Neither microbial metabolites nor organic solutions were found to have a significant effect on the leaching of neptunium. At the same time, for plutonium, a decrease in yield was observed when rocks were treated with organic solutions were compared to groundwater treatment without them. Plutonium leaching is significantly affected by rock composition. In rocks with a low clay fraction content, its yield can reach 40%. At the same time, microbial metabolites can increase americium (Am) desorption from rocks with a low clay fraction content. Additionally, particle size analysis was performed using a step-by-step filtration approach, aiming to evaluate the risks that are associated with colloidal phase formation. It was shown that microbiological stimulation resulted in particle enlargement, substantially diminishing the presence of actinides in the form of dissolved or sub-50 nm nanoparticles. This outcome significantly reduced the potential for colloidal and pseudocolloidal transfer, thereby lowering associated risks.

Keywords: in situ bioremediation; actinides; aquifers; organic substates; milk whey; acetate; sugar; denitrification; desorption

Citation: Myasnikov, I.; Artemiev, G.; Lavrinovich, E.; Kazinskaya, I.; Novikov, A.; Safonov, A. Simple and Complex Substrates (Sugar, Acetate and Milk Whey) for In Situ Bioremediation of Groundwater with Nitrate and Actinide Contamination. *Hydrology* **2023**, *10*, 175. https://doi.org/10.3390/hydrology10080175

Academic Editor: Meysam Vadiati

Received: 2 July 2023
Revised: 10 August 2023
Accepted: 15 August 2023
Published: 18 August 2023

Copyright: © 2023 by the authors. Licensee MDPI, Basel, Switzerland. This article is an open access article distributed under the terms and conditions of the Creative Commons Attribution (CC BY) license (https://creativecommons.org/licenses/by/4.0/).

1. Introduction

The contamination of groundwater with actinides due to accidents and unsafe storage of radioactive waste poses a long-term uncontrolled migration risk and the potential for the entry of such pollutants into water supply systems [1–3]. Low-level radioactive waste repositories with open-type storage are considered to be a significant part of the groundwater contamination problem [4–6]. Such storage facilities were used in the uranium enrichment process and nuclear fuel reprocessing plants in Russia and other countries during the mid-20th century. Currently, these storage facilities are either undergoing preservation or have already been preserved, and the areas have been remediated [7–9]. However, after more than 50 years of operation, the protective properties of engineering barriers under the influence of technological and natural factors are gradually decreasing, and this results in increased permeability and subsequent contamination of aquifers with soluble waste components. For uranium mining, processing plants, and sublimation plants, the contamination of groundwater with macrocomponents, such as nitrate, sulfate, carbonate, and ammonium ions, as well as microcomponents, including radioactive elements (Ra, Th, and U) and stable elements (As, Cr, Fe, Mn, Ni, V, Zn, etc.), poses a significant issue [4,10,11]. In the case of open-type radioactive waste storage basins, besides nitrates and sulfates, the migration of soluble forms of long-lived radionuclides, such as actinides (U, Np, Pu, and Am) and fission products (Cl, Cs, Se, Sr, and Tc), into groundwater poses a greater risk. It is important to note that the solubility and the migration ability of many elements in higher oxidation states (U, Np, Pu, Se, and Tc) increase in the presence of nitrate and carbonate ions, certain organic complexing agents, iron-bearing clay, and organic colloidal particles [11–17]. The distribution of radioactive and chemical pollutants in groundwater, in addition to their chemical properties, depends on the porosity of aquifers, the mineral composition of surrounding rocks, and the characteristics of hydrodynamic regimes within aquifers. One of the factors influencing the migration of pollutants in groundwater is the biogeochemical one, which is driven by the activity of subsurface microflora capable of being activated by bioavailable elements present in the waste (C, Fe, N, S) [18–22]. In the case of complex contamination, one suitable method for removing nitrogen, sulfur, and actinides is in situ bioremediation, which involves the injection of inexpensive soluble substrates in order to activate the microbial community [23–26]. Microbial processes facilitate the removal of nitrate ions through denitrification (i.e., the reduction of nitrate ions to molecular nitrogen) and the removal of ammonium through nitrification (i.e., the two-step oxidation of ammonium in the presence of oxygen). This technology has proven successful, and it has been actively used worldwide for several decades [5,27,28]. A range of inexpensive soluble organic compounds, such as methanol, acetate [29,30], or more complex substrates, including vegetable oil [31,32], are used as substrates for bioremediation. The choice of substrate depends on the microbial composition of the groundwater and hydrogeological parameters. Hydrophobic complex substrates are used when groundwater flow rates are high [33].

In the case of bioremediation, an important task is to assess the impact of the substrate on the behavior of actinides. It is known that actinides could form soluble complexes with organic substances, which can increase their migration activity [2,3]. Complexes of uranium with isosaccharinic acid and other products of microbial metabolism have been well studied [34]. Some soluble substrates could lead to the desorption of actinides from rocks due to soluble complex formation. In our previous work, it was determined that bioremediation could reduce the risk of colloidal and pseudocolloidal transport of actinides through the aggregation of colloids by microbial metabolites during in situ bioremediation of groundwater near the B2 storage facility at the Siberian Chemical Combine [12]. In this case, milk whey, sugar, and acetate were all used in laboratory and field experiments. Their addition during in situ bioremediation led to the removal of nitrate ions and the formation of reducing conditions in the formation [24] that were favorable for the immobilization of actinides.

This study aims to select the most optimal substrate for nitrate removal in groundwater samples from the horizon near the B2 storage facility at the Siberian Chemical Combine in order to assess the possible impact of these substrates on actinide immobilization and remobilization as well as the risk of forming colloidal phases. The obtained results will serve to develop recommendations for industrial stakeholders to enhance the monitoring and the control of radioactive waste repositories and contaminated areas, thereby ensuring better environmental protection and safety measures in handling radioactive materials.

2. Materials and Methods

2.1. Water Samples

The main flow of groundwater in this aquifer is located at a depth of 10–20 m and depends on the local content of the clay phase of 10–30 m/year. When unloading the aquifer into the nearby river, soluble components of waste are possible. The study used groundwater samples taken from a highly polluted area (sample 1) and from an uncontaminated area (sample 2) near the B2 storage facility at the Siberian Chemical Combine at a 15 m depth (Table 1).

Table 1. Physico-chemical parameters of water samples.

Well	1	2
Sampling depth, m	15	15
TOS Salinity, mg/L	3952.0	109.0
pH	6.58	6.41
Eh	65	−30
Oxidizability, mg O/L	13.10	5.11
Fe(total)	0.25	2.38
Na^+	604.0	3.41
K^+	3.09	0.59
Ca^{2+}	316.60	15.39
Mg^{2+}	63.20	2.76
NH_4^+	7.64	<0.5
NO_3^-	2517.0	0.77
SO_4^{2-}	72.40	0.84
Cl^-	4.52	2.26
HCO_3^-	331.0	67.10
NO_2^-	<0.2	<0.2

Sample 1 was taken at a distance of 20 m from the storage in the direction of the groundwater flow. Sample 2 was taken at a distance of 60 m from the other side of the storage away from the direction of the groundwater flow. The technogenic impact was caused by an increase in the concentration of soluble waste components: nitrate ions, hydrocarbonate, and ions Na, K, Ca, Mg, and NH_4. A decrease in iron content was noted, which was probably due to its oxidation in an oxidizing environment as well as the formation of low-soluble phases. The composition of microbial communities has been described in our previous work [24]. It was found that the samples contained a significant group of bacteria capable of being nitrate reducers (denitrifiers) and sulfate reducers as well as iron reducers.

2.2. Sandy-Loam Samples

Rock (sandy loam) samples with different clay fraction contents from contaminated and background areas were used. The loam samples were taken at different distances from the repository at a 15 m depth. Samples s2, s3, and s4 were all taken near the repository, and they have a greater technogenic impact. Sample S2 was taken at the point of water sample 1, and sample s1 was taken at the point of water sample 2 in the uncontaminated zone. Sample s3 was taken at a distance of 30 m from the repository, and sample s4 was taken at a distance of 40 m from the repository. The choice of loam sample selection points was dictated by the importance of evaluating the technogenic factor in changing their sorption parameters in relation to actinides. All of the samples had a typical mineralogical composition, with a predominance of quartz from 36% (n5) to 53% (n4) and CMP from 10% (n5) to 23% (n2). Up to 25% albite was detected in the sample with a low KSP content. Goethite phase was detected in most of the samples, with a maximum content of up to 4% in sample s1. Samples s1, s3, s4, and s5 contained siderite up to 4% (s1,s5). Sample n2, taken near the repository, contained calcite, which was most likely due to the high calcium content from the repository. It is possible that anthropogenic influences led to a decrease in both goethite and siderite content and to an increase in both smectite and illite content (Tables 2 and 3).

Table 2. Elemental composition of sandy loams, mass%.

Sample	Na_2O	MgO	Al_2O_3	SiO_2	K_2O	CaO	TiO_2	MnO	Fe_2O_3	P_2O_5	S
s1	1.68	1.14	8.64	79.32	1.69	1.46	0.39	0.038	2.75	0.07	0.09
s2	3.18	2.80	7.78	74.20	1.98	4.45	0.01	0.001	2.15	0.05	2.78
s3	1.11	0.98	11.81	77.51	2.82	0.44	0.600	0.11	4.41	0.10	<0.02
s4	1.20	0.96	11.69	79.03	3.00	0.46	0.570	0.06	2.84	0.08	<0.02
s5	1.2	1.4	13.4	75.7	2.9	1.5	0.78	0.11	2.9	0.1	<0.02

Table 3. Mineral Composition of sandy loams, %.

Mineral Phase	Samples				
	s1	s2	s3	s4	s5
Quarz	50	42	49	53	36
Siderite	4	-	3	1	4
Calcite	-	4	-	-	-
Potassium feldspar	22	20	20	20	10
Albite	3	6	-	-	25
Amphibole	-	-	-	-	2
Goethite	4	2	3	1	-
Smectite	8	13	5	5	10
Kaolinite	3	3	5	5	5
Illite	5	10	5	5	5
Chlorite	-	-	-	-	3

At the same time, the total iron content in the sample was insignificantly different from the others. This could be a consequence of the dissolution of crystalline phases and the formation of both X-ray amorphous phases of iron oxide and the penetration of this element into the interlayer space of clay minerals. In the remaining samples, the clay mineral content ranged from 15% to 20%.

2.3. Methods

Substrate screening was carried out using 20 mL of water samples in 30 mL hermetically sealed penicillin vials, where dissolved sterilized organic additives were added at a concentration of 2–4 g/L, depending on the sample. Sterilized nitrate ions solution was added to sample 1 to a final concentration of 1000 mg/L. The experiment was carried out for 28 days by measuring the content of nitrate and nitrite ions in the water phase at a temperature of 12 °C, which was close to natural conditions. The gas phase was represented by argon. The rate of nitrate removal and nitrite formation was defined as the difference in their concentrations at the start and the end point, divided for an experimental time.

Sorption experiments were carried out with in sealed penicillin vials, which were preliminarily purged with argon. Radionuclide solutions (10^{-7} M of ^{233}U, ^{237}Np, and ^{239}Pu) were injected with an insulin syringe. The phase ratio was 0.1 g/mL. Radionuclides were added to air-dried rock samples (1 g weight) in nitrogen-oxide form with a maximum concentration in a minimum volume (0.01 mL) of solution. After addition, samples were moistened and stirred thoroughly for 60 days [14,35]. The samples were then air-dried and stored for 6 months in sealed tubes in a refrigerator. Solutions of a mixture of sugar, acetate, and whey (2 g/L) were used as the leaching aqueous phase. The volume of the aqueous phase was 5 mL. The establishment of equilibrium was monitored by conducting measurements on 1 mL aliquots of the solution at specific time intervals, including 15 min, 75 min, 6 days, and 34 days.

Desorption experiments. Radioisotopes ^{239}Pu, ^{237}Np, and ^{152}Eu in the form of nitric acid were introduced into 1 g air-dried rock samples, with the maximum concentration in a minimal volume (0.01 mL) of solution. After the introduction, the samples were moistened and subjected to thorough agitation for a duration of 60 days.

Subsequently, the samples were dried to an air-dried state and stored for 6 months in sealed vials in a refrigerator. A solution composed of a mixture of sugar, acetate, and whey (at a specific concentration) was utilized as the leaching aqueous phase. The volume of the aqueous phase was 5 mL. After removing the supernatant liquid, the rocks were treated with a 0.04 M $NH_2OH{\cdot}Cl$ solution in 25% CH_3COOH with a volume of 5 mL, followed by intensive agitation, and they were then left in contact for 10 days. Subsequently, the aqueous phase of the rocks was opened with a mixture of hydrofluoric and sulfuric acid in order to calculate the balance of the radioisotope content after radiochemical analysis.

The size of particles in the samples was determined by step-by-step filtration with syringe-mounted Vladipor (Russia) filters that were 2.4, 1.2, 0.8, 0.4, 0.22, 0.1, and 0.05 μm in diameter [36]. The concentration of the ions and radionuclides was determined in each filtrate.

Analysis of the loam samples composition was carried out by X-ray phase analysis on a Panalytical Aeris powder X-ray diffractometer with Cu Kα anode at 40 kV and 15 mA at the Shared Use Center, Institute of Physical Chemistry and Electrochemistry, Russian Academy of Sciences. For XRD analysis after drying, the samples were ground in a corundum mortar to a powder state. The survey was carried out on a PANalytical AERIS device (Malvern, United Kingdom). The sample was taken from 2° to 65° 2θ in 0.002 increments and a ¼ rad slit. The obtained results were interpreted using the HighScore Plus program, PDF2 database.

Anion and cation concentrations were measured using a CGE capillary electrophoresis system (Capel-105M, LUMEX Instruments, Sankt Peterburg, Russia).

The determination of Eh and pH values were carried out using an ANION-4100 pH meter/ionomer (Novisibirsk, Russia) with an electrode combination.

Radiochemical analytic. The equilibrium for neptunium was achieved at the slowest rate, taking 6 days, whereas europium and plutonium reached equilibrium within 75 min. Based on these observations, all desorption experiments were conducted with a desorbent solution incubation period of 6 days. The plutonium content was determined by alpha spectrometry, neptunium was determined by the luminescent method [36,37], and europium was determined by gamma spectrometry. In this study, europium was used as a model for studying americium. Therefore, the results will be provided as americium studies.

3. Results

3.1. Screening of Substrates for Nitrate Removal (Accumulation of Nitrite Ions) in Samples of Groundwater

Table 4 presents the results of the substrate screening for denitrifier activation in groundwater samples from wells 1 and 2. A sample from well 2, which had the lowest contamination, was supplemented with sodium nitrate at a concentration of 1 g/L. It was found that milk whey (MW), glucose (Gl), and sodium acetate (Ac) based on nitrate removal rates were the most optimal substrates for activating denitrification processes in both samples. These substrates resulted in low nitrite accumulation rates (columns 3, 5). Previously, using milk whey at various concentrations in field experiments showed high effectivity of nitrate removal in situ in our research [24] and in other research [37,38] due to its complex composition and the presence of bioavailable elements, including potassium, phosphorus, etc.

Table 4. Rates of nitrate removal/nitrite accumulation in groundwater samples 1 and 2, mg/L/day.

Substrates	Sample 1		Sample 2	
	NO_3^- Removal	NO_2^- Accumulation	NO_3^- Removal	NO_2^- Accumulation
Hydrogen	89.1	0.2	102.6	0.5
Methanol	106.2	5.0	41.8	7.9
Ethanol	96.5	2.9	92.2	8.4
Glycerin	45.7	2.9	55.8	3.9
Oxalate	33.4	8.6	32.4	11.4
Acetate	115.1	4.3	160.6	6.8
Lactate	86.5	4.3	39.5	3.9
Glucose	96.5	2.1	104.1	6.7
Sucrose	136.1	5.0	106.5	7.5
Milk whey	232.8	0.2	136.7	0.15
Brewing waste	222.1	16.7	94.0	0.56
Sunflower oil	78.6	16.0	56.8	1.1

3.2. Distribution of Radionuclide Forms in the Solution/Sediment System in A Sample of Natural Water

The distribution of actinides between the solid and liquid phases in samples before and after microbial processes are presented in Figure 1. At the beginning of the experiments, more than 85% of uranium was in the liquid phase, and the maximum removal effect was achieved by milk whey (MW), with a removal efficiency of over 85% from the solution. This can be attributed to the high intensity of the microbial processes and the presence of phosphates in MW, which could create low-solubility precipitates with uranium. The removal efficiency of uranium using acetate and sucrose reached 70–75%. For the other actinides, the maximum removal efficiency from the solution was also observed when MW was added, specifically 95% for neptunium, 98% for plutonium, and 92% for americium. The removal efficiency of americium and plutonium through microbial activation using acetate and sucrose was above 90%, while for neptunium, the removal efficiency was over 65% when sucrose was added, and was around 78% when acetate was added. It is important to note that in sample 2 with a nitrate content of 3.2 g/L, before microbial processes, more than 85% of uranium, 78% of neptunium, 55% of americium, and 45% of plutonium were present in the solution. This could be related to hydrolysis processes of actinides and their binding with clay and iron colloidal particles in water. Previously, we described the binding process of natural colloids during the flow of microbial processes in a sample of groundwater with the stimulation of microbial processes by molecular hydrogen [12].

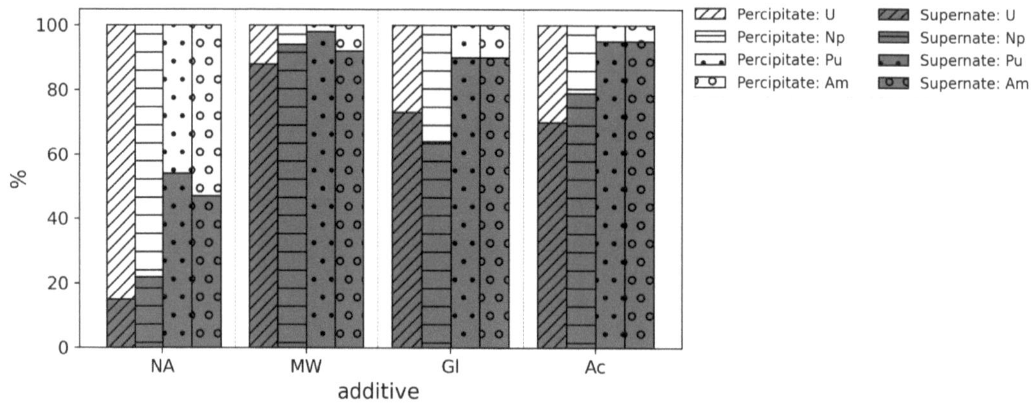

Figure 1. Actinide distribution between solid and liquid phase before and after microbial processes in sample 2 stimulated by milk whey (MW), glucose (Gl), acetate (Ac), and without stimulation (NA).

3.3. *Actinides Size Distribution in a Sample of Natural Water after Substrates Addition*

At Figure 2A–D, the distribution of actinides between the solid and liquid phases before and after microbial processes in sample 2 are shown. In the samples without microbial activation (NA), more than 60% of uranium (Figure 2A), 45% of neptunium (Figure 2B), 30% of plutonium (Figure 2C), and 50% of americium (Figure 2D) were dissolved or in the form of nanoparticles smaller than 50 nm.

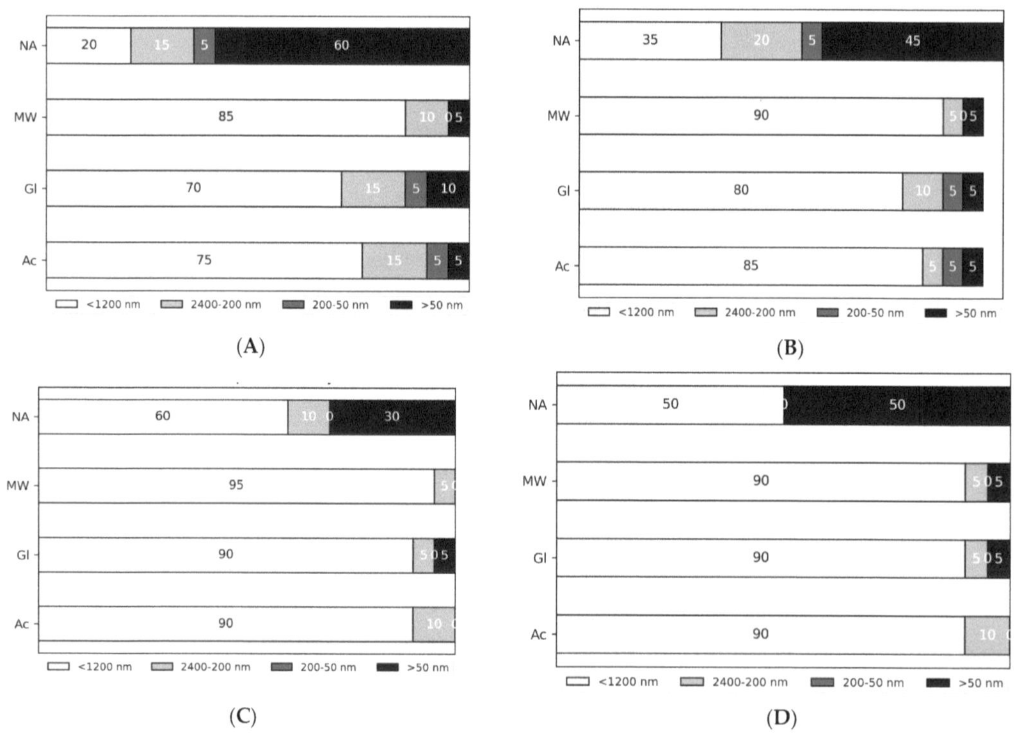

Figure 2. (**A**) Uranium content in particles by size distribution in sample 2 stimulated by milk whey (MW), glucose (Gl), acetate (Ac), and without stimulation (NA). (**B**) Neptunium content in particles

by size distribution in sample 2 stimulated by milk whey (MW), glucose (Gl), acetate (Ac), and without stimulation (NA). (**C**) Plutonium content in particles by size distribution in sample 2 stimulated by milk whey (MW), glucose (Gl), acetate (Ac), and without stimulation (NA). (**D**) Americium content in particles by size distribution in sample 2 stimulated by milk whey (MW), glucose (Gl), acetate (Ac), and without stimulation (NA).

It was found that when MW was added, more than 85% of uranium was present in particles larger than 1200 nm. When acetate and glucose were added, the content of particles larger than 1200 nm was 70–75%. Moreover, the content of dissolved and colloidal forms smaller than 50 nm was no more than 5% when MW and acetate were added, and it did not exceed 10% when glucose was added (Figure 2A). After microbiological stimulation in the case of neptunium, the formation of particles larger than 1200 nm was observed, contributing to over 80% of the total. The contribution of dissolved and small colloidal forms was below 5% (Figure 2B). In the case of plutonium and americium, the proportion of particles larger than 1200 nm was more than 90%, while particles smaller than 50 nm accounted for less than 50% (Figure 2C,D).

3.4. Actinides Sorption on Rock Samples in the Presence of Milk Whey, Sugar, Acetate, and Metabolites

Figure 3 presents the distribution coefficients of U, Np, Pu, and Am in rocks obtained from wells B2-32 and B2-38 in the presence of milk whey (MW), glucose (Gl), acetate (Ac), metabolites (Mb), and without additives (NA). For all of the radionuclides in each media, a slightly higher distribution coefficient was observed for sample 1 compared to sample 2. It should be mentioned that sample 1 contained a higher concentration of iron-bearing minerals, such as siderite and goethite (see Table 2), and this could contribute to better retention of radionuclides by the rock. Furthermore, the highest distribution coefficient was consistently found in the presence of milk whey, while the lowest was observed for the glucose. The highest distribution coefficient was noted for americium, with a comparable value for neptunium, while the coefficients for uranium and plutonium exceeded them by nearly threefold.

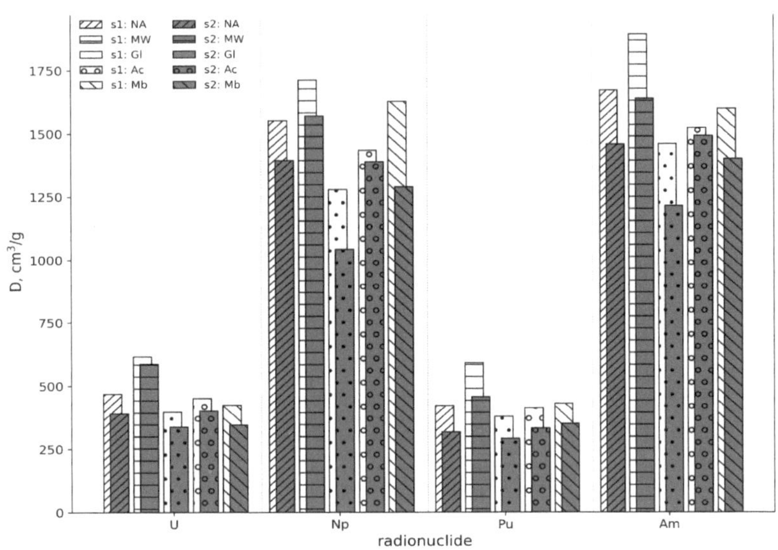

Figure 3. The distribution coefficient of Np, Pu, and Am in soils in the presence of organic additives and metabolites, cm^3/g NA, no additives, milk whey MW, glucose Gl, sodium acetate, Ac, metabolites Mb.

3.5. Desorption of Radionuclides from Rock Samples in the Presence of Substrates

The leaching was measured (Eq) into two different medium water and metabolites, and, after that, leaching by hydroxylamine (HA) was carried out. Figures 4–6 represent the leaching results of Pu, Np, and Am, respectively, into metabolites (right-hatch pattern) and water (left-hatch pattern), followed by the subsequent leaching in the presence of hydroxylamine (white-colored bars).

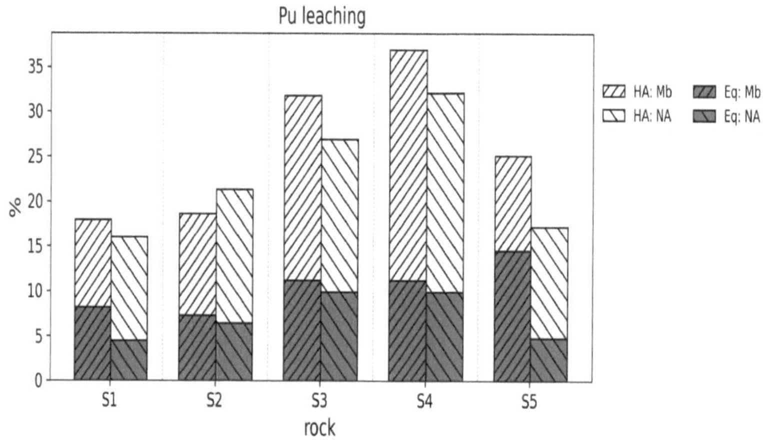

Figure 4. Total amount of plutonium desorption from the rock in the presence of metabolites (right-hatch pattern, white and grey bars) and without additives (left-hatch pattern, white and grey)—white bar's for natural water (Eq) and by hydroxylamine (left-hatch pattern).

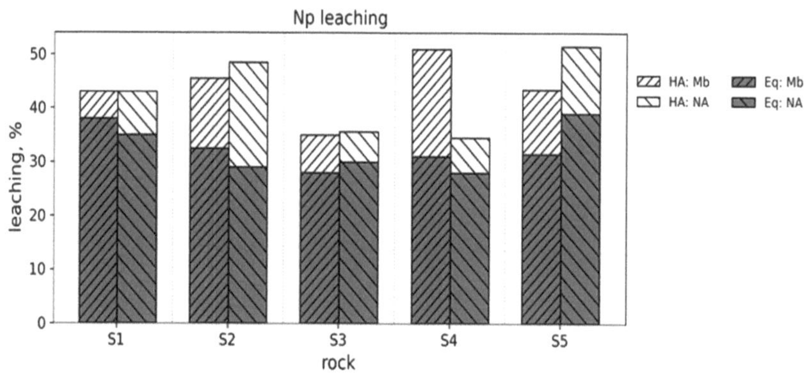

Figure 5. Total amount of neptunium desorption from the rock in the presence of metabolites (right-hatch pattern, white and grey bars) and without additives (left-hatch pattern, white and grey bars)—white bars for natural water (Eq) and by hydroxylamine (left-hatch pattern).

For Pu, Np, and Am, leaching percentages ranged from 16% to 37%, 14% to 70%, and 35% to 51.5% for all of the samples. The lower contribution of hydroxylamine to overall leaching in the case of Np was observed. Plutonium demonstrated the highest overall desorption in samples 3 and 4, while europium showed this trend in samples 2 and 5. Notably, plutonium exhibited comparable desorption values with both metabolites and water. In the case of americium, leaching with HA was generally higher, particularly in samples 2 and 5, which displayed maximum leaching. The leaching percentages of radionuclides into metabolites indicated a preferential interaction between the radionuclides and organic components that were present in the samples. As previously mentioned, minerals such as siderite and goethite could be crucial to the affinity of the rock with radionuclides. Sample 1 contained the highest fraction of siderite and goethite, while sample 3 had a value close to this but still lower. Furthermore, sample 5 showed a high concentration of siderite similar to sample 1, though it did not contain goethite. In addition, sample 4 had the lowest presence of siderite and goethite, while sample 2 did not contain siderite but did have a slightly higher amount of goethite.

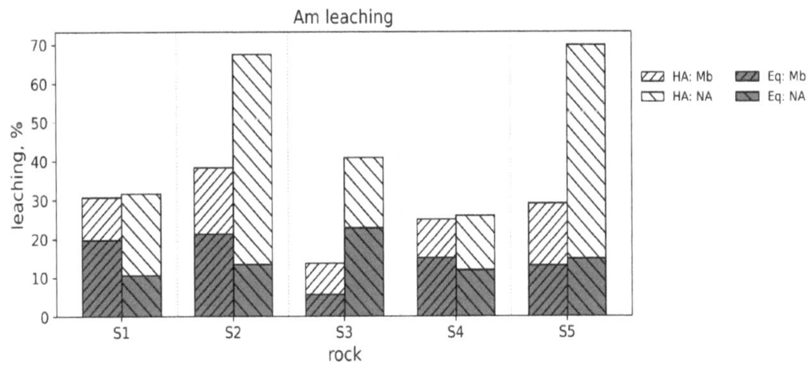

Figure 6. Total amount of americium desorption from the rock in the presence of metabolites (right-hatch pattern, white and grey bars) and without additives (left-hatch pattern, white and grey bars)—white bars for natural water (Eq) and by hydroxylamine (left-hatch pattern).

4. Conclusions

The comparison of various substrates for nitrate removal from groundwater and the evaluation of the impact of these substrates on americium, plutonium, and neptunium mobility, as well as the risks associated with colloidal phase formation, was conducted in this study.

The highest efficiency in activation of microbial communities was shown by complex substrates (milk whey) and monosubstrates (glucose, sugar, acetate). In laboratory experiments, these substrates do not affect the degree of leaching of all radionuclides. Moreover, the formation of large actinide-containing particles observed in the solution can lead to the sedimentation of clay and iron-containing particles, as we have previously shown in [12]. This can reduce the risk of its colloidal and pseudocolloidal transfer.

Based on the results of laboratory experiments, it was found that:

1. The mobility of neptunium was more significant than americium and plutonium, and the leaching by natural water was in a range from 28 to 39%.
2. For samples with high aluminum and sulfur content, a significant reduction in americium leaching was observed after pre-treatment of microbial activation. In the presence of hydroxylamine (HA), this difference reached up to 40% of the total.
3. In the case of the plutonium in the samples with a high aluminum content, a minor increase in leaching with pre-treatment was observed in the presence of HA. The proportion of plutonium leached without pre-treatment in this case constituted 70% to 80% of the total.
4. The occurrence of active microbial processes on rocks may be associated with the reduction of iron and the conversion of highly crystalline ferruginous phases into less crystalline ones with a higher sorption capacity with respect to actinides.

These results emphasize the importance of considering rock composition when developing strategies to modify the mobility of actinides using bioremediation techniques.

The interactions between hydroxylamine and rocks, specifically with iron-containing minerals as well as manganese oxides due to their solvation, reduces the effectiveness of the immobilization of actinides. Moreover, the specific phase of the mineral, whether crystalline or amorphous, during the interactions is important [2,11,20,39,40]. Additionally, to effectively retain plutonium and americium, the capacity of aluminosilicates is expected to be more important. Furthermore, it is worth mentioning that the impact of metabolites on the retention of americium also depends on the rock type. In some rocks, the use of metabolites leads to an increased retention of americium. Thus, it is shown that the method of in situ groundwater bioremediation can be safely applicable in case of its complex pollution with nitrogen compounds and actinides. Examples of such cases are aquifers near the surface of liquid waste storages in radiochemical plants at a stage of their conservation or in a stage of long-term protection in the post-conservation period. Moving forward, we intend to assess actinide behavior under varying pH conditions, focusing on the impact of alkaline and acidic risks during bioremediation. Additionally, we hope to study the role of microbial iron mineral transformation in actinide immobilization. The goal of these studies is to enhance our understanding of actinide interactions, which will lead to more effective bioremediation strategies for contaminated sites.

Author Contributions: Conceptualization, A.N. and A.S.; formal analysis, I.M., G.A. and A.S.; investigation, I.M., E.L., I.K. and A.S.; methodology, I.M., G.A., A.N. and A.S.; project administration, A.N.; software, G.A.; supervision, A.N. and A.S.; writing—original draft, I.M., A.N. and A.S. All authors have read and agreed to the published version of the manuscript.

Funding: This research was funded by Russian Science Foundation, grant number No. 23-24-00381, https://rscf.ru/en/project/23-24-00381/.

Data Availability Statement: Not applicable.

Conflicts of Interest: The authors declare no conflict of interest.

References

1. Rawson, S.A.; Walton, J.C.; Baca, R.G. Migration of actinides from a transuranic waste disposal site in the vadose zone. *Radiochim. Acta* **1991**, *52*, 477–486. [CrossRef]
2. Maher, K.; Bargar, J.R.; Brown, G.E., Jr. Environmental speciation of actinides. *Inorg. Chem.* **2013**, *52*, 3510–3532. [CrossRef] [PubMed]
3. Choppin, G. Actinide speciation in the environment. *J. Radioanal. Nucl. Chem.* **2007**, *273*, 695–703. [CrossRef]
4. Safonov, A.V.; Boguslavsky, A.E.; Gaskova, O.L.; Boldyrev, K.A.; Shvartseva, O.S.; Khvashchevskaya, A.A.; Popova, N.M. Biogeochemical modelling of uranium immobilization and aquifer remediation strategies near NCCP sludge storage facilities. *Appl. Sci.* **2021**, *11*, 2875. [CrossRef]
5. Nash, K.L.; Jensen, M.P.; Schmidt, M.A. In-situ mineralization of actinides for groundwater cleanup: Laboratory demonstration with soil from the Fernald Environmental Management Project. In *Science and Technology for Disposal of Radioactive Tank Wastes*, 1st ed.; Schulz, W.W., Lombardo, N.J., Eds.; Springer: New York, NY, USA, 1998; pp. 507–518.
6. Buesseler, K.O.; Dai, M.; Repeta, D.J. *Speciation, Mobility and Fate of Actinides in the Groundwater at the Hanford Site*; Woods Hole Oceanographic Institution: Falmouth, MA, USA, 2003.
7. Razvorotneva, L.I.; Boguslavskii, A.E.; Markovich, T.I. Geochemical aspects of environmentally safe conservation of liquid radioactive waste. *Radiochemistry* **2016**, *58*, 317–322. [CrossRef]
8. Pronkin, N.S.; Sharafutdinov, R.B.; Kovalevich, O.M.; Smetnik, A.A.; Levin, A.G.; Kabakchi, S.A.; Masanov, O.L. Hazard classification of holding ponds for liquid radioactive wastes. *AT Energy* **2003**, *94*, 398–404. [CrossRef]
9. Bayliss, C.; Langley, K. *Nuclear Decommissioning, Waste Management, and Environmental Site Remediation*; Elsevier: Amsterdam, The Netherlands, 2003.
10. Boguslavsky, A.E.; Gaskova, O.L.; Naymushina, O.S.; Popova, N.M.; Safonov, A.V. Environmental monitoring of low-level radioactive waste disposal in electrochemical plant facilities in Zelenogorsk, Russia. *Appl. Geochem.* **2020**, *119*, 104598.
11. Safonov, A.; Popova, N.; Boldyrev, K.; Lavrinovich, E.; Boeva, N.; Artemiev, G.; Kuzovkina, E.; Emelyanov, A.; Myasnikov, I.; Zakharova, E.; et al. The microbial impact on U, Pu, Np, and Am immobilization on aquifer sandy rocks, collected at the deep LRW injection site. *J. Geochem. Explor.* **2022**, *240*, 107052. [CrossRef]
12. Safonov, A.; Lavrinovich, E.; Emel'yanov, A.; Boldyrev, K.; Kuryakov, V.; Rodygina, N.; Zakharova, E.; Novikov, A. Risk of colloidal and pseudo-colloidal transport of actinides in nitrate contaminated groundwater near a radioactive waste repository after bioremediation. *Sci. Rep.* **2022**, *12*, 4557. [CrossRef]

13. Baumer, T.; Hellebrandt, S.; Maulden, E.; Pearce, C.I.; Emerson, H.P.; Zavarin, M.; Kersting, A.B. Pu distribution among mixed waste components at the Hanford legacy site, USA and implications to long-term migration. *Appl. Geochem.* **2022**, *141*, 105304. [CrossRef]
14. Novikov, A.; Vlasova, I.; Safonov, A.; Ermolaev, V.; Zakharova, E.; Kalmykov, S. Speciation of actinides in groundwater samples collected near deep nuclear waste repositories. *J. Environ. Radioact.* **2018**, *192*, 334–341. [CrossRef] [PubMed]
15. Kalmykov, S.N.; Zakharova, E.V.; Novikov, A.P.; Myasoedov, B.F.; Utsunomiya, S. Effect of redox conditions on actinide speciation and partitioning with colloidal matter. In *Actinide Nanoparticle Research*; Springer: Berlin/Heidelberg, Germany, 2011; pp. 361–375.
16. Mibus, J.; Sachs, S.; Pfingsten, W.; Nebelung, C.; Bernhard, G. Migration of uranium (IV)/(VI) in the presence of humic acids in quartz sand: A laboratory column study. *J. Contam. Hydrol.* **2007**, *89*, 199–217. [CrossRef] [PubMed]
17. Santschi, P.H.; Kimberly, A.R.; Laodong, G. Organic nature of colloidal actinides transported in surface water environments. *Environ. Sci. Technol.* **2022**, *36*, 3711–3719. [CrossRef] [PubMed]
18. Andryushchenko, N.D.; Safonov, A.V.; Babich, T.L.; Ivanov, P.V.; Konevnik, Y.V.; Kondrashova, A.A.; Proshin, I.M.; Zakharova, E.V. Sorption characteristics of materials of the filtration barrier in upper aquifers contaminated with radionuclides. *Radiochemistry* **2017**, *59*, 414–424. [CrossRef]
19. Safonov, A.V.; Andryushchenko, N.D.; Ivanov, P.V.; Boldyrev, K.A.; Babich, T.L.; German, K.E.; Zakharova, E.V. Biogenic Factors of Radionuclide Immobilization on Sandy Rocks of Upper Aquifers. *Radiochemistry* **2019**, *61*, 99–108. [CrossRef]
20. Safonov, A.V.; Boguslavskii, A.E.; Boldyrev, K.A.; Zayceva, L.V. Biogenic Factors of Formation of Geochemical Uranium Anomalies near the Sludge Storage of the Novosibirsk Chemical Concentrate Plant. *Geochem. Int.* **2019**, *57*, 709–715. [CrossRef]
21. Gillow, J.B.; Dunn, M.; Francis, A.J.; Lucero, D.A.; Papenguth, H.W. The potential of subterranean microbes in facilitating actinide migration at the Grimsel Test Site and Waste Isolation Pilot Plant. *Radiochim. Acta* **2022**, *88*, 769–775. [CrossRef]
22. Banaszak, J.; Rittmann, B.; Reed, D. Subsurface interactions of actinide species and microorganisms: Implications for the bioremediation of actinide-organic mixtures. *J. Radioanal. Nucl. Chem.* **1999**, *241*, 385–435. [CrossRef]
23. Nash, K.; Jensen, M.; Schmidt, M. Actinide immobilization in the subsurface environment by in-situ treatment with a hydrolytically unstable organophosphorus complexant: Uranyl uptake by calcium phytate. *J. Alloys Compd.* **1998**, *271-273*, 257–261. [CrossRef]
24. Safonov, A.V.; Babich, T.L.; Sokolova, D.S.; Grouzdev, D.S.; Tourova, T.P.; Poltaraus, A.B.; Zakharova, E.V.; Merkel, A.Y.; Novikov, A.P.; Nazina, T.N. Microbial Community and in situ Bioremediation of Groundwater by Nitrate Removal in the Zone of a Radioactive Waste Surface Repository. *Front. Microbiol.* **2018**, *9*, 1985. [CrossRef]
25. Renshaw, J.C.; Butchins, L.J.C.; Livens, F.R.; May, I.; Charnock, J.M.; Lloyd, J.R. Bioreduction of Uranium: Environmental Implications of a Pentavalent Intermediate. *Environ. Sci. Technol.* **2005**, *39*, 5657–5660. [CrossRef] [PubMed]
26. Newsome, L.; Morris, K.; Lloyd, J.R. The biogeochemistry and bioremediation of uranium and other priority radionuclides. *Chem. Geol.* **2013**, *363*, 164–184. [CrossRef]
27. Kanematsu, H.; Barry, D.M. Environmental problems: Soil and underground water treatment and bioremedi-ation. In *Biofilm and Materials Science*; Springer: Cham, Switzerland, 2015; pp. 117–123.
28. Xu, M.; Wu, W.-M.; Wu, L.; He, Z.; Van Nostrand, J.D.; Deng, Y.; Luo, J.; Carley, J.; Ginder-Vogel, M.; Gentry, T.J.; et al. Responses of microbial community functional structures to pilot-scale uranium in situ bioremediation. *ISME J.* **2010**, *4*, 1060–1070. [CrossRef] [PubMed]
29. Madden, A.S.; Palumbo, A.V.; Ravel, B.; Vishnivetskaya, T.A.; Phelps, T.J.; Schadt, C.W.; Brandt, C.C. Donor-dependent extent of uranium reduction for bioremediation of contaminated sediment mi-crocosms. *J. Environ. Qual.* **2009**, *38*, 53–60. [CrossRef]
30. Anderson, R.T.; Vrionis, H.A.; Ortiz-Bernad, I.; Resch, C.T.; Long, P.E.; Dayvault, R.; Karp, K.; Marutzky, S.; Metzler, D.R.; Peacock, A.; et al. Stimulating the In Situ Activity of *Geobacter* Species To Remove Uranium from the Groundwater of a Uranium-Contaminated Aquifer. *Appl. Environ. Microbiol.* **2003**, *69*, 5884–5891. [CrossRef] [PubMed]
31. Watson, D.B.; Wu, W.-M.; Mehlhorn, T.; Tang, G.; Earles, J.; Lowe, K.; Gihring, T.M.; Zhang, G.; Phillips, J.; Boyanov, M.I.; et al. In Situ Bioremediation of Uranium with Emulsified Vegetable Oil as the Electron Donor. *Environ. Sci. Technol.* **2013**, *47*, 6440–6448. [CrossRef] [PubMed]
32. Hall, S. *Groundwater Restoration at Uranium In-Situ Recovery Mines, South Texas Coastal Plain*; US Geological Survey: Denver, CO, USA, 2009. [CrossRef]
33. Williamson, A.J.; Binet, M.; Sergeant, C. Radionuclide biogeochemistry: From bioremediation toward the treatment of aqueous radioactive effluents. *Crit. Rev. Biotechnol.* **2023**, 1–19. [CrossRef]
34. Kobayashi, T.; Sasaki, T.; Kitamura, A. Thermodynamic interpretation of uranium(IV/VI) solubility in the presence of α-isosaccharinic acid. *J. Chem. Thermodyn.* **2019**, *138*, 151–158. [CrossRef]
35. Novikov, A.P.; Goryachenkova, T.A.; Sobakin, P.I.; Kazinskaya, I.E.; Ryleeva, V.S. Speciation of plutonium and americium in the soils affected by Kraton-3 accidental underground nuclear explosion in Yakutia (Russia). *J. Radioanal. Nucl. Chem.* **2015**, *307*, 691–697. [CrossRef]
36. Novikov, A.P.; Safonov, A.V.; Babich, T.L.; Boldyrev, K.A.; Kryuchkov, D.V.; Lavrinovich, E.A.; Kuzovkina, E.V.; Emel'yanov, A.M.; Goryachenkova, T.A. Biotransformation of Neptunium in Model Groundwaters. *Geochem. Int.* **2020**, *58*, 182–188. [CrossRef]
37. Margalef-Marti, R.; Carrey, R.; Soler, A.; Otero, N. Evaluating the potential use of a dairy industry residue to induce denitrification in polluted water bodies: A flow-through experiment. *J. Environ. Manag.* **2019**, *245*, 86–94. [CrossRef]

38. Banerjee, S.; Kundu, A.; Dhak, P. Bioremediation of uranium from waste effluents using novel biosorbents: A review. *J. Radioanal. Nucl. Chem.* **2022**, *331*, 2409–2435. [CrossRef]
39. Del Nero, M.; Froideval, A.; Gaillard, C.; Mignot, G.; Barillon, R.; Munier, I.; Ozgümüs, A. Mechanisms of uranyl sorption. *Geol. Soc. London, Spéc. Publ.* **2004**, *236*, 545–560. [CrossRef]
40. Neu, M.P.; Boukhalfa, H.; Merroun, M.L. Biomineralization and biotransformations of actinide materials. *MRS Bull.* **2010**, *35*, 849–857. [CrossRef]

Disclaimer/Publisher's Note: The statements, opinions and data contained in all publications are solely those of the individual author(s) and contributor(s) and not of MDPI and/or the editor(s). MDPI and/or the editor(s) disclaim responsibility for any injury to people or property resulting from any ideas, methods, instructions or products referred to in the content.

hydrology

Article

Hydropedological Characterization of a Coal Mining Waste Deposition Area Affected by Self-Burning

Jorge Espinha Marques [1,*], Aracelis Narayan [1], Patrícia Santos [1], Joana Ribeiro [2,3], Sara C. Antunes [4,5], Armindo Melo [6,7], Fernando Rocha [8], Deolinda Flores [1] and Catarina Mansilha [6,7]

1 Institute of Earth Sciences, Department of Geosciences, Environment and Spatial Planning, Faculty of Sciences, University of Porto, Rua do Campo Alegre, s/n, 4169-007 Porto, Portugal; up201912339@edu.fc.up.pt (A.N.); patricia.santos@fc.up.pt (P.S.); dflores@fc.up.pt (D.F.)
2 Institute Dom Luiz, Rua Sílvio Lima, 3030-790 Coimbra, Portugal; joana.ribeiro@uc.pt
3 University of Coimbra, Department of Earth Sciences, Rua Sílvio Lima, 3030-790 Coimbra, Portugal
4 Interdisciplinary Centre of Marine and Environmental Research, Av. General Norton de Matos, s/n, 4450-208 Matosinhos, Portugal; scantunes@fc.up.pt
5 Department of Biology, Faculty of Sciences, University of Porto, Rua do Campo Alegre, s/n, 4169-007 Porto, Portugal
6 National Institute of Health Doutor Ricardo Jorge, Department of Environmental Health, Rua Alexandre Herculano, 321, 4000-055 Porto, Portugal; armindo.melo@insa.min-saude.pt (A.M.); catarina.mansilha@insa.min-saude.pt (C.M.)
7 LAQV/REQUIMTE, University of Porto, Praça do Coronel Pacheco, 15, 4050-453 Porto, Portugal
8 GeoBioTec, Department of Geosciences, University of Aveiro, Campus de Santiago, 3810-193 Aveiro, Portugal; tavares.rocha@ua.pt
* Correspondence: jespinha@fc.up.pt

Abstract: Coal mining often produces severe environmental effects, including impacts on the soil system and, specifically, on hydropedological conditions that control the leaching of significant ions and Potentially Toxic Elements (PTEs). The research objective is to assess changes in the hydropedological conditions in an area with a coal mining waste pile that underwent self-burning. An integrative approach was implemented, starting with the definition of hydropedological zoning based on field observations of soil formation factors (namely, parent material, relief, biological activity, anthropic influence, and time). The soil profile in each hydropedological zone was characterized regarding morphological features. The upper mineral horizons were sampled and characterized in terms of mineralogy and PTE geochemistry. Field measurements of unsaturated hydraulic conductivity, soil water content, and hydrophobicity were performed. Afterwards, the hydrogeochemistry of leachates was determined, and the soil leaching potential was evaluated. The research outcomes express substantial differences regarding the hydropedological zones: development of different soil profiles, diverse mineralogy and PTE geochemistry, higher unsaturated hydraulic conductivity and leaching of major ions, and PTEs in soils affected by coal mining activities. Finally, a Principal Component Analysis confirmed the existence of significant contrasts according to hydropedological zoning.

Keywords: hydropedology; coal mining; soil leaching; soil and water pollution; coal waste self-burning

Citation: Espinha Marques, J.; Narayan, A.; Santos, P.; Ribeiro, J.; Antunes, S.C.; Melo, A.; Rocha, F.; Flores, D.; Mansilha, C. Hydropedological Characterization of a Coal Mining Waste Deposition Area Affected by Self-Burning. *Hydrology* **2024**, *11*, 62. https://doi.org/10.3390/hydrology11050062

Academic Editor: Meysam Vadiati

Received: 21 March 2024
Revised: 19 April 2024
Accepted: 20 April 2024
Published: 25 April 2024

Copyright: © 2024 by the authors. Licensee MDPI, Basel, Switzerland. This article is an open access article distributed under the terms and conditions of the Creative Commons Attribution (CC BY) license (https://creativecommons.org/licenses/by/4.0/).

1. Introduction

Mining activities, especially the operation and disposal of mining residues, usually produce severe environmental and social impacts, including changes in land use originating ecosystem disturbance, loss of biodiversity, as well as soil and groundwater and surface water bodies degradation, e.g., [1,2]. In many countries, current mining activities follow regulations that require and allow environmental protection [3]. However, past mining activities created environmental impacts that persist nowadays. Indeed, the exploitation of mineral resources may induce profound transformations in the local hydrological processes as well as interrupt the regular pedological evolution, with long-term impacts on habitats and landscapes at both local and regional scales [4]. As a result, the hydrological and pedological features of the soil system often change dramatically. For example, mining frequently produces deforestation, which may disturb rainfall interception, overland flow, infiltration, and percolation. In contrast, the disposal of mining wastes in waste piles forms new soils—Technosols—frequently with different mineralogical and geochemical characteristics. In these environmental conditions, mining areas may develop new hydropedological characteristics, namely, the volume and velocity of water circulating in the upper unsaturated zone and the leaching potential of significant ions and Potentially Toxic Elements (PTEs). In this context, to understand how this type of environment functions, it is very useful to adopt a hydropedological perspective, which is the basis of an integrative scientific approach focusing on hydropedological zoning.

Coal mining activities have the potential to accumulate and release high concentrations of PTEs into nearby soils and water, contributing to environmental degradation [5–8]. Residues from coal mining pose environmental hazards and can potentially lead to human health issues [9,10], as several PTEs are frequent and abundant in coals [11]. Coal mining presents some specific hazards when compared with the exploitation of metals [5,12–14]. Spontaneous combustion and self-burning of coal in seams, stockpiles, and mining waste deposits is known worldwide as the cause of economic losses and environmental impacts, principally related to the emission of gases to the atmosphere and the alteration of the mode of occurrence of toxic compounds (including inorganic and organic compounds). Additionally, water percolation and acid drainage in coaly materials may also promote the leaching of compounds naturally occurring in coals, including PTEs and polycyclic aromatic hydrocarbons.

Coal mining in the Douro Coalfield (NW Portugal) occurred from the end of the 18th century until 1994. The thin and elongated shape of the coalfield explains the existence of various underground mines which stand out among others for their dimension and importance, the São Pedro da Cova and Pejão mines. The legacy of the coal mining in the Douro Coalfield is materialized today by the existence of dozens of waste piles. Five of them underwent self-burning after ignition caused by wildfires.

Previous studies investigated the waste piles in the Douro Coalfield [15–20] including: (i) the characterization of waste materials and identification of potential environmental impacts of coal waste piles; (ii) the identification of changes in mining waste caused by self-burning and the products generated by the combustion process in waste piles, as well as the associated potential impacts on the environment and human health; and (iii) the integrated monitoring of the combustion process ongoing in the São Pedro da Cova waste pile. Recently, the characterization of the residues deposited in the Fojo waste pile (in the Pejão area) was assessed by [21].

This research is part of a multidisciplinary project encompassing other subjects, namely ecotoxicology and environmental geochemistry (the latest is ongoing). The ecotoxicological study assessed the effects of soil leachates in seed germination and individual and subindividual parameters in *Lactuca sativa* [22], as well as the toxicity effects of soil leachates in aquatic species (*Allivibrio fischeri*, *Lemna minor*, and *Daphnia magna*)—[23]. In the ecotoxicological study, pedological zoning was defined, and the hydropedological zoning established in the present work (see Sections 2 and 3) is a development and upgrade of this preliminary zoning.

The research objective is to assess changes in the hydropedological conditions in an area with a coal mining waste pile that underwent self-burning. The assessment was based on an integrative approach encompassing hydrology, geology, pedology, mineralogy, geochemistry, and hydrogeochemistry and started with the definition of the hydropedological zoning, considering field observations related to soil formation factors (namely, parent material, relief, biological activity, anthropic influence, and time)—as described by [24,25]—and their relation with local hydrological conditions and processes (specifically, rainfall interception, infiltration, percolation, overland flow, and interflow). One of the hydropedological zones represents conditions prior to coal mining influence, while others reflect anthropogenic influence, that is, the deposition of coal mining waste and, to a lesser extent, intensive forestry.

The study began with the observation of local hydrological conditions and processes, including rainfall interception, infiltration, percolation, overland flow, and interflow. Then, the soil profile was described in each hydropedological zone regarding morphological features, and the upper mineral horizons were sampled and characterized in terms of mineralogy and PTE geochemistry. Field measurements of unsaturated hydraulic conductivity (K), soil water content, and hydrophobicity were performed. Afterwards, the hydrogeochemistry of leachates was determined and the soil leaching potential was evaluated.

The research results revealed that coal mining waste disposal in the Fojo area originated new soil types with hydropedological characteristics different from the pre-depositional setting.

2. Materials and Methods

2.1. Study Area

The study site encompasses the Fojo coal mine waste pile, and the surrounding area and is located in the Pejão mining area, Douro Coalfield, NW Portugal (Figure 1). In the Pejão area, coal mining was carried out from 1920 until 1994 and originated a number of waste piles [15]. The ignition and consequent self-burning of the Fojo waste pile was caused by a wildfire in October 2017. Afterwards, between 2017 and 2019, an operation to control and extinguish the coal waste self-burning was conducted, comprising the remobilization of the coal waste and the application of water mixed with a cooling accelerator agent.

Figure 1. Fojo coal mine waste pile and the surrounding area and sampling sites (satellite image from *Google Earth*).

Due to the self-burning process and the fire control and extinction operation, significant changes occurred in the local pedological and hydrological conditions. These alterations led to the development of new environmental features, allowing the definition of a hydropedological zoning comprising the waste pile and its surrounding area (Figure 1): Uphill Soil (US), Unburned Coal Waste Pile (UCW), Burned Coal Waste Pile (BCW), Mixed Burned Coal Waste (MBW), Downhill Soil (DS). Further descriptions of these hydropedological zones are provided in Section 3.1.

2.2. Field Methods

The hydropedological characterization of the Fojo coal mine waste pile and the surrounding area was carried out from May 2021 (first field survey) until July 2023 (last hydropedological field measurements). The hydropedological zoning of the study area was defined in accordance with the soil mapping of [26] and the World Reference Base for Soil Resources [27]. In each zone, the morphological description of the soil profile followed the FAO's guidelines [28], considering the following characteristics: (i) the formation of soil horizons, in the case of normal pedogenesis as well in the case of the influence of mine waste deposition and self-burning; (ii) the depth and thickness of the soil horizons; (iii) the type of soil horizon boundaries; (iv) the type of aggregation—structure; (v) soil texture; (vi) soil colour—described using the Munsell colour system; (vii) accumulation of humified organic matter; (viii) soil porosity; and (ix) biological activity.

Twenty-nine sites were selected in the study area for soil sampling and field measurement of unsaturated hydraulic conductivity (K), soil water content, and hydrophobicity (Figure 1). In all soil types (as detailed in Section 3), the uppermost mineral horizon was sampled to a depth of 20 cm. In the BCW zone, samples were taken from the cover layer and the waste itself, in both cases, to a depth of 20 cm. Unsaturated hydraulic conductivity (K) was assessed using a mini disk infiltrometer—e.g., [29–31]. All tests were performed with a suction rate of −1 cm. The volumetric water content was measured with a capacitance probe, specifically the ThetaProbe model ML3. Soil hydrophobicity was measured using the Water Drop Penetration Time (WDPT), using the procedure outlined by [32]. The hydropedological field measurements were carried out in March 2022 (wet season) and in July 2023 (dry season) in the A horizon of the US and DS zones, as well as in the C horizon of the UCW and the MBW zones and the C1 and C2 horizons of the BCW zone. An additional sample (BW10) from the deepest part of the burned waste pile was also collected close to the NE limit of the BCW hydropedological zone to provide a preliminary insight into the conditions prevailing in this environment.

Water from interflow in the self-burning waste pile, which may occur after periods of heavy rainfall, was sampled in March 2022.

2.3. Laboratory Methods

The soil geochemistry concerning PTEs was determined at the Bureau Veritas Mineral Laboratories (Vancouver, BC, Canada) via inductively coupled plasma mass spectrometry (ICP-MS) after ignition at 550 °C and acid digestion using an acid solution of (2:2:1:1) H_2O-HF-$HClO_4$-HNO_3.

The soil mineralogy was characterized by X-ray diffraction (XRD). First, samples were dried at about 50 °C and then disaggregated. The sedimentation method was applied to separate the fractions with particle size under 2 µm. For the fractions under 63 µm and 2 µm, the mineralogical analysis was carried out with a Panalytical X'Pert-Pro MPD, Kα Cu (λ = 1.5405 Å) radiation on random-oriented powders (total sample) and oriented aggregates (<2 µm). The mineralogical composition was assessed using (hkl) peaks (on random powder mounts) for non-clay minerals and (001) peaks (on oriented aggregates) for clay minerals; the mineral phases were identified through the criteria recommended by [33,34] and the Joint Committee for Powder Diffraction Standards. The semi-quantification of the mineralogical determinations was performed using the procedures described by [35,36].

The leaching of major ions, PTEs, and Fe in soils was assessed using the USGS Field Leach Test—USGS FLT, [37]. This leaching method is simple to apply and time-effective, allowing the simulation, prediction, and characterization of the water–soil geochemical interaction during the percolation of rainwater in the upper soil profile. The pH, electrical conductivity (EC), and alkalinity were measured in unfiltered leachate subsamples. Leachates were then filtered using 0.45 μm pore-size nitrocellulose membrane filters and a glass vacuum filtration apparatus. Filtrates were collected and preserved for analysis, which was performed according to procedures outlined in [38,39].

The values of pH and EC were determined using a Crison MultiMeter MM 41. Total alkalinity and bicarbonate (HCO_3^-) were analysed by titration. Total organic carbon (TOC) was analysed using a Shimadzu TOC-V (TOC-ASI-V, Shimadzu Corporation, Kyoto, Japan); Potentially Toxic Elements (Cr, Mn, Ni, Cu, Zn, As, Al, Cd, and Pb), and Fe, were analysed in a Varian AA240 Atomic Absorption Spectrometer (Varian Inc., Palo Alto, CA, USA) and a Continuous Segmented Flow Instrument (CSF) (San-Plus Skalar, Skalar Analytical, Breda, The Netherlands), respectively. The major inorganic ions (Na^+, K^+, Mg^{2+}, Ca^{2+}, Cl^-, NO_3^-, and SO_4^{2-}) were analysed by ion chromatography, CI (DionexTM system DX-120/ICS-1000, Dionex Corporation, Sunnyvale, CA, USA).

As for water from interflow in the self-burning waste pile, the same parameters and components were analysed as in the case of soil leachates.

2.4. Statistical Analysis

A Principal Component Analysis (PCA) was performed to relate soil types in each hydropedological zone with the composition of soil leachates. Before conducting the multivariate analysis, the data underwent standardization to address the high variability in parameter values. Additionally, redundant variables were eliminated to streamline the analysis. A Principal Component Analysis (PCA) was then carried out using the software CANOCO for Windows version 4.5®.

3. Results and Discussion

The research outcomes reveal significant contrasts in the studied hydropedological features, as described in the following subsections. Indeed, the results highlighted apparent differences among the hydropedological zones defined in the study area, which are mainly driven by soil-forming factors: the type of parent material (in situ metasedimentary rock, unburned coal mining waste, burned coal mining waste, as well as a mixture of all these types of materials), topography both of the coal mining waste pile and the surrounding area, biological influence (type of vegetation cover), time, and anthropogenic influence (including the spatial distribution of coal mine waste accumulation and the self-burning process).

The observed hydropedological contrasts encompass the morphological characteristics of the soil profile, soil mineralogy and geochemistry, unsaturated hydraulic conductivity, soil water content, soil hydrophobicity, and hydrogeochemistry of soil leachates. These features are crucial for understanding the water–soil interaction and, consequently, comprehending the susceptibility of the coal waste pile and the surrounding soils to leaching.

Soil leaching also depends on other hydrological factors that influence the water available for infiltration and percolation, such as the volume of precipitation, interception, and evapotranspiration. Factors related to water movement in the porous media, namely K and hydrophobicity, are also relevant.

3.1. Hydropedological Zoning and Soil Morphology

In the study area, soils were classified as Regosols and Technosols, according to the criteria of the World Reference Base for Soil Resources [27]. Before the self-burning period (2017–2019), pedogenesis was driven by the regular functioning of the soil formation factors. However, self-burning disrupted this process in part of the waste pile, forming a distinct Technosol. Based on the environmental features after the self-burning process, hydropedological zoning was established, comprising the waste pile and its surroundings, in which every hydropedological zone corresponds to a specific environment (Figures 1 and 2):

Figure 2. Aspects of the Fojo coal mining waste pile and the surrounding area: (**a**) vegetation cover in the Uphill Soil; (**b**) soil profile in the Uphill Soil; (**c**) Unburned Coal Waste; (**d**) Mixed Burned Coal Waste; (**e**) sampling point of the Burned Waste layer (BW), and Cover Layer (CL); (**f**) Burned Waste layer; (**g**) Downhill Soil; (**h**) application of the mini disk infiltrometer in the Unburned Coal Waste.

(i) Uphill Soil (US): with a soil type representing the geological and hydropedological conditions before the deposition of the coal mining waste. This zone is situated in the highest part of the study area and is free from the influence of coal mining waste. The parent material consists of carboniferous metasedimentary rocks. The vegetation cover consists mainly of *Eucalyptus globulus*, *Acacia melanoxylon*, *Acacia dealbata*, and *Pinus pinaster*.

The soil profile is O-Ah-C-R. The analytical results and field measurements refer to the Ah horizon.

(ii) Unburned Coal Waste Pile (UCW): with a soil type composed of ordinary coal mining waste, in which self-burning did not take place. The vegetation cover is recent and consists mainly of *Acacia dealbata* and *Eucalyptus globulus*. The prevailing soil profile is O-A-C. The accumulation of humified organic matter in A horizon is very incipient and does not occur in part of the hydropedological zone. The analytical results and field measurements refer to the C horizon.

(iii) Burned Coal Waste Pile (BCW): with a soil type composed of coal mining waste influenced by self-burning (BW), covered by a 30–40 cm protective layer consisting of mixed material from the C horizon and especially the R horizon of nearby soils with several metasedimentary parent materials. The protective layer is referred to as the Cover Layer (CL). Vegetation is almost absent. The soil profile is C1–C2, with the C1 horizon corresponding to the CL protective layer and the C2 horizon to the BW layer. The analytical results and field measurements refer to both horizon C1 and horizon C2.

(iv) Mixed Burned Coal Waste (MBW): with a soil type consisting of a mixture of US, UCW, and BW material. The vegetation cover consists of *Eucalyptus globulus* planted in 2020. The soil profile is C-R. The R horizon consists of carboniferous metasedimentary rocks. The analytical results and field measurements refer to the C horizon.

(v) Downhill Soil (DS): with a soil type where the prevailing geological and pedological conditions are the ones before the deposition of the coal mining waste. However, since this zone is situated in the lowest part of the study, this soil type is influenced by the waste located uphill, namely, through the input of coal waste debris and water from overland flow and interflow. The vegetation cover consists mainly of *Eucalyptus globulus*. The soil profile is O-Ap-C-R. The analytical results and field measurements refer to the Ap horizon.

Regosols correspond to soils located uphill and downhill from the Fojo coal mining waste pile. In contrast, Technosols correspond to soils with all or part of the parent material composed of coal mining waste (Figures 1 and 2). According to the available soil mapping [26] and the criteria of the World Reference Base for Soil Resources [27], soil types from US and DS hydropedological zones have Regosol features, while soil types from UCW, BCW, and MBW hydropedological zones have Technosol features. Tables S1–S5 present selected morphological features of each soil type in the study area according to the FAO guidelines for soil description [28].

3.2. Hydropedological Field Measurements

Field tests were conducted at each soil sampling point (Figures 1 and 2). Unsaturated hydraulic conductivity (K) measurements were carried out using a mini disk infiltrometer (applying a suction rate of −1 cm), while the volumetric water content was measured using a capacitance probe, and the WDPT was applied to evaluate hydrophobicity. The results concerning K, volumetric water content, and hydrophobicity are presented in Table 1.

The mini disk infiltrometer test results indicate that, in all soils, the magnitude of K remains consistent between the wet season and the dry season, suggesting that the volumetric water content is not a key factor controlling this hydropedological feature. As for hydrophobicity, all soils are wettable throughout the year except for the soil from the US zone, which is severely water-repellent. In the Ah horizon (US zone), measured K is 0.00 cm/s in the wet and dry seasons, while most of the remaining K mean values are around 10^{-4} cm/s. In the UCW zone (C horizon), K values range from 1.39×10^{-4} cm/s to 1.79×10^{-3} cm/s. In the BCW zone, K values range from 3.05×10^{-5} cm/s to 1.90×10^{-3} cm/s in the C1 horizon (CL material) and from 1.76×10^{-4} cm/s to 5.50×10^{-3} cm/s in the C2 horizon (BW material). In the MBW zone (C horizon), K values range from 2.42×10^{-4} cm/s to 1.83×10^{-3} cm/s. In the DS zone (Ap horizon), K values range from 3.74×10^{-5} cm/s to 5.94×10^{-4} cm/s. Finally, the BW10 sample, composed of material from the deepest part of the burned waste pile, presents K in the same order of magnitude as the lowest values measured in the C2 horizon of the BCW zone (that is, 10^{-4} cm/s).

Table 1. Unsaturated hydraulic conductivity, volumetric water content, and hydrophobicity measured in soils from the Fojo coal mine waste pile and the surrounding area.

Hydropedological Zone/Soil Sample	Soil Horizon/ Number of Measurements	March 2022		July 2023		Hydrophobicity
		K (cm/s)	VWC (%)	K (cm/s)	VWC (%)	
US	Ah (n = 5)	0.00	8.5	0.00	6.5	Severely water-repellent
UCW	C (n = 5)	8.59×10^{-4}	15.6	7.47×10^{-4}	12.1	Wettable
MBW	C (n = 6)	8.12×10^{-4}	24.4	6.97×10^{-4}	10.3	Wettable
BCW	C1 (n = 9)	7.45×10^{-4}	19.2	5.90×10^{-4}	8.8	Wettable
	C2 (n = 9)	1.10×10^{-3}	34.2	1.10×10^{-3}	14.4	Wettable
DS	Ap (n = 3)	1.65×10^{-4}	25.3	6.78×10^{-4}	7.2	Wettable
BW10	C2 (n = 1)	9.19×10^{-4}	15.3	7.83×10^{-4}	5.4	Wettable

Mean values of unsaturated hydraulic conductivity (K) and volumetric water content (VWC); mini disk suction rate: −1 cm; hydrophobicity classification according to [32].

3.3. Soil Mineralogy

The analytical results highlight distinct mineralogical signatures within each hydropedological zone in the fine-earth and clay fractions.

The overall mineralogical composition of the fine-earth fraction (Table 2) is silicate, with an absolute predominance of Quartz, followed by phyllosilicates (essentially Muscovite and Kaolinite) and feldspars (potassic and calc-sodic). Iron oxides (Hematite and, in some samples, Goethite) and Titanium oxides (Anatase) are present in almost all the samples, as well as sulphates, namely, Jarosite and Alunite. As for carbonates, only Siderite was detected, which is relatively frequent but in small quantities.

The fine-earth fraction also reveals some mineralogical differences among hydropedological zones. The US soil is characterized by its almost monomineralic composition, given the predominance of Quartz (85.8%). The UCW soil is differentiated by some accessory minerals, namely, the higher values of Chlorite, Siderite, and Pyrite. The MBW soil shows an indistinct composition which reflects the mixture of US, UCW, and BCW material. In the case of the CL layer (C1 horizon of the BCW soil), Quartz is much less abundant than in the US soil (58%), offset by greater quantities of Muscovite, Kaolinite, and Chlorite; as accessory minerals, Goethite (instead of Hematite) and Siderite stand out. In addition to containing less quartz (55.8%), the BW layer (C2 horizon of the BCW soil) is noticeably enriched in Hematite, Jarosite, and Alunite. The presence of Siderite and Pyrite in the UW soil is related to the reducing environment in the coal mine rock massif, while the presence of Jarosite and Alunite in the BW layer reflects the oxidizing environment in the waste pile affected by self-burning. The DS soil, despite having a similar amount of Quartz to the US soil (82,5%), is characterized by some accessory minerals, namely because it is the only one in which the ubiquitous presence, although always discreet, of Opal C/CT, Zeolites, and Pyrophyllite has been identified. The fine-earth mineralogical composition of the BW10 sample is quite similar to that of the BW material.

Regarding the overall mineralogical composition of the clay fraction (Table 3), Illite is dominant in all soils, followed by Kaolinite. Smectite and Chlorite are also common, while Pyrophyllite is somewhat less abundant.

The clay fraction results also show distinctive aspects. The US soil is characterized by the very significant presence of Smectite at the expense of a relative decrease in Illite. The UCW soil shows a relative enrichment in Chlorite and Pyrophyllite. In addition to Illite and Kaolinite, the MBW soil is also characterized by the presence of Smectite. In the BCW soil, the CL layer is essentially distinguished by some of the accessory minerals, namely the ubiquitous presence of Smectite, Chlorite, and Pyrophyllite. An almost monomineralic composition characterizes the BW layer, such is the predominance of Illite.

However, the BW10 sample has more diversified clay mineralogy, including Smectite, Chlorite, and Pyrophyllite, like the DS soil, which is enriched in these minerals and depleted in Illite.

Table 2. Mean mineralogical composition (%) of the fine-earth fraction of soils from the Fojo coal mine waste pile and its surrounding area.

Minerals	Hydropedological Zones						BW10 (n = 1)
	US (n = 2)	UW (n = 2)	MBW (n = 2)	BCW		DS (n = 2)	
				CL (n = 5)	BW (n = 5)		
Quartz	85.8	52.8	69.0	58.0	55.5	82.5	55.0
K Feldspar	1.3	3.3	1.3	5.5	5.2	3.5	5.5
Plagioclase	4.3	7.3	5.0	5.6	5.8	3.0	4.5
Opal C/CT	0.0	0.0	0.0	0.0	0.0	traces	0.0
Zeolites	0.0	0.0	0.0	0.0	0.0	traces	0.0
Muscovite	4.5	13.0	9.5	12.9	12.4	4.3	13.5
Kaolinite	2.0	3.5	2.5	4.1	2.5	1.0	1.5
Chlorite	0.0	2.8	0.0	1.0	0.0	0.0	0.0
Pyrophyllite	0.0	0.0	0.0	0.0	0.0	traces	0.0
Hematite	0.0	1.5	3.8	0.9	6.5	0.0	8.0
Goethite	0.0	1.0	1.0	1.4	0.0	0.0	0.0
Anatase	2.3	4.5	2.5	3.7	3.6	2.0	3.5
Siderite	0.0	2.5	1.3	1.5	0.8	0.8	0.5
Pyrite	0.0	4.0	0.0	0.0	0.0	0.0	0.0
Jarosite	0.0	2.5	2.8	2.8	4.8	2.3	5.0
Alunite	0.0	2.0	1.5	2.8	3.3	1.0	3.5
Gypsum/Anhydrite	0.0	0.0	0.0	0.6	0.0	0.0	0.0

Table 3. Mean mineralogical composition (%) of the clay fraction of soils from the Fojo coal mine waste pile and its surrounding area.

Minerals	Hydropedological Zones						BW10 (n = 1)
	US (n = 2)	UW (n = 2)	MBW (n = 2)	BCW		DS (n = 2)	
				CL (n = 5)	BW (n = 5)		
Smectite	27.5	0.0	2.8	7.8	0.0	10.0	9.0
Chlorite	0.0	7.0	0.0	5.4	0.0	8.5	6.0
Illite	53.5	72.0	83.5	65.0	88.0	55.0	65.0
Pyrophyllite	0.0	5.0	0.0	1.8	0.0	15.5	10.0
Kaolinite	19.0	16.0	14.0	20.0	12.0	11.0	10.0

3.4. Hydrogeochemistry of Soil Leachates and Interflow

The results of the soil leaching tests showed that the water–soil interaction concerning major ions, Fe, and PTEs differs considerably according to the hydropedological zone. Similar contrasts were also observed in the pH and EC of leachates. In March 2022, after a rainy period, interflow was observed at the base of the waste pile, and a water sample was collected to carry out a hydrogeochemical characterisation similar to that of the leachates. Mean values of pH, EC, TOC, and major ions in leachates are presented in Table 4, whereas Fe and PTE mean contents are presented in Table 5 (which also includes soil geochemistry of PTEs for comparison purposes). The hydrogeochemical features are illustrated in Figures 3 and 4.

Table 4. Major ion content, pH, EC, and TOC of soil leachates and interflow from the Fojo coal mine waste pile and its surrounding area.

Hydropedological Zone/Soil Sample/Interflow		pH	EC	TOC	HCO_3^-	Cl^-	SO_4^{2-}	Na^+	K^+	Ca^{2+}	Mg^{2+}	Σ Major Ions
US (n = 5)		5.7	11	6.15	2.95	1.14	0.78	1.76	0.52	0.62	0.17	7.94
UW (n = 5)		4.7	67	3.33	1.05	0.58	20.27	0.42	0.88	4.05	1.67	28.92
MBW (n = 6)		4.1	65	0.90	0.05	0.35	19.49	0.89	0.77	5.02	1.26	27.83
BCW	CL (n = 9)	5.0	23	1.04	1.64	0.60	7.11	0.75	0.60	1.30	0.88	12.85
	BW (n = 9)	4.2	61	0.67	<0.05	0.34	19.25	0.73	0.70	2.90	1.42	25.35
DS (n = 3)		4.7	49	1.72	0.71	0.60	17.61	1.23	0.96	2.00	2.26	25.37
BW10 (n = 1)		4.7	28	0.62	<0.05	0.37	8.98	0.57	1.16	1.81	0.61	13.50
Interflow (n = 1)		3.7	2690	12.20	<0.05	10.20	2635.00	83.00	11.00	301.50	206.50	3247.20

Mean values; EC (electrical conductivity) in µS/cm; major ion and Total Organic Carbon (TOC) content in mg/L. Detection limits of analytical methods (LD) for leachates: 0.05 mg/L for HCO_3^-; 0.03 mg/L for Cl^-, SO_4^{2-}, Na^+, K^+, Ca^{2+}, Mg^{2+}.

Table 5. Potentially Toxic Elements and Fe content of soil leachates, soil, and interflow from the Fojo coal mine waste pile and its surrounding area.

Hydropedological Zone/Soil Sample/ Interflow			Fe	Al	Mn	As	Cd	Ni	Cu	Cr	Zn	Pb
US (n = 5)		Leachate	77.78	64.36	1.00	1.17	n.d.	0.91	1.70	0.38	n.d.	0.20
		Soil	1.97	5.71	18.20	17.16	0.01	9.52	10.16	49.60	14.32	22,61
UW (n = 5)		Leachate	66.51	97.19	170.45	0.48	0.04	7.58	2.52	n.d.	35.60	1.05
		Soil	4.55	8.37	230.00	53.62	0.17	29.44	51.52	87.40	74.44	68,37
MBW (n = 6)		Leachate	21.81	441.88	62.42	0.23	0.15	9.72	9.42	0.36	8.00	2.76
		Soil	5.46	8.53	135.50	52.98	0.12	20.82	49.20	89.33	66.65	47,76
BCW	CL (n = 9)	Leachate	77.55	43.02	59.88	0.21	0.03	5.59	1.83	0.06	0.22	1.03
		Soil	5.59	12.53	83.78	39.04	0.07	33.37	58.17	100.89	37.68	46.16
	BW (n = 9)	Leachate	32.23	276.57	84.18	0.32	n.d.	8.92	2.43	n.d.	39.78	n.d.
		Soil	5.94	9.37	134.67	58.94	0.15	22.13	51.79	90.22	68.86	52.55
DS (n = 3)		Leachate	24.27	154.93	385.57	0.57	0.07	9.20	4.70	n.d.	7.33	n.d.
		Soil	3.01	9.12	334.33	35.90	0.02	22.57	24.97	82.67	42.10	31,75
BW10 (n = 1)		Leachate	77.28	29.63	78.70	<LD	n.d.	3.83	n.d.	n.d.	126.00	n.d.
		Soil	6.44	9.33	196.00	79.60	0.12	20.80	41.3	102.00	64.2	62.58
Interflow (n = 1)			2948.00	1002.00	14,600.00	2.70	12.00	137.20	110.00	6.45	n.a.	n.d.

Mean values; concentrations in leachates and interflow in µg/L; concentrations in soil in mg/kg; n.a.—not analysed; n.d.—not detected. Detection limits of analytical methods (LD) for leachates: 1 µg/L for Cd, As, Pb, Ni, Cu, Zn, and Cr; 15 µg/L for Fe and Al.

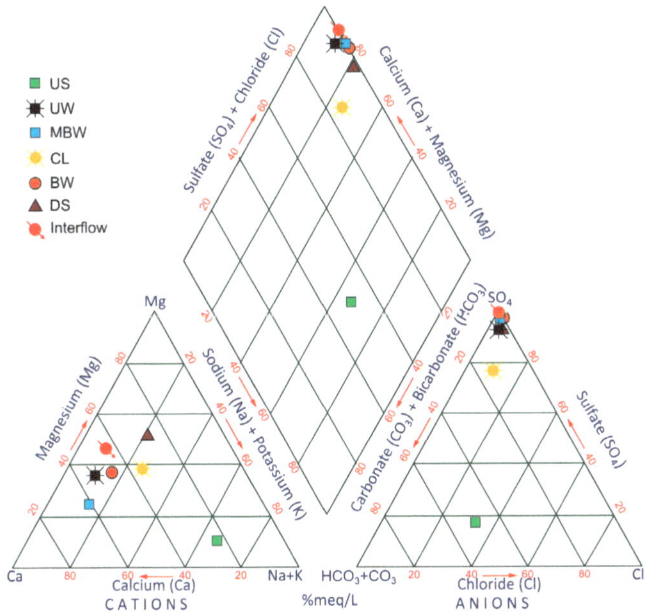

Figure 3. Piper diagram of soil leachates and interflow from the Fojo coal mine waste pile and the surrounding area.

The pH values are higher in samples without coal mining influence, specifically, 5.7 in US and 5.0 in CL. In the case of the samples with coal mining influence, pH values are more acidic, especially in materials with self-burning: 4.2 in BW and 4.1 in MBW. Soil electrical conductivity also reflects the coal mining influence since the lower values were measured in the US and CL samples (11 µS/cm and 22 µS/cm, respectively). In comparison, the higher values were measured in UCW, BW, and MBW (67 µS/cm, 58 µS/cm, and 65 µS/cm, respectively). In the DS samples, intermediate pH and EC values were measured. The TOC in leachates is higher in soils with humified organic matter, particularly in the US soil, but also in the BCW soil, the BW layer of the BCW soil, and the DS soil.

Regarding major ion content, US and CL leachates show much lower mineralisation than the UCW, BW, MBW, and DS leachates, which is consistent with the EC results. The hydrogeochemical facies of leachates are Na-HCO_3 and Mg-SO_4 for US and CL, respectively, Ca-SO_4 in the case of UCW, BW, and MBW, and Mg-Ca-SO_4 in the case of DS. The BW10 leachate features are analogous to BW, except for the lower SO_4 content.

The results also revealed distinct distribution patterns of PTEs (Al, Mn, As, Cd, Ni, Cu, Cr, Zn, Pb) and Fe according to the hydropedological zone. The concentration of PTEs and Fe in leachates is represented in Figure 4, showing higher values in soil samples influenced by coal mining (UCW, MBW, and BW) when compared with US and CL. Indeed, the UW and BW leachates have similar total concentrations of PTEs and Fe, with more significant differences observed in Al, Fe, Mn, and Zn. Compared to the BCW material, the BW10 sample contains more Fe (consistent with pyrite weathering in an environment with more intense self-burning) and Zn. However, this sample is relatively depleted in Al and, to a lesser extent, in Mn, Ni, and Cu.

The comparison of soil leachate hydrogeochemistry with soil geochemistry in Table 5 also highlights a different leaching potential in PTEs and Fe. In the case of Mn and Zn, higher hydrogeochemical concentrations correspond to higher geochemical concentrations, while in the case of Fe, Al, As, Cd, Ni, Cu, Cr, and Pb, this relation is not apparent.

Figure 4. Mean contents of major ions, Potentially Toxic Elements, and Fe of soil leachates from the Fojo coal mine waste pile and the surrounding area.

The hydrogeochemistry of soil leachates reveals that the water–soil interaction in the upper unsaturated zone is more intense in the soils affected by coal mining, resulting in higher concentrations of readily soluble major ions and PTEs. Also, the DS soils show signs of coal mining influence (namely, the relatively high concentration of SO_4, Ca, Mg, Al, Mn, Ni, Cu, and Zn). This influence results from overland flow and interflow in the coal mine waste pile, which has probably been going on for decades since the waste was first deposited but has intensified since the self-burning event. Overland flow in the BCW and MBW hydropedological zones (Figure 1) was observed during heavy precipitation events. This process promoted the transport of solid particles and dissolved chemical compounds into the DS zone. Moreover, water from interflow emerging at the base of the waste pile, with a high concentration of major ions, PTEs, and Fe (see Tables 4 and 5), also reaches the DS material, contributing to the change in its geochemical and hydrogeochemical signature.

The hydrogeochemistry of interflow results from a longer and deeper flow path, in which the fluid becomes progressively more acidic due to the weathering of pyrite, promoting the interaction with the mine waste, resulting in a much more intense leaching of major ions, PTEs, and Fe. The pH and EC values are 3.7 and 2690 µS/cm, respectively. The mineralisation is two orders of magnitude higher than in the soil leachates, and the hydrogeochemical facies is Ca-SO_4. Likewise, the concentrations of Fe and several PTEs in interflow are more than one order of magnitude higher than in soil leachates, namely, in the case of Al, Mn, As, Cd, Ni, Cu, and Cr.

3.5. Hydropedological Setting and Soil Leaching

Field observations and measurements, along with mineralogy, geochemistry, and hydrogeochemistry results, pointed out that coal mining produced severe changes in the hydropedological conditions of the study area. The US hydropedological zone represents the conditions prior to the mining waste disposal, characterized by the least favourable setting for soil leaching. In this case, the rainfall interception caused by the abundant vegetation cover, which has existed in this location for decades, if not centuries, decreases the volume of water that reaches the ground surface and may originate infiltration. Additionally, the infiltration capacity of the Ah horizon is also reduced by its hydrophobic nature (originated by soil organic matter, e.g., [32]), which seems to persist during most of the year (see Table 1), increasing overland flow e.g., [40,41]. Under these conditions, the unsaturated hydraulic conductivity in the Ah horizon is very low or even negligible, and the water infiltration and percolation in the upper soil horizons primarily take place during the wettest events of the year, originating hydropedological conditions less favourable to leaching. Water percolation is also limited at greater depth, particularly from the R horizon downwards, due to the relatively low permeability of the fractured metasedimentary rock. Another relevant feature of the US zone is that the pedological evolution and the soil leaching have been going on for centuries or even longer. Consequently, the topsoil has become geochemically depleted in leachable compounds and enriched in minerals resistant to leaching, thus explaining the lower concentration of major ions, PTEs, and Fe in the soil leachates.

Moreover, the conditions in the hydropedological zones affected by coal mining, as well as in the DS zone, are more favourable to water infiltration/percolation in the upper horizons and may induce solute transport into the deepest part of the unsaturated zone and the unconfined aquifer. In the UCW and MBW hydropedological zones, the vegetation cover is recent and less dense than in the US zone, and the A horizon is incipient and very shallow (in part of the UCW zone) or absent (in the remaining UCW zone and all of the MBW zone). In these circumstances, not only is the interception much less effective than in the US zone, but the soil is much less hydrophobic as a result of the scarcity or lack of humified organic matter, allowing for higher infiltration and percolation in a more permeable porous medium, especially during the wet season, when most of the precipitation occurs, and evapotranspiration is lower. This situation is even more extreme in the BCW zone, where the vegetation is almost absent, the soil is hydrophilic, and K is one order of magnitude higher in the C2 horizon (see Table 1). In addition, the more permeable porous medium in the burned coal waste only gives way to the fractured medium at a greater depth than in the remaining hydropedological zones, originating a longer flow path with more favourable conditions for leaching.

Also, the mineralogical composition of the UCW and MBW soils reflects the incipient pedogenetic evolution and is more diversified and abundant in leachable compounds. In the case of the BW layer of the BCW soil, the self-burning process induced important mineralogical transformations, including oxidising processes leading to the destruction of Pyrite and Siderite followed by the formation of Hematite, Jarosite, and Alunite. As a result of the weathering of Pyrite, the water percolating through the BW material becomes progressively acidic and more able to promote leaching, resulting in highly mineralized interflow water.

The DS soil, which, before the coal mine waste deposition, was similar to the US, has been under the influence of the coal mine waste pile for decades. This influence includes the transport of solid particles (with size from clay to gravel), especially during overland flow in the BCW and MBW hydropedological zones, as well as dissolved chemical components from interflow through the deepest part of the burned waste pile in an acidic environment.

The hydrogeochemical signature of this influence includes higher contents of SO_4^{2-}, Ca^{2+}, Mg^{2+}, Al, Mn, Ni, Cu, and Zn than the US soil.

As confirmed by PCA statistical analysis, the chemical composition of soil leachates reflects the hydropedological zoning. The PCA graph (Figure 5) clearly differentiates soil samples from the studied hydropedological zones. The US samples are clustered regarding the pH of soil and leachates, TOC, HCO_3, Cl, Na, Fe, and As. Soils with self-burning influence (BW layer and MBW) are clustered in terms of SO_4, Mg, Ca, Ni, Mn, Al, Cu, Zn, and Pb. On the other hand, the UCW samples appear dispersed (possibly because the coal waste pile is composed of somewhat heterogeneous material with different petrological features and weathering), while the CL samples are clustered close to the central part of the graphic. Nevertheless, self-burning, together with the waste mobilization during the fire control operation, seems to cause a process of homogenization with regard to leaching in the BW layer. A similar effect is observed in the MBW due to soil mobilization for planting eucalyptus. Finally, the DS samples define a boundary between the US and the burned material (BW and MBW).

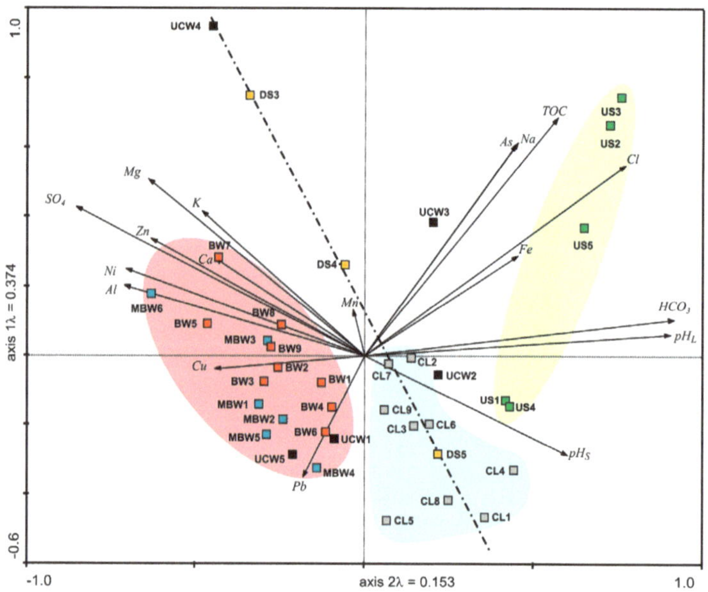

Figure 5. PCA analysis of chemical parameters in soil leachates from the Fojo waste pile and the surrounding area; Uphill Soil (US) cluster in green; Burned Waste layer (BW) and Mixed Burned waste (MBW) cluster in red; Cover Layer (CL) cluster in blue; the dashed line marks the boundary defined by the Downhill Soil (DS) samples; pH_L-leachates pH; pH_S-soil pH.

4. Concluding Remarks

The environmental impact of coal mining on soils and water bodies is a well-established fact. In coal mining areas, landscapes and ecosystems often undergo dramatic transformations, accompanied by changes in hydrological and pedological features and processes. The local water cycle is disturbed concerning rainfall interception, overland flow, infiltration, and percolation. Additionally, the local soil system is altered by waste disposal in piles, which are often large in volume and area, originating new soil types classified as Technosols.

A hydropedological perspective (in the form of the definition of hydropedological zoning) is valuable for understanding the water–soil interaction in this environment. It also provides a basis for an integrative scientific approach for assessing changes in processes and features in the upper unsaturated zone.

The research results showed clear contrasts between the hydropedological zones in all aspects studied, from soil morphology to soil leaching. Indeed, the hydropedological conditions favour rainfall infiltration, percolation, and leaching of major ions and PTEs in the upper soil horizons of hydropedological zones with mining influence compared to the US zone, which represents the pre-waste deposition setting. In addition, the leaching potential in the US zone is expected to be lower because of the soil mineralogy and the percolation into the deepest part of the soil profile (C horizon and, especially, R horizon), which takes place in a fractured medium corresponding to a material of lower permeability and lower specific surface than in the case of UCW, BCW, and MBW zones. Also, based on the morphological characteristics of the soil profile, hydropedological field measurements, and the structure of the unsaturated zone, it is to be expected that the transport of pollutants to greater depths and eventually to the unconfined aquifer will be more effective in the UCW, MBW, and BCW hydropedological zones.

The leachates from soils with mining influence are more acidic, especially those with self-burning, and have higher major ion content (with Ca-SO_4 hydrogeochemical facies) and higher PTE content. The influence of self-burning makes it possible to distinguish the soil in the UCW zone from the BW layer of the BCW zone. The BW layer is characterized by a higher K value, a somewhat different mineralogical composition (both in the fine soil fraction and the clay fraction), leachates with a slightly more acidic pH, and different concentrations of some PTEs.

Although, before the waste deposition, the DS soil was similar to the US soil, it presently reveals an apparent mining influence, as evidenced by leachates with a composition closer to that of mining-influenced soils.

In summary, the hydropedological setting in the mining-influenced hydropedological zones corresponds to soils more susceptible to major ion and PTE leaching and with a greater ability to disperse pollutants in the environment, namely in groundwater.

The study of the Fojo mine waste pile illustrates the interest in applying hydropedological concepts, methods, and techniques to assess the environmental impacts of coal mining. Such an approach can be of great value in decision-making related to the environmental management of coal mining areas and can be applied to the exploitation of many other geological resources.

Supplementary Materials: The following supporting information can be downloaded at: https://www.mdpi.com/article/10.3390/hydrology11050062/s1, Table S1: Soil morphological characteristics of the Uphill Soil (US) hydropedological zone; Table S2: Soil morphological characteristics of the Unburned Coal Waste Pile (UCW) hydropedological zone; Table S3: Soil morphological characteristics of the Mixed Burned Coal Waste (MBW) hydropedological zone; Table S4: Soil morphological characteristics of the Burned Coal Waste Pile (BCW) hydropedological zone; Table S5: Soil morphological characteristics of the Downhill Soil (DS) hydropedological zone.

Author Contributions: Conceptualization, J.E.M. and A.N.; methodology, J.E.M., A.N., J.R., S.C.A., F.R. and C.M.; validation, J.E.M., A.N., J.R., F.R., S.C.A., D.F. and C.M.; formal analysis, J.E.M., A.N., P.S., J.R., F.R., D.F. and C.M.; investigation, J.E.M., A.N., P.S., J.R., S.C.A., A.M., F.R., D.F. and C.M.; resources, J.E.M., F.R., D.F. and C.M.; writing—original draft preparation, J.E.M., A.N., J.R., F.R., D.F. and C.M.; writing—review and editing, J.E.M., A.N., P.S., J.R., S.C.A., A.M., F.R., D.F. and C.M.; supervision, J.E.M., D.F. and C.M.; project administration, D.F.; funding acquisition, D.F. All authors have read and agreed to the published version of the manuscript.

Funding: This research was supported by the "SHS: Soil health surrounding former mining areas: characterization, risk analysis, and intervention" Project, financed by NORTE-45-2020-75-SISTEMA DE APOIO À INVESTIGAÇÃO CIENTÍFICA E TECNOLÓGICA—"PROJETOS ESTRUTURADOS DE I&D"—HORIZONTE EUROPA, Ref. NORTE-01-0145-FEDER-000056 and framed within the ICT activities (projects UIDB/04683/2020, https://doi.org/10.54499/UIDB/04683/2020 and UIDP/04683/2020, https://doi.org/10.54499/UIDP/04683/2020). The research was also supported by GeoBioTec Research Centre (UIDB/04035/2020).

Data Availability Statement: The original contributions presented in the study are included in the article and supplementary materials, further inquiries can be directed to the corresponding author.

Acknowledgments: The authors thank the anonymous reviewers for their insightful comments and suggestions, which helped improve and clarify this manuscript.

Conflicts of Interest: The authors declare no conflicts of interest.

References

1. Sonter, L.; Moran, C.J.; Barrett, D.J.; Soares-Filho, B. Processes of land use change in mining regions. *J. Clean. Prod.* **2014**, *84*, 494–501. [CrossRef]
2. Mononen, T.; Kivinen, S.; Kotilainen, J.M.; Leino, J. *Social and Environmental Impacts of Mining Activities in the EU*; PE 729.156—May; Policy Department for Citizens' Rights and Constitutional Affairs; Directorate-General for Internal Policies; European Union: Brussels, Belgium, 2022; 81p.
3. Gankhuyag, U.; Gregoire, F. *Managing Mining for Sustainable Development: A Sourcebook*; UNDP and UN Environment: Bangkok, Thailand, 2018; 111p.
4. Yang, J.; Wei, H.; Quan, Z.; Xu, R.; Wang, Z.; He, H. A global meta-analysis of coal mining studies provides insights into the hydrologic cycle at watershed scale. *J. Hydrol.* **2023**, *617*, 129023. [CrossRef]
5. Suárez-Ruiz, I.; Crelling, J.C. (Eds.) *Applied Coal Petrology. The Role of Petrology in Coal Utilization*; Elsevier: Amsterdam, The Netherland, 2008; 388p.
6. Suárez-Ruiz, I.; Flores, D.; Mendonça Filho, J.G.; Hackley, P.C. Review and update of the applications of organic petrology: Part 2, geological and multidisciplinary applications. *Int. J. Coal Geol.* **2012**, *98*, 73–94. [CrossRef]
7. Marove, C.A.; Sotozono, R.; Tangviroon, P.; Tabelin, C.B.; Igarashi, T. Assessment of Soil, Sediment and Water Contaminations around Open-Pit Coal Mines in Moatize, Tete Province, Mozambique. *Environ. Adv.* **2022**, *8*, 100215. [CrossRef]
8. Liu, X.; Shi, H.; Bai, Z.; Zhou, W.; Liu, K.; Wang, M.; He, Y. Heavy Metal Concentrations of Soils near the Large Opencast Coal Mine Pits in China. *Chemosphere* **2020**, *244*, 125360. [CrossRef]
9. Finkelman, R.B. Potential Health Impacts of Burning Coal Beds and Waste Banks. *Int. J. Coal Geol.* **2004**, *59*, 19–24. [CrossRef]
10. Li, H.; Ji, H. Chemical Speciation, Vertical Profile and Human Health Risk Assessment of Heavy Metals in Soils from Coal-Mine Brownfield, Beijing, China. *J. Geochem. Explor.* **2017**, *183*, 22–32. [CrossRef]
11. Finkelman, R.B.; Palmer, C.A.; Wang, P. Quantification of the Modes of Occurrence of 42 Elements in Coal. *Int. J. Coal Geol.* **2018**, *185*, 138–160. [CrossRef]
12. Stracher, G.B.; Prakash, A.; Sokol, E.V. (Eds.) *Coal and Peat Fires: A Global Perspective*; Elsevier: Amsterdam, The Netherland, 2011; Volume 1, 380p.
13. Stracher, G.B.; Prakash, A.; Sokol, E.V. (Eds.) *Coal and Peat Fires: A Global Perspective, Volume 3—Case Studies*; Elsevier: Amsterdam, The Netherland, 2015; Volume 3, 786p.
14. Stracher, G.B.; Prakash, A.; Sokol, E.V. (Eds.) *Coal and Peat Fires: A Global Perspective, Volume 5—Case Studies—Advances in Field and Laboratory Research*; Elsevier: Amsterdam, The Netherland, 2019; Volume 5, 542p.
15. Ribeiro, J.; Ferreira da Silva, E.; Flores, D. Burning of coal waste piles from Douro Coalfield (Portugal): Petrological, geochemical and mineralogical characterization. *Int. J. Coal Geol.* **2010**, *81*, 359–372. [CrossRef]
16. Espinha Marques, J.; Martins, V.; Santos, P.; Ribeiro, J.; Mansilha, C.; Rocha, F.; Flores, D. Changes induced by self-burning in Technosols from a coal mine waste pile: A hydropedological approach. *Geosci. J.* **2021**, *11*, 195. [CrossRef]
17. Mansilha, C.; Melo, A.; Flores, D.; Ribeiro, J.; Rocha, J.; Martins, V.; Santos, P.; Espinha Marques, J. Irrigation with coal mining effluents: Sustainability and water quality considerations (S. Pedro da Cova, N Portugal). *Water J.* **2021**, *13*, 2157. [CrossRef]
18. Celebi, E.; Ribeiro, J. Prediction of acid production potential of self-combusted coal mining wastes from Douro Coalfield (Portugal) with integration of mineralogical and chemical data. *Int. J. Coal Geol.* **2023**, *265*, 104152. [CrossRef]
19. Santos, P.; Ribeiro, J.; Espinha Marques, J.; Flores, D. Environmental and health risk assessment of soil adjacent to a self-burning waste pile from an abandoned coal mine in Northern Portugal. *Environments* **2023**, *10*, 53. [CrossRef]
20. Teodoro, A.; Santos, P.; Espinha Marques, J.; Ribeiro, J.; Mansilha, C.; Melo, A.; Duarte, L.; Rodrigues, C.; Flores, D. An integrated multi-approach to environmental monitoring of a self-burning coal waste pile: The São Pedro da Cova mine (Porto, Portugal) study. *Environments* **2021**, *8*, 48. [CrossRef]
21. Ribeiro, J.; Suárez-Ruiz, I.; Flores, D. Coal related fires in Portugal: New occurrences and new insights on the characterization of thermally affected and non-affected coal waste piles. *Int. J. Coal Geol.* **2022**, *252*, 103941. [CrossRef]
22. Diogo, B.; Narayan, A.; Mansilha, C.; Espinha Marques, J.; Flores, D.; Antunes, S. Phytotoxicity of coal waste elutriates (Douro Coalfield, North Portugal) in Lactuca sativa. *Environ. Sci. Pollut. Res.* **2023**, *30*, 107650–107660. [CrossRef] [PubMed]
23. Narayan, A.; Diogo, B.S.; Mansilha, C.; Espinha Marques, J.; Flores, D.; Antunes, S.C. Assessment of ecotoxicological effects of Fojo coal mine waste elutriate in aquatic species (Douro Coalfield, North Portugal). *Front. Toxicol.* **2024**, *6*, 1334169. [CrossRef] [PubMed]
24. Jenny, H. Factors of Soil Formation. *Soil Sci.* **1941**, *52*, 415. [CrossRef]
25. Amundson, R. Factors of soil formation in the 21st century. *Geoderma* **2021**, *391*, 114960. [CrossRef]

26. Agroconsultores and Geometral. *Carta de Solos e Carta de Aptidão da Terra de Entre Douro e Minho, Escala 1:100 000*; Direcção Regional da Agricultura e Entre Douro e Minho: Braga, Portugal, 1995.
27. FAO—Food and Agriculture Organization of the United Nations; IUSS Working Group WRB. *World Reference Base for Soil Resources 2014, Update 2015—International Soil Classification System for Naming Soils and Creating Legends for Soil Maps*; World Soil Resources Reports No. 106; Food and Agriculture Organization of the United Nations (FAO): Rome, Italy, 2015; 192p.
28. FAO—Food and Agriculture Organization of the United Nations. *Guidelines for Soil Description*, 4th ed.; Food and Agriculture Organization of the United Nations (FAO): Rome, Italy, 2006; pp. 67–71.
29. Zhang, R. Determination of soil sorptivity and hydraulic conductivity from the disk infiltrometer. *Soil Sci. Soc. Am. J.* **1997**, *61*, 1024–1030. [CrossRef]
30. Reynolds, W.D. Unsaturated Hydraulic Properties: Field Tension Infiltrometer. In *Soil Sampling and Methods of Analysis*, 2nd ed.; Carter, M.R., Gregorich, E.G., Eds.; CRC Press: Boca Raton, FL, USA, 2008; pp. 1–22.
31. METER Group. *Mini Disk Infiltrometer*; Meter Group, Inc.: Court Pullman, WA, USA, 2020; pp. 1–15.
32. Dekker, L.W.; Ritsema, C.J.; Oostindie, K.; Moore, D.; Wesseling, J.G. Methods for determining soil water repellency on field-moist samples. *Water Resour. Res.* **2009**, *45*, W00D33. [CrossRef]
33. Schultz, L.G. *Quantitative Interpretation of Mineralogical Composition from X-ray and Chemical Data for the Pierre Shale*; United States Geological Survey Professional Paper: Washington, DC, USA, 1964; Volume 391-C, pp. 1–31.
34. Brindley, G.W.; Brown, G. *Crystal Structures of Clay Minerals and Their X-ray Identification*; Monograph No. 5; Mineralogical Society of Great Britain and Ireland: London, UK, 1980; pp. 1–495.
35. Galhano, C.; Rocha, F.; Gomes, C. Geostatistical analysis of the influence of textural, mineralogical and geochemical parameters on the geotechnical behavior of the "Clays Aveiro" formation (Portugal). *Clay Miner.* **1999**, *34*, 109–116. [CrossRef]
36. Oliveira, A.; Rocha, F.; Rodrigues, A.; Jouanneau, J.; Dias, A.; Weber, O.; Gomes, C. Clay minerals from the sedimentary cover from the Northwest Iberian shelf. *Prog. Oceanogr.* **2002**, *52*, 233–247. [CrossRef]
37. Hageman, P.L. U.S. Geological Survey field leach test for assessing water reactivity and leaching potential of mine wastes, soils, and other geologic and environmental materials. In *U.S. Geological Survey Techniques and Methods*; USGS: Fairfax County, VA, USA, 2007; Book 5, Chapter D3; p. 14.
38. American Public Health Association; American Water Works Association; Water Environment Federation. *Standard Methods for the Examination of Water and Wastewater (SMEWW)*, 23rd ed.; Rice, E.W., Baird, R.B., Eaton, A.D., Eds.; American Public Health Association; American Water Works Association; Water Environment Federation: Washington, DC, USA, 2017; p. 168.
39. Rodier, J.; Legube, B. *Eaux Naturelles, Eaux Residuaires, Eau de Mer*, 10th ed.; De L'Eau, L., Ed.; DUNOD: Malakoff, France, 2016; p. 1824.
40. Leighton-Boyce, G.; Doerr, S.H.; Shakesby, R.A.; Walsh, R.P.D. Quantifying the impact of soil water repellency on overland flow generation and erosion: A new approach using rainfall simulation and wetting agent on in situ soil. *Hydrol. Process. Int. J.* **2007**, *21*, 2337–2345. [CrossRef]
41. Butzen, V.; Seeger, M.; Marruedo, A.; Jonge, L.; Wengel, R., Ries, J.B., Casper, M.C. Water repellency under coniferous and deciduous forest—Experimental assessment and impact on overland flow. *Catena* **2015**, *133*, 255–265. [CrossRef]

Disclaimer/Publisher's Note: The statements, opinions and data contained in all publications are solely those of the individual author(s) and contributor(s) and not of MDPI and/or the editor(s). MDPI and/or the editor(s) disclaim responsibility for any injury to people or property resulting from any ideas, methods, instructions or products referred to in the content.

hydrology

Article

Assessment and Mitigation of Groundwater Contamination from Phosphate Mining in Tunisia: Geochemical and Radiological Analysis

Younes Hamed [1,2], Matteo Gentilucci [3], Naziha Mokadem [2], Rayan Khalil [4], Yosra Ayadi [5], Riheb Hadji [6,*] and Elimame Elaloui [1]

1 Laboratory for the Application of Materials to the Environment, Water and Energy, University Campus, Gafsa 2112, Tunisia; hamedhydro.tn@gmail.com (Y.H.); limam_aloui@yahoo.fr (E.E.)
2 Faculty of Science of Gafsa, University of Gafsa, Gafsa 2112, Tunisia; mokademnaziha86@gmail.com
3 School of Science and Technology, Geology Division, University of Camerino, Via Gentile III da Varano, 7, 62032 Camerino, Italy; matteo.gentilucci@unicam.it
4 Faculty of Earth Sciences, King Abdulaziz University, Jeddah 21589, Saudi Arabia; rekhalil@kau.edu.sa
5 Department of Chemistry, Faculty of Science of Gabes, University of Gabes, Gabes 6029, Tunisia; ayadiyosraa@yahoo.fr
6 Laboratory of Applied Research in Engineering Geology, Geotechnics, Water Sciences, and Environment, Setif 1 University, Setif 19000, Algeria
* Correspondence: hadjirihab@yahoo.fr

Abstract: Groundwater contamination in the Mediterranean Basin is a severe problem that has a significant impact on environmental ecosystems and human health. The unconventional uranium and the potentially toxic elements (PTEs) of phosphate rocks are the principal contaminants in the phosphate mining industry in Tunisia. Phosphogypsum (PG) results from the valorization of phosphate to fertilizers and phosphoric acid. PG stocks can be used in cement production, brick manufacturing, and soil amendments in desertic land, and can be resolved by using nanomaterial adsorbents. In the flat area of the study area, the increase in radioactivity (^{40}K) is due to abusive fertilizer use. Geochemical and radiological analyses in the northern part of Tunisia and its karst shallow aquifer indicate significant contamination levels. The northern part exhibits moderate contamination, whereas the karst shallow aquifer shows higher contamination levels, particularly with elevated nitrate concentrations. In the phosphate basin, both washing phosphate and phosphogypsum reveal high levels of radioactive elements, with the latter showing especially high concentrations of radium. The shallow aquifer in this region has moderate contamination levels, while the deep geothermal aquifer also shows noticeable contamination but to a lesser degree compared to the shallow aquifer. The shallow groundwater is characterized by a higher value of radioactivity than the groundwater due to the contamination impact from the phosphate industry and the cumulative radioactivity disintegration. Finally, the nanoparticles and the electrostatic adsorption can decrease the PTEs and radionuclides from the contaminated water in the study area. Moreover, other key issues for advancing research on groundwater contamination are proposed in this study. It is time to valorize this PG and the other mines of (Fe, Pb, and Zn) in the socioeconomic sector in Tunisia and to minimize the environmental impact of the industrial sector's extraction on groundwater and human health in the study area.

Keywords: groundwater contamination; phosphate mining; uranium; potential toxic elements; phosphogypsum contamination

Citation: Hamed, Y.; Gentilucci, M.; Mokadem, N.; Khalil, R.; Ayadi, Y.; Hadji, R.; Elaloui, E. Assessment and Mitigation of Groundwater Contamination from Phosphate Mining in Tunisia: Geochemical and Radiological Analysis. *Hydrology* **2024**, *11*, 84. https://doi.org/10.3390/hydrology11060084

Academic Editor: Meysam Vadiati

Received: 29 March 2024
Revised: 12 June 2024
Accepted: 14 June 2024
Published: 17 June 2024

Copyright: © 2024 by the authors. Licensee MDPI, Basel, Switzerland. This article is an open access article distributed under the terms and conditions of the Creative Commons Attribution (CC BY) license (https://creativecommons.org/licenses/by/4.0/).

1. Introduction

One-third of the global population depends on groundwater for drinking water [1–3]. In Tunisia, groundwater is a major source of fresh water for around 12 million people for domestic, agricultural, industrial, and tourism uses. However, urbanization, agricultural practices, industries, touristic activities, and climate change all pose significant threats to groundwater quantity and quality. Radionuclides, potentially toxic elements, organic and inorganic contaminants, pesticides, fertilizers, nanoparticles, hydrocarbon, microplastic, and other emerging contaminants, are threats to human health, environmental services, and sustainable socioeconomic development [4–12].

The geoscientific literature has discussed the potential health impacts of ionizing radiation emitted by radionuclides, which are components of naturally occurring radioactive materials. These radionuclides are often found in petroleum and gas products, as well as their by-products, including crude oil, scale, sludge, polluted water, and sand [13]. They are also prevalent in phosphate mining areas and processing facilities [14]. Naturally occurring radioactive materials radionuclides in phosphate mining areas can lead to both external and internal exposure to gamma radiation, as well as β and α particles emitted from ^{238}U, ^{232}Th, and their decay products, alongside ^{40}K. Most radionuclides found in phosphate rocks are part of the ^{238}U and ^{232}Th natural decay series, in addition to K-40. This exposure poses a significant risk to human health, necessitating rigorous monitoring and management of these radioactive materials.

It is estimated that 20% of global aquifers are overexploited, leading to subsidence, saltwater intrusion, and increased treatment costs [15]. In addition, nitrate, phosphate, sulfate, fertilizers, pesticides, heavy metals, and radionuclides in air, soil, and bedrock enter groundwater by surface runoff, erosion, desorption, and dissolution. Radionuclides and PTEs from phosphate rocks and phosphate/stock from other mines are transported by wind and seep into groundwater [16].

However, widespread pollution and increasing dependence on groundwater have resulted in a decline in quality and quantity. Groundwater reserves in the majority of the world's basins are depleting at a rapid rate, such as in North Africa (Morocco, Algeria and Tunisia) (20 mm-8m/y), Libya (25 mm-4m/y), Egypt (45 mm-10m/y), the Middle East (Saudia Arabia = 50 mm-6m/y), Iraq (20 mm-10m/y) northwestern India (40 mm/y), North China (22 mm/y), and the high plains of the USA (27.6 mm/y) [17–27].

In the study area, the phosphatic rock extraction started in the mid-1950s after the phosphate mine was discovered in 1899 in the Gafsa basin (Metlaoui region–Thelja Gorge) by the Company of Phosphate of Gafsa (CPG). The annual production of exported phosphate varied from 6 to 9 million tons, placing Tunisia fifth among world phosphate producers before 2010 [28,29]. The Tunisian area (Gafsa-El Kef basin) is one of the most productive phosphate regions in North Africa; it is ranked as the third country in phosphate production (phosphate, phosphoric acid, and fertilizers) during this period. In this report, we study the impact of the phosphate industry on the environment and human health in the study area.

2. Study Area

In the study area, the precipitation and surface water are limited due to the arid climate of North Africa. Groundwater is a particularly important resource in this arid and semi-arid land. The study area is located in the southern Mediterranean Basin, North Africa. It is limited to the north and the east by the Mediterranean Sea, to the west by the Algerian territory (Atlas Monts of North Africa), and to the south by the Saharan platform (Figure 1). The great desert of North Africa (Algeria, Tunisia, and Libya) is characterized by climatic fluctuation and irregularity with a precipitation rate varying between 50 mm/y and 1500 mm/y. The climatic and meteorological parameters vary from the north to the south. The north is characterized by ($1000 < P < 1500$ mm/y; $-5 < T < 20$ °C; $1000 < ETP < 1200$ mm/y and $865 <$ solar radiation < 1045 Kwh/m^2/y). However, the south is characterized by ($50 < P < 350$ mm/y; $-10 < T < 45$ °C; $1750 < ETP < 2500$ mm/y and $870 <$ solar radiation < 1075 Kwh/m^2/y). The study area is characterized by an arid to semi-arid climate in the northern part of North Africa–Southern Mediterranean Basin [15,28].

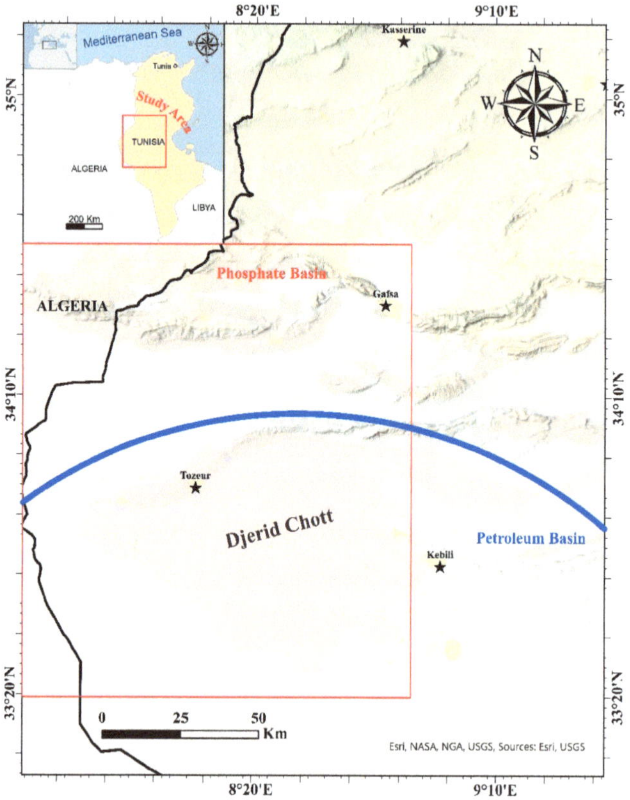

Figure 1. Geographic location of the study area.

3. Socioeconomic of the Study Area

Tunisia's economy has long been heavily dependent on extracting and exporting phosphate resources. In addition, the country's petroleum industry plays a significant role in contributing to its gross domestic product (GDP), (Figure 1). Many of Tunisia's social and economic activities, as well as the development of important cities like Gabès and Gafsa, are deeply intertwined with these extractive industries. The phosphate ore mines that are exploited in Tunisia are primarily located in the southwestern regions of the country. This geographical concentration of phosphate resources has been a key factor in shaping the economic landscape and regional development patterns. During the last half-century, anthropogenic activities increased to boost the Tunisian economy. These Eocene (Ypresian) rocks (47 to 55 million years) are very rich in uranium and some other rare elements (toxic and non-toxic: Cd, F, Li, Cr, Cu, As, Pb, Zn, Fe, Sr, Sb, V, and Eu) [9–11,30]. The uranium content in this rock depends on climatic, tectonic, and eustatic paleo conditions. The uranium distribution in the different phosphate layers in the study area provides information about the sedimentation and oxidoreduction conditions, and the determination of the activity ratios of $^{234}U/^{238}U$ in the phosphate layers can give information about the paleoclimatic and paleogeomorphology conditions. The natural radionuclide uranium has three isotopes in the natural environment U-234 (0.0054%), U-235 (0.72%), and U-238 (99.27%) [31]. Of these isotopes, U-238 is the most important natural radionuclide due to its long half-life (4.5×10^6 years) and chemical and radiation toxicities [32,33].

4. Lithological Proprieties

Based on data from cores collected from different stations in the study area (Tables 1 and 2). In the north part of the study area (Table 1), we could distinguish two different lithological levels [5–7]:

1- Alluvial sands and gravel (riverbed) are characterized by homogeneous fine-to-medium sands intercalated by laminates of silts, with a predominance of silt and clay (approximately between 60 and 66%), and a thickness of 0.5 to 2.5 m;
2- Limestones are characterized by carbonate and dolomite (fractured and karstified rocks) with a thickness of 2 to 50 m.

Table 1. Characteristics of the lithological levels of the study area (north of Tunisia).

Lithology	Mineralogy	Porosity (%)	Density (kg/m^3)	Silt (%) (<0.01 mm)	Thickness (m)
Alluvial sands and gravel	Quartz (97–99%) Silt and organic matter (1–3%)	36–42	1.68	0.72–1.85	0.5–2.5
Limestones	Carbonate (87–94%) Dolomite (6–13%)	25–32	1.71	0.8–2.4	2–50

Table 2. Characteristics of the lithological levels of the study area (south of Tunisia).

Lithology	Mineralogy	Porosity (%)	Density (kg/m^3)	Silt (%) (<0.01 mm)	Thickness (m)
Aeolian deposits	Quartz (97–99%) Silt and organic matter (1–3%)	33–37	1.74	0.8–2.5	0.5–2
Alluvial sands	Quartz (85–92%) Silt and organic matter (8–15%)	30–38	1.71	1.4–10.4	2–3

In the southern part of the study area (Table 2), we could distinguish two different lithological levels [34]:

1- Aeolian deposits are mainly composed of fine–medium-grained sands, with a predominance of silts (55–65%) and clays (15%), and have a low level of organic matter. They have a variable thickness from one region to another, ranging from 0.5 to 2 m. Over the last ten years, this thickness has increased year by year, essentially due to the phenomenon of continuous desertification driven by sandstorms from south to north;
2- Alluvial sands (riverbeds) are characterized by homogeneous fine-to-medium sands intercalated by laminates of silts, with a predominance of silts (62–64%) and clays (approximately 16%), with a thickness ranging from 2 to 3 m.

5. Hydrology and Hydrogeology Setting

In the north, the Majerda drainage network is dendritic and perennial due to the natural recharge (precipitation and springs) and anthropogenic activities (wastewaters from the Fe, Pb, and Zn mines). In the southern part, the study area is characterized by non-perennial wadies due to the arid climate, but the industrial sector (washing phosphate and phosphogypsum valorization) creates a new stage by discharging wastewater directly into the drainage network. This situation causes groundwater contamination through the alluvial deposits and the major faults. The hydrogeology part is based on data from the lithostratigraphic data from piezometers and exploitation boreholes from the multi-aquifer system in the study area (10 < depth < 2150 m) (Figure 2). Several aquifer horizons can be distinguished in the study area [11,23]:

1- Alluvial aquifer, Pliocene, and Pleistocene–Holocene sands (shallow aquifer);
2- Mio-Plio-Quaternary aquifer (semi-confined aquifer);
3- Upper Cretaceous limestone/dolomite (semi-confined aquifer);
4- Lower Cretaceous sand (confined aquifer), contains fossil water (40–45 million years) and geothermal water (35–92 °C).

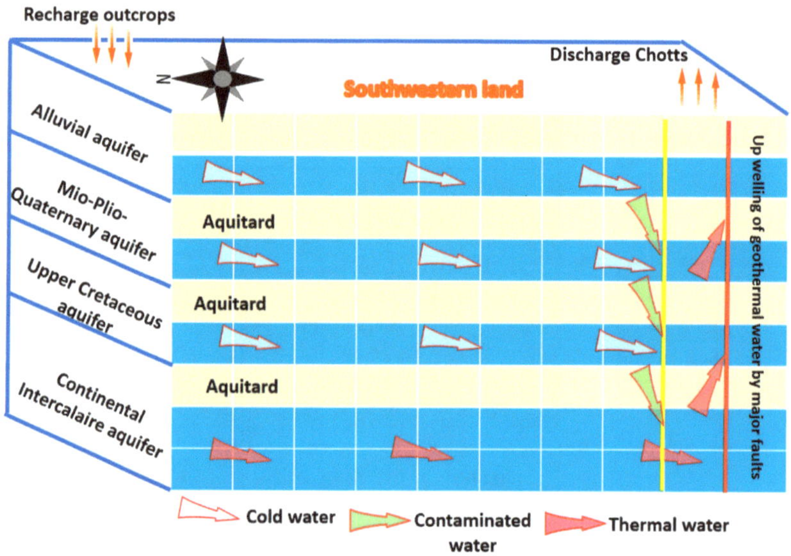

Figure 2. A mathematical conceptual model showing the hydrodynamic of the multi-aquifer system in the southern part of the study area.

The groundwater contained in these aquifers is mainly recharged by precipitation and perennial drainage networks. Numerous hydraulic exchanges take place in these layers and directly influence the hydrogeological conditions, ensuring their hydraulic intercommunications (major faults impact).

6. Material and Methods

During spring, summer, and autumn 2023, 105 sediment samples, 50 air locations, and 33 deep wells in this study were analyzed using a TERRA detector (pocket-type instrument) in the field, known for its high sensitivity to climatic variations such as temperature and humidity, as well as anthropogenic activities. This state-of-the-art detector is capable of accurately detecting and quantifying the cumulative levels of alpha and beta particles, as well as gamma rays, providing precise measurements of the diverse forms of total radioactivity (U-Th-Ra and R) present in the analyzed samples. It features high-resolution capabilities for sediment samples and includes a user-friendly display screen. The samples were further dried at 105 °C for 24 h to ensure the complete removal of moisture. Quality control (QC) procedures were applied using homemade control samples and reference samples provided by the IAEA (Environmental Monitoring Laboratory, Department of Energy, Ukraine). An IAEA reference multi-standard radionuclide source was utilized to calibrate the energy and efficiency of the detector. The sediment sampling was based on the collection of sediment cores at depths of up to 2 m and cutting was conducted every 10 cm for physicochemical and radiometric analyses. The groundwater wells are based on data sampling from the exploitation boreholes in the study area.

The starting materials were washed with distilled water, dried at 110 °C overnight, and sieved to give a homogeneous sample, and we followed the international protocol in this geochemical analysis in the LAM3E laboratory, Chemistry Department, Faculty of Sciences of Gafsa, Tunisia, and the radio-analysis laboratory of Bizerte Faculty, Tunisia. Analyses were performed with a broad-energy germanium detector (Canberra), with a relative efficiency of 43 percent. The U-238 activity was determined using four gamma peaks of Bi-214 and two peaks of Pb-214 and Rn-222. U-235 was determined using the photopeak at 143 KeV, and Th-232 was determined by the 911 KeV peak of Ac-228. Calibration of the detection system was performed using a certified multi-gamma source and was controlled using reference materials from the International Atomic Energy Agency (AIEA), IAEA 327, and IAEA 375.

7. Radiochemical Determination Methods

7.1. Determination of Po and Th

The sample solution passed through the TEVA columns, which were conditioned with 10 mL of 3 M HNO_3. The columns were rinsed with 6 mL 3 M HNO_3. These effluents were retained as U portions and evaporated to dryness; 20 mL of 8 M HNO_3 was used to elute polonium. After this, the thorium was eluted with 20 mL of 10 M HCl. The Po portion solution was retained for the following micro-co-precipitation, and the Th solution was evaporated to dryness. For radioactivity analysis, Po and Th alpha sources were prepared via the CuS micro-co-precipitation method and counted using alpha spectrometry for 24 h.

7.2. Determination of U

The UTEVA columns were pre-conditioned with 15 mL of 8 M HNO_3, and the dried U samples were dissolved in 8 M HNO_3. The sample solutions were loaded into UTEVA columns and rinsed with 15 mL of 8 M HNO_3. Next, the columns were washed with 15 mL of 4 M HCl, converting the columns to chloride form and removing any residual thorium or plutonium. After this, the U was eluted with 20 mL of 0.5 M HCl; then, 1 mL of 10% sodium sulfate Na_2SO_4 and 5 mL of concentrated HNO_3 were added. The solution was then evaporated to dryness. The U alpha source was obtained using the NdF3 micro-co-precipitation method, and radioactivity was measured via alpha spectrometry [35].

7.3. Radioactivity Measurements of Ra

Groundwater was used directly without chemical separation for the determination of Ra isotopes by gamma spectroscopy. Before gamma counting, groundwater samples were stored for a few weeks, reaching a complete equilibrium between radium and radon. Approximately 2.5 L of sample aliquot was sealed in a sample container and was counted in a high-purity germanium detector for 48 h, a long enough time to minimize the counting errors. In the direct approach, 226Ra was determined by measuring the peak at 186.2 keV. The radioactivity of 228Ra was obtained by counting its daughter 228Ac (t1/2 = 6.15 h) [36].

8. Results and Discussions

8.1. Radionuclides Transfer Mode and Environmental Impact

8.1.1. Environmental Contamination Measurements

Extensive measurements were conducted using in situ gamma spectrometry techniques to accurately estimate the inventories of radionuclides in contaminated soils and groundwaters within the study area. This approach is an efficient alternative to the traditional, time-consuming methods of air, soil, and groundwater sampling followed by laboratory analysis. The surface gamma spectrometry investigation depth ranged from 0.2 to 2.5 m, depending on the specific soils and radionuclides under consideration.

8.1.2. Gamma Spectrometry Data

The gamma spectrometry data provided detailed insights into the spatial distribution of radionuclides within the study area. The results revealed significant variability in radionuclide concentrations across different sites. The specific radionuclides measured included uranium (U), thorium (Th), and radium (Ra); the average concentrations are detailed in Table 1.

8.1.3. Mathematical Modeling of Radionuclide Transport

To further understand the transport and distribution of radionuclides, we adopted mathematical models for solute transport in groundwater, as developed by references [23,37–42]. These models are essential for demonstrating the system's functioning in a steady state, particularly concerning the flow speed and the transport of radionuclides in the unsaturated zone. This transport is influenced by artificial recharge from phosphate wastewater, phosphogypsum (PG), and other mining fluids within the multi-layer aquifer system.

8.1.4. Modeling Approach and Findings

The transfer of radionuclides was modeled under the assumption that the medium is laterally isotropic and homogeneous, and vertically anisotropic and heterogeneous. The solute transfer occurs primarily through mass effect. Initially, a coupled 1D and saturated 3D model was explored to simplify the unsaturated model. However, due to the inconclusive nature of the unsaturated 1D approach, an unsaturated 2D model was subsequently constructed to provide a more realistic representation of groundwater flow and radionuclide migration. The results of this modeling are shown in Figures 3 and 4a.

Figure 3. The origin of groundwater contamination in southwestern Tunisia (**a**) in the wadi; (**b**) in the Sebkha.

8.1.5. Radionuclide Distribution Patterns

Figures 3 and 4a illustrate the spatial variability in radionuclide distribution and the predicted flow patterns of groundwater. The data indicate that radionuclide concentrations are higher in areas closer to mining activities and artificial recharge points. This spatial distribution highlights the significant impact of anthropogenic activities on environmental contamination levels.

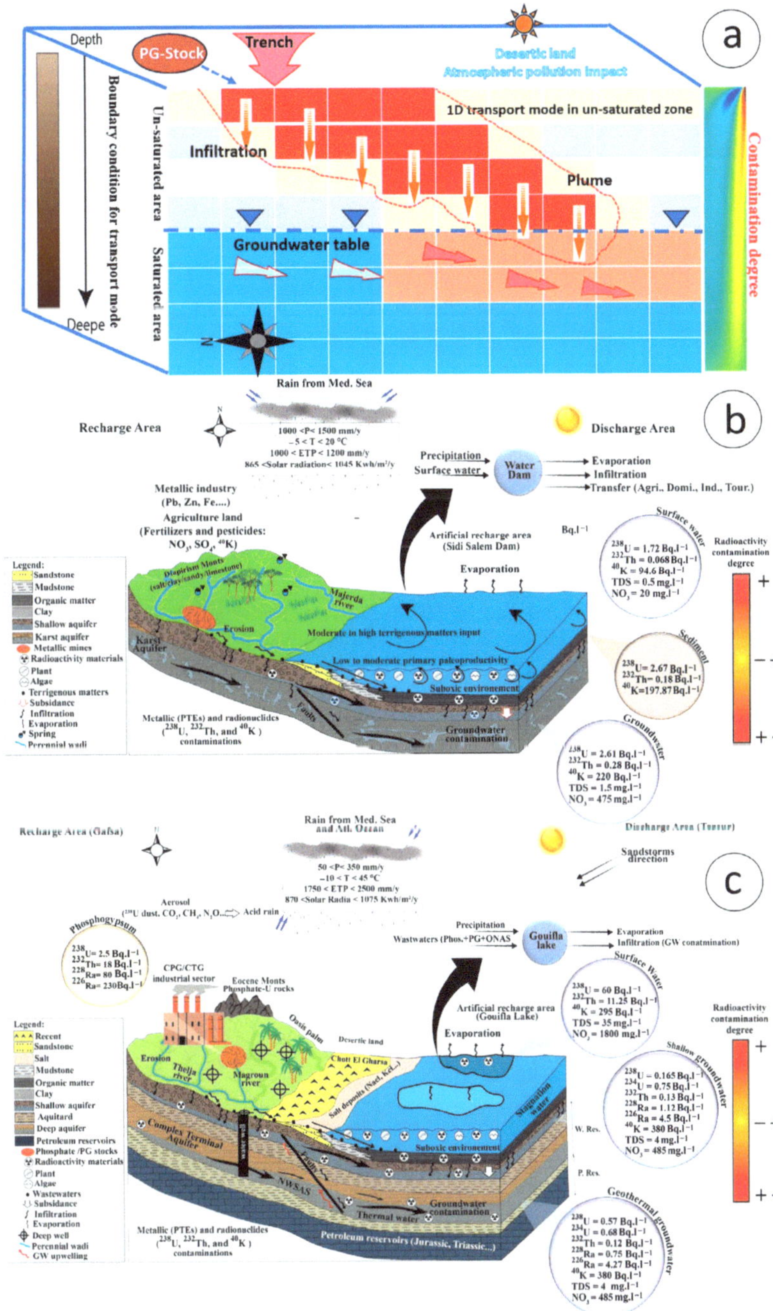

Figure 4. (**a**) Model of radionuclide propagation through groundwater transfer in southwestern Tunisia. (**b**) The conceptual model shows the dam surface water and karst groundwater contamination in northern Tunisia. (**c**) The conceptual model shows the dam surface water and karst groundwater contamination in northern Tunisia [12].

8.1.6. Health and Ecological Risk Assessment

A preliminary health and ecological risk assessment was conducted based on the measured and modeled radionuclide concentrations. The assessment identified potential risks to human health and the ecosystem, particularly in areas with elevated radionuclide levels. These findings underscore the importance of continuous monitoring and the implementation of mitigation strategies to protect public health and the environment.

The deterioration of the ecosystem and human health is in relation to the water resources contamination by the phosphate extraction in the study area. The rate of gamma radioactivity in the deep groundwater ranges between 0.07 and 0.31 mSv/y, and this water contains higher radioactivity compared to shallow groundwater (Table 3). The increasing radioactivity in shallow wells is due to the direct contamination by dust aerosols (atmospheric pollution) and soil (pedological pollution) from phosphate-U extraction and phosphogypsum-U solid stock in the M'dhila region. But far from this region (Gafsa North, Sidi Boubaker, and Oum Laksab), this water is not contaminated. The increasing radioactivity in deep groundwater of the CI transboundary aquifer (thermal fossil deep groundwater from the North-Western Sahara Aquifer System) is due to the water–rock interaction and the dissolution impact of the radionuclides in the aquifer (long residence time). The concentration of all the samples measured is higher than the guidance level (1 Bq/L) for controlling radionuclide health risks from drinking water, as proposed by the WHO The high concentrations of radioactivity have been found near the industrial area (phosphate-U of the Gafsa basin), in the deep geothermal water in the Tozeur region (GWs 18, 19, 20), and the deep geothermal water in the Sidi Ahmed Zarroug region (GW 17- Gafsa region). In the shallow aquifers, the radionuclide concentrations are U-238 = 0.165 Bq/L, U-234 = 0.75 Bq/L, Th-228= 0.12 Bq/L, Ra-228 = 1.12 Bq/L, and Ra = 4.5 Bq/L. Nevertheless, these concentrations in the deep aquifers are U-238 = 0.57 Bq/L, U-234 = 0.68 Bq/L, Th-228 = 0.12 Bq/L, Ra-228 = 0.75 Bq/L, and Ra = 4.27 Bq/L (Table 3).

Table 3. Concentrations ^{238}U, ^{228}Th, ^{226} Ra (Bq/L) and gamma radioactivity (mSv/y) in shallow and deep groundwater samples in southern Tunisia.

Region	Sample ID	Gamma Radioactivity mSv/y	^{238}U	^{234}U	^{228}Th	^{228}Ra	^{226}Ra
M Dilla1	* GW1	0.18	0.19	0.83	0.24	2.05	7.95
M Dilla2	* GW2	0.16	0.2	0.75	0.22	2.00	7.98
M Dilla3	* GW3	0.22	0.22	1.22	0.1	0.84	4.48
El Guettar1	* GW4	0.24	0.22	1.23	0.11	0.88	4.56
El Guettar2	* GW5	0.19	0.20	1.09	0.1	0.78	4.49
El Guettar3	* GW6	0.16	0.21	0.76	0.24	2.04	8.02
Gafsa North1	* GW7	0.09	0.12	0.67	0.12	1.24	7.87
Gafsa North2	* GW8	0.06	0.09	0.5	0.09	1.17	7.32
Gafsa North3	* GW9	0.07	0.08	0.48	0.085	1.15	7.28
Gafsa North4	* GW10	0.08	0.11	0.52	0.11	1.18	7.64
Metalouil1	* GW11	0.14	0.2	0.61	0.11	1.21	6.98
Metlaoui2	* GW12	0.15	0.19	0.58	0.12	1.24	7.09
Metlaoui3	* GW13	0.18	0.23	0.66	0.14	0.96	1.87
Moulares1	** GW14	0.08	0.1	0.32	0.13	0.52	0.89
Moulares2	** GW15	0.07	0.11	0.33	0.16	0.66	1.05
Moulares3	** GW16	0.09	0.12	0.42	0.18	0.79	2.07
S. Ahmed Zarroug	** GW17	0.24	1.05	0.52	0.21	1.02	7.56

Table 3. *Cont.*

Region	Sample ID	Gamma Radioactivity mSv/y	^{238}U	^{234}U	^{228}Th	^{228}Ra	^{226}Ra
Tozeur1	** GW18	0.23	0.82	0.47	0.19	0.81	5.21
Tozeur2	** GW19	0.26	0.98	0.41	0.2	0.72	4.45
Tozeur3	** GW20	0.24	0.87	0.52	0.09	0.55	7.02
Sidi Boubaker1	** GW21	0.10	0.09	0.31	0.04	0.42	2.63
Sidi Boubaker2	** GW22	0.09	0.1	0.34	0.05	0.33	2.91
Oum Laksab1	* GW23	0.07	0.08	0.31	0.02	0.29	2.78
Oum Laksab2	* GW24	0.07	0.05	0.38	0.02	0.29	2.78
Oum Laksab3	* GW25	0.08	0.04	0.35	0.01	0.31	3.21
Gabès1	* GW26	0.26	0.29	1.04	0.13	0.2	0.98
Gabès2	** GW27	0.29	0.32	0.98	0.12	0.14	1.02
Gabès3	** GW28	0.28	0.24	0.78	0.1	0.16	1.12
Gabès4	** GW29	0.31	0.33	1.05	0.11	0.12	ND *
Jerba1	* GW30	0.06	0.04	0.28	0.02	0.24	3.01
Jerba2	** GW31	0.24	0.23	0.66	0.11	0.14	1.14
Jerba3	** GW32	0.22	0.31	0.74	0.14	0.42	2.02
Jerba4	** GW33	0.3	0.33	0.81	0.16	0.36	ND *
---	Mean-shallow	0.13	0.165	0.75	0.12	1.12	4.5
---	Mean-Deep	0.202	0.57	0.68	0.12	0.75	4.27

* GW: shallow groundwater, ** GW: deep groundwater, ND *: not detected.

Adding contributions from all radionuclides present in the study area, the total annual effective dose HE ($mSvy^{-1}$) to an individual was found for the studied samples using Equation (1):

$$HE = \sum(R_a \times A_i \cdot C_f) \quad (1)$$

Ra is the radioactivity concentration of radionuclides in the water samples (BqL^{-1}), Ai is the annual intake (ly^{-1}), and Cf is the ingested dose conversion factor for radionuclides ($mSvy^{-1}$), which varies with the type of radionuclide, as well as the age and gender of the individuals ingesting the toxic radionuclides.

8.2. Groundwater Contamination

Groundwater contamination poses significant risks to the ecosystem, human health, environmental quality, and socioeconomic development in the study area. Our study focused on the transboundary region of the Tuniso-Algerian basin, examining both the northern and southern parts of this area.

8.2.1. Northern Transboundary Part (Tuniso-Algerian Basin)

In the northern part of the basin, the shallow karst aquifers are particularly vulnerable to surface contamination from perennial drainage networks and agricultural activities. The use of surface water from dams and irrigation with water from shallow aquifers, both contaminated with potentially toxic elements and radionuclides, can lead to the accumulation of these toxic elements in crops such as cereals and vegetables, posing direct health risks to humans.

The geochemical and radiological analyses revealed the following concentrations in surface water from the Sidi Salem Dam:

^{238}U: 1.72 Bq/L, ^{232}Th: 0.068 Bq/L, ^{40}K: 94.6 Bq/L, TDS: 0.5 mg/L and NO_3: 20 mg/L.
In contrast, the shallow karst aquifer exhibited higher contamination levels:
^{238}U: 2.61 Bq/L, ^{232}Th: 0.28 Bq/L, ^{40}K: 220 Bq/L, TDS: 1.5 mg/L and NO_3: 475 mg/L.

These findings indicate that groundwater has higher radioactivity levels compared to surface water, likely due to cumulative effects and radioactive decay. The excessive use of fertilizers and pesticides in agricultural areas contributes to elevated nitrate levels (NO_3 = 475 mg/L) and potassium-40 (^{40}K = 220 Bq/L).

Our conceptual model, which integrates geological, hydrological, hydrogeological, geochemical, and radiological data, illustrates the relationship between surface water and groundwater in the study area (Figure 4b).

8.2.2. Southern Transboundary Part (Tuniso-Algerian Basin)

In the southern part of the basin, the shallow and deep aquifers, including the Complex Terminal and North-Western Sahara Aquifer System, are also susceptible to surface contamination from perennial drainage networks such as the Thelja and Magroun wadis. Contamination in these areas arises from domestic use and irrigation with water from shallow and deep aquifers containing potentially toxic elements and radionuclides. These contaminants accumulate in fruits (e.g., dates, pistachios, olives) and vegetables, increasing health risks to humans.

A major source of contamination in this region is the phosphate–uranium rock extraction industry in the Gafsa industrial sectors. Activities such as extraction, washing, and processing into phosphogypsum, fertilizers, and phosphoric acid contribute to atmospheric, pedologic, hydrologic, and hydrogeologic pollution.

The geochemical and radiological analyses from washing phosphate show ^{238}U = 60 Bq/L, ^{232}Th = 11.25 Bq/L, ^{40}K = 180 Bq/L, TDS = 2100 mg/L, and NO_3 = 524 mg/L; from PG, the measurements are ^{238}U = 2.5 Bq/L, ^{232}Th = 18 Bq/L, ^{228}Ra = 80 Bq/L, ^{226}Ra = 230 Bq/L. Shallow groundwater shows ^{238}U = 0.165 Bq/L, ^{234}U = 0.75 Bq/L, ^{232}Th = 0.13 Bq/L, ^{228}Ra = 1.12 Bq/L, ^{226}Ra = 4.5 Bq/L, ^{40}k = 380 Bq/L, TDS = 4 g/L and NO_3 = 485 mg/L. Deep geothermal aquifer results are ^{238}U = 0.57 Bq/L, ^{234}U = 0.68 Bq/L, ^{232}Th = 0.12 Bq/L, ^{228}Ra = 0.75 Bq/L, ^{226}Ra = 4.27 Bq/L, ^{40}k = 380 Bq/L, TDS = 4 g/L and NO_3 = 485 mg/L. The shallow groundwater is characterized by a higher value of radioactivity than the groundwater from the contamination impact from the phosphate industry and the cumulative and radioactivity disintegration impacts. The excessive use of fertilizers and pesticides in agricultural areas causes an increase in NO_3 (485 m/L) and ^{40}K (380 Bq/L). The following conceptual model refers to the combination of all the geological, hydrological, hydrogeological, geochemical, and radiological data showing the relationship between the surface water and the groundwater in the study area (Figure 4c). The artificial lake of the Gouilfa area is characterized by the stagnation of surface wastewater from washing phosphate, phosphogypsum stock, and human wastewater (ONAS). In this place, we detected a high rate of gamma radioactivity and a high value of organic matter. The transfer of this contaminated water to the deep groundwater caused the deterioration of the deep groundwater used in the domestic and irrigation sectors in the study area.

Uranium is a radiotoxic and chemotoxic element with an average abundance of 2.7 mg/kg in the Earth's crust. It generally occurs in minerals such as carnotite, uraninite, and autunite in rocks and ores. Uranium concentrations as high as 15 mg/kg and 120 mg/kg are found in volcanic rock (granite) and phosphate rock, respectively [43]. Uranium is 500 times more abundant than gold, and 40 times more than silver [44]. There are three naturally occurring radioactive isotopes of uranium: (^{238}U-99.27%, $t_{1/2} = 4.468 \times 10^9$ years), (^{235}U-0.711%, $t_{1/2} = 7 \times 10^8$ years), and (^{234}U-0.005%, $t_{1/2} = 24.5 \times 10^4$ years).

The radiotoxicity transfer from sediments to water resources in the study area depends on the pH, Eh, and the form of the father radionuclides (^{238}U). Uranium exists in various natural oxidation states, such as +3, +4, +5, and +6. In an aquatic environment, the predominate states of uranium are tetravalent (U-IV) and hexavalent (U-VI), with the chemical forms UO_2 (uranous oxide) and UO_2^{2+} (uranyl ion), respectively. U-IV is sparsely soluble

(10–6 m/kg at 25 °C) and immobile, whereas (U-VI) is soluble in water, mobile, and toxic to the ecosystem and the environment [45,46]. In nature, other soluble forms of uranium are UO_2F_2, UO_2 $(NO_3)_2$, UO_2Cl_2, UF_6, uranyl acetates, uranyl sulfates, and uranyl carbonates [47]. However, oxidation-reduction and geochemical processes like mineral dissolution and desorption from uranium-hosting minerals (Fe-Mn (oxy/hydroxides), silicates, and phosphate rocks) are mainly responsible for uranium release in groundwater [48]. High uranium concentrations in deep groundwater are predominantly due to U-VI species that leach from host rocks under oxic to suboxic conditions [49]. This radionuclide can be redeposited onto sandstone (Complex Terminal "Segui and Beglia sandy Fms." and NWSAS "Sidi Aich sandy Fm.") under reducing environments. The geothermal deep water of NWSAS can be a good catalyst of uranium mineral dissolution but the clay mineral and the organic matter of the aquitards in the study area can be the inhibitors of radioactivity propagation in groundwater depending on their thickness [12]. In this sandy geothermal transboundary aquifer, the uranium is favorably fractionated in high-temperature meta-luminous melts and is dependent on the geothermal gradient (30 °C/Km). The continental sandstone of the Sidi Aich Fm. of NWSAS is radiologically stratified by these radionuclides. These radiological groundwater contaminations are observed in many regions of the world like the USA, the Republic of Korea, Australia, Brazil, Pakistan, Switzerland, Finland, Egypt, Algeria, Morocco, KSA, and Kuwait [28,50–55].

8.3. Human Health Impacts

Fertilizers and pesticides used in agricultural areas can increase groundwater contamination. Nitrogen contaminants, such as nitrate, nitrite, and ammonia nitrogen, are prevalent inorganic contaminants. Nitrate is predominantly from anthropogenic sources, including agriculture and domestic wastewater [12,56–58]. Other potentially toxic and non-toxic elements found in groundwater include anions, oxyanions, and cations, such as Cd, Pb, Zn, Cr, F, SO_4, Cl, Na, K, Ca, and Mg. Total dissolved solids (TDSs), which refer to the total amount of inorganic and organic ligands in water, also may be increased in groundwater (water–rock interactions). These contaminants are usually of natural origin, but human activities also can elevate concentrations in groundwater [59,60].

Potentially toxic elements (PTEs), metalloids, radionuclides (^{238}U, ^{236}Th, ^{224}Ra, ^{222}Ra, and ^{40}K), and both natural and anthropogenic aerosols (particle dust, CO_2, CH_4, N_2O...) are risk factors for the natural ecosystem and human health. Exposure to excessive concentrations of these PTEs can lead to severe poisoning, although some of these elements are essential micronutrients at lower doses [8,61]. For example, exposure to hexavalent chromium (Cr^{6+}) can increase the risk of cancer [62]. Arsenic is ranked as a group 1 human carcinogen by the US Environmental Protection Agency (EPA) and the International Agency for Research on Cancer (IARC), and As^{3+} can react with the sulfhydryl (-SH) group of proteins and enzymes, upsetting cellular functions; these elements are persistent in the environment and are subject to moderate bioaccumulation when they enter the food chain [61,62].

The questionnaire was administered considering variables such as gender, age, and residency status, distinguishing between inhabitants of mining regions and those living outside these areas (150 persons). The collected data were then analyzed to determine the prevalence of diseases, with results expressed as percentages for each demographic group and geographic location. This method allowed for a comprehensive assessment of how these factors influence health outcomes in both mining and non-mining communities. Males are affected at a rate of 30% with the most affected age group being over 65 years old. For females, the rate is 70%, with the most affected ages between 45 and 65 years old (Figure 5a–c). The different types of cancers in both sexes (male—30% and female—70%) are breast—52%; prostate—15%; lung—9%; uterus—8%; stomach and colon—6% each; and liver—4%) (Figure 5d).

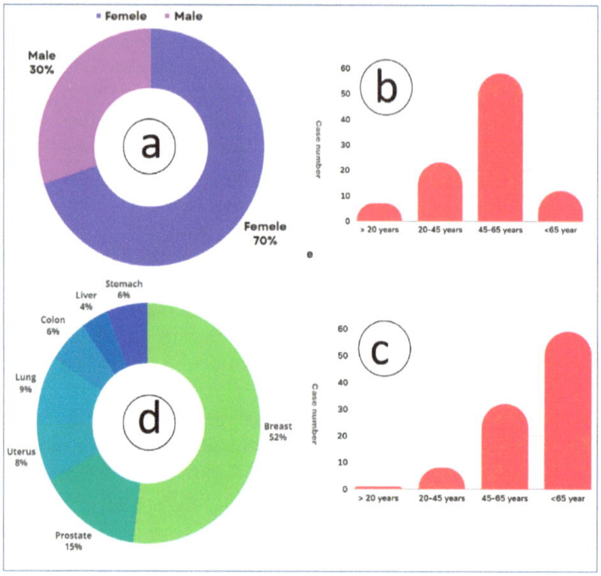

Figure 5. (**a–c**) Human health contamination vs. sex and age. (**d**) The different types of cancer in the study area.

In the study area, the radioactivity impact on human health depends on the adsorption dose (cumulative dose/day and/or per year) from natural and anthropogenic radioactivity. High radiation can damage the DNA in our cells, as well as neurological, respiratory, and reproductive systems, leading to cancer (Figure 6).

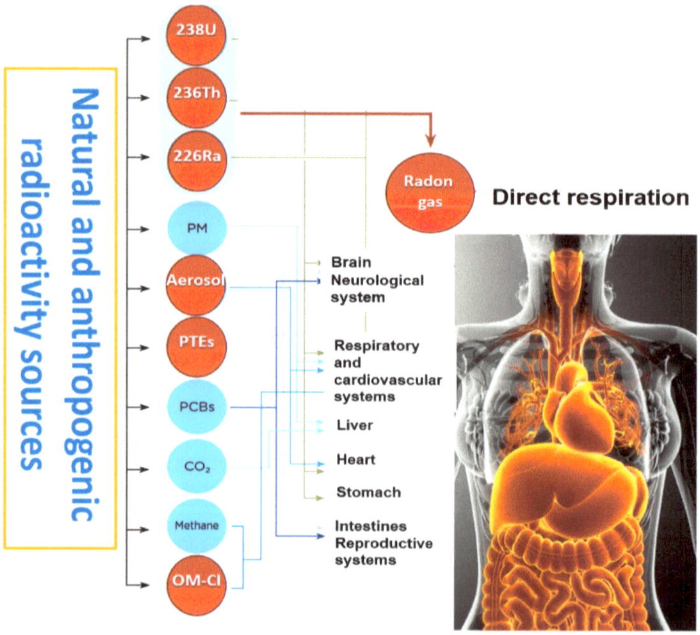

Figure 6. Natural and anthropogenic radioactivity impacts on human health in the study area.

The study area suffers from radionuclide disintegration (^{238}U, ^{236}Th, ^{228}Ra, ^{226}Ra, and ^{222}R), as well as the impact of radon (gas) from the phosphate rock and phosphogypsum on human health and the ecosystem. The effects of radiation can be divided into two categories, showing subtly different relationships between dose and effect. Tissue effects, seen at high doses (>1 Sv), are associated with cell killing in the tissues exposed and show a direct correlation with dose and exposition time. The Gafsa area (phosphate basin) is characterized by a high rate of cancer in Tunisia [12]. Cancer is a public health concern, particularly regarding its stochastic effects on age and sex. Atmospheric, pedologic, hydrologic, and hydrogeological contaminations are the most common sources of the ecosystem and human health deterioration (Figure 7).

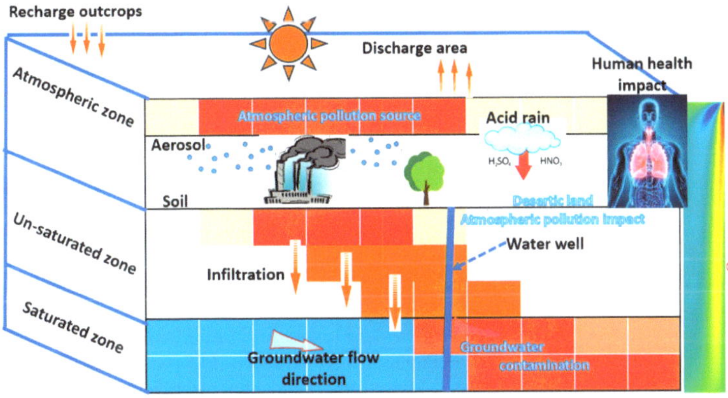

Figure 7. Model of radionuclide propagation via air, soil, and groundwater transfer in the phosphate industry basin.

9. Conclusions and Perspectives

This study highlights the significant impacts of the phosphate industry on groundwater quality in the Gafsa-El Kef basin in Tunisia, emphasizing the presence of radionuclides and potentially toxic elements (PTEs) in both surface and groundwater. The extensive analysis of sediment, air, and groundwater samples using advanced gamma spectrometry and geochemical techniques reveals that uranium, thorium, and potassium isotopes are prevalent in the region, posing considerable health and environmental risks. The results demonstrate that radionuclides from phosphate mining and processing activities are widely distributed in the study area. These contaminants migrate through both surface runoff and subsurface flow, facilitated by the region's geological and hydrological characteristics. The solute transport models indicate that artificial recharge from industrial activities significantly influences radionuclide distribution in the multi-layer aquifer system. This study confirms that groundwater in the region, especially in the shallow karst aquifers and deeper aquifers such as the Complex Terminal and North-Western Sahara Aquifer System, is contaminated with high levels of radionuclides and PTEs. The contamination is more severe in groundwater than in surface water, suggesting cumulative effects and significant groundwater–surface water interactions. The elevated concentrations of uranium, thorium, potassium, and nitrates in groundwater pose direct risks to human health through irrigation and domestic use. The high levels of radioactivity and toxic elements in groundwater used for agricultural purposes could lead to the accumulation of these contaminants in crops, subsequently entering the food chain and affecting human health. The decline in groundwater quality has profound implications for sustainable socioeconomic development in the region. The reliance on contaminated groundwater for agricultural and domestic use threatens the health of local populations and undermines the economic stability dependent on these water resources. Given the significant risks identified, there is a pressing need to rigorously monitor and manage groundwater resources in the study area. Implementing comprehensive groundwater protection strategies, including regular monitoring of radionuclides and PTEs, promoting sustainable agricultural practices, and controlling industrial discharges, is crucial for mitigating health and environmental risks. This study underscores the urgent need to address groundwater contamination in the Gafsa-El Kef basin. Ensuring the sustainability of groundwater resources in this arid and semi-arid region requires coordinated efforts between governmental bodies, industries, and local communities to protect and manage water resources effectively. The findings provide a foundation for future research and policy-making aimed at safeguarding groundwater quality and ensuring the health and well-being of the region's inhabitants.

Author Contributions: Conceptualization, Y.H., M.G. and E.E.; methodology, Y.H.; software, Y.H.; validation, Y.H., M.G., E.E. and R.H.; formal analysis, Y.H.; investigation, Y.H. and E.E.; resources and data curation, R.K. and E.E.; writing—original draft preparation, Y.H.; writing—review and editing, Y.H., N.M., Y.A. and E.E.; visualization, R.H. and E.E.; supervision, R.K. and E.E. All authors have read and agreed to the published version of the manuscript.

Funding: This research received no external funding.

Institutional Review Board Statement: Not applicable.

Informed Consent Statement: Not applicable.

Data Availability Statement: Data are contained within the article.

Acknowledgments: The authors would like to thank the editor and the anonymous reviewers who greatly improved this paper with their relevant critiques; we also thanks Africa and Middle East Earth Scope Consulting and the Laboratory of Applied Research in Engineering Geology, Geotechnics, Water Sciences, and Environment (LAREGGWE). Special thanks to LEGAI, Setif, Algeria, for providing the TERRA material for gamma/beta radioactivity analysis. We are grateful to the staff of the LAM3E lab in Gafsa, Tunisia, and the private Sfax Tina lab, Tunisia. We also acknowledge the contributions of the GeoRS lab members in the Department of Earth and Atmospheric Sciences at the University of Houston, USA. Finally, we thank the staff of the International Association of Water Resources in the Southern Mediterranean Basin for their generous contributions of time and expertise, which greatly enhanced the quality of this manuscript.

Conflicts of Interest: The present paper is an original work; it is the first in the study area, and all the authors declare that they have no conflicts of interest. The authors jointly carried out this research study and they confirm that they are not associated with or involved in any profitable organization or company that has any financial interests.

References

1. Beitz, H.; Schmidt, H.; Herzel, F. Occurrence, Toxicological and Ecotoxicological Significance of Pesticides in Groundwater and Surface Water. In *Pesticides in Ground and Surface Water*; Springer: Berlin/Heidelberg, Germany, 1994; pp. 1–56. [CrossRef]
2. Balbus, J.M.; Boxall, A.B.A.; Fenske, R.A.; McKone, T.E.; Zeise, L. Implications of global climate change for the assessment and management of human health risks of chemicals in the natural environment. *Environ. Toxicol. Chem.* **2013**, *32*, 62–78. [CrossRef] [PubMed]
3. El Gayar, A.; Hamed, Y. Climate change and water resources management in Arab countries. Springer international publishing AG-Euro-Mediterranean and surrounding regions. In *Advances in Science, Technology & Innovation*; Springer International Publishing AG: Cham, Switzerland, 2017.
4. Khleifia, N.; Hannachi, A.; Abbes, N. Studies of uranium recovery from Tunisian wet process phosphoric acid. *Int. J. Innov. Appl. Stud.* **2013**, *3*, 1066–1071.
5. Ayadi, Y.; Mokadem, N.; Besser, H.; Khelifi, F.; Harabi, S.; Hamad, A.; Boyce, A.; Laouar, R.; Hamed, Y. Hydrochemistry and stable isotopes ($\delta^{18}O$ and $\delta^{2}H$) tools applied to the study of karst aquifers in Southern Mediterranean basin (Teboursouk area, NW Tunisia). *J. Afr. Earth Sci.* **2017**, *137*, 208–217. [CrossRef]
6. Latifa, A.Y.; Lhoussaine, B.; Etienne, J.; Moussa, M.; Yassine, A.B.; Ahmed, E.M.; Jana, S.; Barbara, R. Impact of rock-water interactions and recharge on water resources quality of the Agadir-Essaouira basin, southwestern Morocco. *Arab. J. Geosci.* **2017**, *10*, 169. [CrossRef]
7. Ayadi, Y.; Mokadem, N.; Besser, N.; Redhaounia, B.; Khelifi, F.; Harabi, S.; Nasri, T.; Hamed, Y. Statistical and geochemical assessment of groundwater quality in Teboursouk area (Northwestern Tunisian Atlas). *Environ. Earth Sci.* **2018**, *77*, 349. [CrossRef]

8. Khelifi, F.; Besser, H.; Ayadi, Y.; Liu, G.; Yousaf, B.; Harabi, S.; Bedoui, S.; Zeghmi, K.; Hamed, Y. Evaluation of potentially toxic elements' (PTEs) vertical distribution in sediments of Gafsa–Metlaoui mining basin (Southwestern Tunisia) using geochemical and multivariate. *Environ. Earth Sci.* **2019**, *78*, 53. [CrossRef]
9. Khelifi, F.; Caporale, A.G.; Hamed, Y.; Adamo, P. Bioaccessibility of potentially toxic metals in soil, sediments and tailings from a north Africa phosphate-mining area: Insight into human health risk assessment. *J. Environ. Manag.* **2021**, *279*, 111634. [CrossRef] [PubMed]
10. Khelifi, F.; Mokadem, N.; Liu, G.; Yousaf, B.; Zhou, H.; Ncibi, K.; Hamed, Y. Occurrence, contamination evaluation and health risks of trace metals within soil, sediments and tailings in southern Tunisia. *Int. J. Environ. Sci. Technol.* **2021**, *19*, 6127–6140. [CrossRef]
11. Hamed, Y.; Khelifi, F.; Houda, B.; Ben Saad, A.; Ncibi, K.; Hadji, R.; Melki, A. Phosphate mining pollution in southern Tunisia: Environmental, epidemiological, and socioeconomic investigation. *Environ. Dev. Sustain.* **2023**, *25*, 13619–13636. [CrossRef]
12. Hamed, Y.; Ayadi, Y.; Hadji, R.; Ben Saad, A.; Gentilucci, M.; Elaloui, E. Environmental Radioactivity, Ecotoxicology (^{238}U, ^{232}Th and ^{40}K) and Potentially Toxic Elements in Water and Sediments from North Africa Dams. *Sustainability* **2024**, *16*, 490. [CrossRef]
13. Hassan, N.M.; Mansour, N.A.; Fayez-Hassan, M.; Sedqy, E. Assessment of natural radioactivity in fertilizers and phosphate ores in Egypt. *J. Taibah Univ. Sci.* **2016**, *10*, 296–306. [CrossRef]
14. Boumala, D.; Mavon, C.; Belafrites, A.; Tedjani, A.; Groetz, J.E. Evaluation of radionuclide concentrations and external gamma radiation levels in phosphate ores and fertilizers commonly used in Algeria. *J. Radioanal. Nucl. Chem.* **2018**, *317*, 501–510. [CrossRef]
15. Besser, H.; Hamed, Y. Environmental impacts of land management on the sustainability of natural resources in Oriental Erg Tunisia, North Africa. *Environ. Dev. Sustain.* **2021**, *23*, 11677–11705. [CrossRef]
16. Zhang, F.; Wang, Y.; Liao, X. Recognition method for the health risks of potentially toxic elements in a headwater catchment. *Sci. Total Environ.* **2022**, *839*, 156287. [CrossRef] [PubMed]
17. Arabi, A.M.; El Ahmed, N.K.; Din, K.S. Natural radionuclides and dose estimation in natural water resources from Elba protective area, Egypt. *Radiat. Prot. Dosim.* **2006**, *121*, 284–292. [CrossRef] [PubMed]
18. Benmarce, K.; Hadji, R.; Zahri, F.; Khanchoul, K.; Chouabi, A.; Zighmi, K.; Hamed, Y. Hydrochemical and geothermometry characterization for a geothermal system in semiarid dry climate: The case study of Hamma spring (Northeast Algeria). *J. Afr. Earth Sci.* **2021**, *182*, 104285. [CrossRef]
19. Dragović, S.D.; Janković-Mandić, L.J.; Dragović, R.M.; Đorđević, M.M.; Đokić, M.M. Spatial distribution of the 226Ra activity concentrations in well and spring waters in Serbia and their relation to geological formations. *J. Geochem. Explor.* **2012**, *112*, 206–211. [CrossRef]
20. Scanlon, B.R.; Faunt, C.C.; Longuevergne, L.; Reedy, R.C.; Alley, W.M.; McGuire, V.L.; McMahon, P.B. Groundwater depletion and sustainability of irrigation in the US High Plains and Central Valley. *Proc. Natl. Acad. Sci. USA* **2012**, *109*, 9320–9325. [CrossRef]
21. Feng, W.; Zhong, M.; Lemoine, J.M.; Biancale, R.; Hsu, H.T.; Xia, J. Evaluation of groundwater depletion in North China using the Gravity Recovery and Climate Experiment (GRACE) data and ground-based measurements. *Water Resour. Res.* **2013**, *49*, 2110–2118. [CrossRef]
22. Famiglietti, J.S. The global groundwater crisis. *Nat. Clim. Chang.* **2014**, *4*, 945–948. [CrossRef]
23. Hamed, Y.; Ahmadi, R.; Demdoum, A.; Bouri, S.; Gargouri, I.; Ben Dhia, H.; Al-Gamal, S.; Laouar, R.; Choura, A. Use of geochemical, isotopic, and age tracer data to develop models of groundwater flow: A case study of Gafsa mining basin-Southern Tunisia. *J. Afr. Earth Sci.* **2014**, *100*, 418–436. [CrossRef]
24. Hamed, Y.; Hadji, R.; Redhaounia, B.; Zighmi, K.; Bâali, F.; El Gayar, A. Climate impact on surface and groundwater in North Africa: A global synthesis of findings and recommendations. *Euro-Mediterr. J. Environ. Integr.* **2018**, *3*, 25. [CrossRef]
25. Post, V.E.A.; Vassolo, S.I.; Tiberghien, C.; Baranyikwa, D.; Miburo, D. Weathering and evaporation controls on dissolved uranium concentrations in groundwater—A case study from northern Burundi. *Sci. Total Environ.* **2017**, *607–608*, 281–293. [CrossRef] [PubMed]
26. Hadji, R.; Raïs, K.; Gadri, L.; Chouabi, A.; Hamed, Y. Slope failures characteristics and slope movement susceptibility assessment using GIS in a medium scale: A case study from OuledDriss and Machroha municipalities, Northeastern of Algeria. *Arab. J. Sci. Eng.* **2017**, *42*, 281–300. [CrossRef]
27. Arif, K.; Khalid, P. Hydrostratigraphy and hydrogeophysical studies to delineate fresh and saline aquifer boundaries in Lesser Cholistan of Pakistan. *PLoS ONE* **2023**, *18*, e0292035. [CrossRef]
28. Mokadem, N.; Hamed, Y.; Ben Saad, A.; Gargouri, I. Atmospheric pollution in North Africa (ecosystems–atmosphere interactions): A case study in the mining basin of El Guettar–M'Dilla (southwestern Tunisia). *Arab. J. Geosci.* **2014**, *7*, 2071–2079. [CrossRef]
29. Marzougui, S.; Sdiri, A.; Rekhiss, F. Heavy metals' mobility from phosphate washing effluents discharged in the Gafsa area (southwestern Tunisia). *Arab. J. Geosci.* **2016**, *9*, 599. [CrossRef]
30. Colman, A.S.; Holland, H.D. *The Global Diagenetic Flux of Phosphorus from Marine Sediments to the Oceans: Redox Sensitivity and the Control of Atmospheric Oxygen Levels*; SEPM Society for Sedimentary Geology: Claremore, OK, USA, 2000. [CrossRef]
31. Selvakumar, R.; Ramadoss, G.; Menon, M.P.; Rajendran, K.; Thavamani, P.; Naidu, R.; Megharaj, M. Challenges and complexities in remediation of uranium contaminated soils: A review. *J. Environ. Radioact.* **2018**, *192*, 592–603. [CrossRef] [PubMed]

32. Veríssimo, C.U.V.; Santos, R.V.; Parente, C.V.; de Oliveira, C.G.; Cavalcanti, J.A.D.; Neto, J.D.A.N. The Itataia phosphate-uranium deposit (Ceará, Brazil) new petrographic, geochemistry and isotope studies. *J. S. Am. Earth Sci.* **2016**, *70*, 115–144. [CrossRef]
33. Gao, S.; Tang, G.; Hua, D.; Xiong, R.; Han, J.; Jiang, S.; Zhang, Q.; Huang, C. Stimuli-responsive bio-based polymeric systems and their applications. *J. Mater. Chem. B* **2019**, *7*, 709–729. [CrossRef]
34. Hamed, Y.; Awad, S.; Ben Sâad, A. Nitrate contamination in groundwater in the Sidi Aïch-Gafsa oases region, Southern Tunisia. *Environ. Earth Sci.* **2013**, *70*, 2335–2348. [CrossRef]
35. Khaing, H.; Thakur, P. Rapid sequential separation method for 210Po and actinides in air flter samples. *J. Radioanal. Nucl. Chem.* **2017**, *314*, 1383–1392. [CrossRef]
36. Nelson, A.W.; Eitrheim, E.S.; Knight, A.W.; May, D.; Mehrhoff, M.A.; Shannon, R.; Litman, R.; Burnett, W.C.; Forbes, T.Z.; Schultz, M.K. Understanding the radioactive ingrowth and decay of naturally occurring radioactive materials in the environment: An analysis of produced fluids from the Marcellus Shale. *Environ. Health Perspect.* **2015**, *123*, 689–696. [CrossRef] [PubMed]
37. Bugai, D.; Dewière, L. *Global Model for the 90Sr Transport in the Aquifer at Chernobyl Pilot Site: Model Calibration and Sensitivity Analyses*; IRSN: Fontenay-aux-Roses, France, 2004.
38. Dewière, L.; Bugai, D.; Grenier, C.; Kashparov, V.; Ahamdach, N. 90Sr migration to the geosphere from a waste burial in the Chernobyl exclusion zone. *J. Env. Radioact.* **2004**, *74*, 139–150. [CrossRef] [PubMed]
39. Roux, C. Etude des Mécanismes de Transfert des Radionucléides en Aval de la Fosse T22 du Site Expérimental de Tchernobyl. Doctoral Dissertation, Aix-Marseille University, Marseille, France, 2013.
40. Jenifer, M.A.; Jha, M.K. Comprehensive risk assessment of groundwater contamination in a weathered hard-rock aquifer system of India. *J. Clean. Prod.* **2018**, *201*, 853–868. [CrossRef]
41. Yuan, Y.; Xiang, M.; Liu, C.; Theng, B.K.G. Chronic impact of an accidental wastewater spill from a smelter, China: A study of health risk of heavy metal(loid)s via vegetable intake. *Ecotoxicol. Environ. Saf.* **2019**, *182*, 109401. [CrossRef] [PubMed]
42. Njuguna, S.M.; Makokha, V.A.; Yan, X.; Gituru, R.W.; Wang, Q.; Wang, J. Health risk assessment by consumption of vegetables irrigated with reclaimed wastewater: A case study in Thika (Kenya). *J. Environ. Manag.* **2019**, *231*, 576–581. [CrossRef] [PubMed]
43. Langmuir, D. *Aqueous Environmental Geochemistry*; Prentice Hall: Upper Saddle River, NJ, USA, 1997; p. 600.
44. De Souza, A.L.; Cotrim, M.E.B.; Pires, M.A.F. An overview of spectrometric techniques and sample preparation for the determination of impurities in uranium nuclear fuel grade. *Microchem. J.* **2013**, *106*, 194–201. [CrossRef]
45. Kertes, A.S.; Guillaumont, R. Solubility of UO_2. A comparative review. *Nucl. Chem. Waste Manag.* **1985**, *5*, 215–219.
46. Bi, Y.; Hyun, S.P.; Kukkadapu, R.K.; Hayes, K.F. Oxidative dissolution of UO_2 in a simulated groundwater containing synthetic nanocrystalline mackinawite. *Geochim. Cosmochim. Acta* **2013**, *102*, 175–190. [CrossRef]
47. Durakoviæ, A. Medical effects of internal contamination with uranium. *Croat. Med. J.* **1999**, *40*, 49–66.
48. Alam, M.S.; Cheng, T. Uranium release from sediment to groundwater: Influence of water chemistry and insights into release mechanisms. *J. Contam. Hydrol.* **2014**, *164*, 72–87. [CrossRef]
49. Ulrich, S.; Gillow, J.; Roberts, S.; Byer, G.; Sueker, J.; Farris, K. Hydrogeochemical and mineralogical factors influencing uranium in background area groundwater wells: Grants, New Mexico. *J. Hydrol. Reg. Stud.* **2019**, *26*, 100636. [CrossRef]
50. Van, K. *Cadmium and Other Minor Elements in World Resources of Phosphate Rock*; The Fertiliser Society Proceedings: York, UK, 1997; No. 400; pp. 15–22, 30–31.
51. Van, K. *World Phosphate Rock Reserves and Resources*; IFDC Publication: Muscle Shoals, AL, USA, 2010; p. 33.
52. U.S. Geological Survey. Mineral Year Book. In *Phosphate Rock*; U.S. Geological Survey: Reston, VA, USA, 2011.
53. NEA; IAEA. *Uranium 2020: Resources, Production and Demand*; Public Online Report; NEA No. 7551; OECD: Paris, France, 2020.
54. Gandhi, T.P.; Sampath, P.V.; Maliyekkal, S.M. A critical review of uranium contamination in groundwater: Treatment and sludge disposal. *Sci. Total Environ.* **2022**, *825*, 153947. [CrossRef]
55. Hidouri, N.; Missaoui, R.; Jraba, A.; Yousaf, B.; Sleimi, N.; Hamed, Y. Assessing the vertical distribution and health risks of trace metal elements in phosphogypsum-enriched agricultural soils of typical peri-urban areas in Southwestern Tunisia. *Water Air Soil. Pollut.* **2023**, *234*, 332. [CrossRef]
56. Serio, F.; Miglietta, P.P.; Lamastra, L.; Ficocelli, S.; Intini, F.; De Leo, F.; De Donno, A. Groundwater nitrate contamination and agricultural land use: A grey water footprint perspective in Southern Apulia Region (Italy). *Sci. Total Environ.* **2018**, *645*, 1425–1431. [CrossRef] [PubMed]
57. He, S.; Wu, J. Hydrogeochemical characteristics, groundwater quality and health risks from hexavalent chromium and nitrate in groundwater of Huanhe Formation in Wuqi County, northwest China. *Expo Health* **2019**, *11*, 125–137. [CrossRef]
58. Li, P.; He, X.; Guo, W. Spatial groundwater quality and potential health risks due to nitrate ingestion through drinking water: A case study in Yan'an City on the Loess Plateau of northwest China. *Hum. Ecol. Risk Assess.* **2019**, *25*, 11–31. [CrossRef]
59. Hamdi, A.; Ben Jamaa, N.; KallelKamoun, I. Potential use of phosphogypsum in paving blocks. *Green Mater.* **2021**, *9*, 97–107. [CrossRef]
60. Garbaya, H.; Jraba, A.; Khadimallah, M.A.; Elaloui, E. The Development of a New Phosphogypsum-Based Construction Material: A Study of the Physicochemical, Mechanical and Thermal Characteristics. *Materials* **2021**, *14*, 7369. [CrossRef]

61. Hashim, M.A.; Mukhopadhyay, S.; Sahu, J.N.; Sengupta, B. Remediation technologies for heavy metal contaminated groundwater. *J. Environ. Manag.* **2011**, *92*, 2355–2388. [CrossRef] [PubMed]
62. He, X.; Li, P. Surface water pollution in the middle Chinese Loess Plateau with special focus on hexavalent chromium (Cr^{6+}): Occurrence, sources, and health risks. *Expo Health* **2020**, *12*, 385–401. [CrossRef]

Disclaimer/Publisher's Note: The statements, opinions and data contained in all publications are solely those of the individual author(s) and contributor(s) and not of MDPI and/or the editor(s). MDPI and/or the editor(s) disclaim responsibility for any injury to people or property resulting from any ideas, methods, instructions or products referred to in the content.

hydrology

MDPI

Article

Assessment of the Impact of Coal Mining on Water Resources in Middelburg, Mpumalanga Province, South Africa: Using Different Water Quality Indices

Mndeni Magagula, Ernestine Atangana * and Paul Oberholster

Centre for Environmental Management, Faculty of Natural and Agricultural Sciences, University of the Free State, Bloemfontein 9300, South Africa; mndenimagagula1@gmail.com (M.M.); OberholsterPJ@ufs.ac.za (P.O.)
* Correspondence: AtanganaE@ufs.ac.za

Abstract: The objective of this study was to assess the water quality status of the surface water and groundwater resources in the Middelburg area, South Africa. The assessment was addressed using combined water quality indices, investigating selected chemical parameters over four different seasons for a period of five years from 2017 to 2021. A combination of the Canadian Council of Ministers of the Environment water quality index and the comprehensive pollution index was used to analyze the water quality status of surface water and groundwater of the town of Middelburg, situated near coal mining activities in Mpumalanga, South Africa. The combination of the indices indicated that some surface water monitoring sites ranged between poor to fair water quality. Groundwater monitoring points also showed a poor to fair ranking. The comprehensive pollution index confirmed that some sites showed very poor water quality in the summer seasons, exceeding expected limits for the period 2017 to 2021. The principal component analysis further showed that both surface water and groundwater sites had high levels of contamination with increased chemical parameters. The results were compared against the different water quality guidelines. In an extensive monitoring program, water management systems must be properly implemented to mitigate impacts on water resources.

Keywords: coal mining; acid mine drainage; water quality; pollution indices; multivariant analysis

Citation: Magagula, M.; Atangana, E.; Oberholster, P. Assessment of the Impact of Coal Mining on Water Resources in Middelburg, Mpumalanga Province, South Africa: Using Different Water Quality Indices. *Hydrology* **2024**, *11*, 113. https://doi.org/10.3390/hydrology11080113

Academic Editor: Meysam Vadiati

Received: 22 June 2024
Revised: 23 July 2024
Accepted: 25 July 2024
Published: 31 July 2024

Copyright: © 2024 by the authors. Licensee MDPI, Basel, Switzerland. This article is an open access article distributed under the terms and conditions of the Creative Commons Attribution (CC BY) license (https://creativecommons.org/licenses/by/4.0/).

1. Introduction

In most developing countries around the world, including South Africa, the mining industry effectively contributes to their socioeconomic development. Over centuries, mining has been an economic driver contributing to the gross domestic product, providing employment and business opportunities. However, the industry also poses a big threat to the natural environment [1,2]. Most mining countries have been victims of water pollution and unrehabilitated land [3]. This has resulted in water quality contamination becoming a significant environmental challenge in many countries around the world. The expansion of coal and gold mining in California has not only contributed to industry growth and success in the United States but also increased detrimental impacts on the environment [4]. In Spain, the Tinto River has experienced endless pollution from mining impacts such as acid mine drainage (AMD) [5]. In China, the Coal Industry Association also showed a reduction in coal mines from approximately 80,000 to 5800; however, most of the closed mines have resulted in environmental challenges and deteriorating water quality [6].

The government of South Africa has set laws and policies through the Department of Environmental Affairs and the Department of Water and Sanitation, previously known as the Department of Water Affairs and Forestry, to ensure that mining does not impact the environment. However, there has not been a significant improvement in the degradation of the environment. The availability and quality of South Africa's water resources are under strain from trends in natural and anthropogenic activities including urbanization, intensive agricultural practices, and significant industrial developments such as mining operations [7]. Since the majority of these rapidly expanding water users are located upstream of sizable rivers and dams, the effects of their effluent discharges and return flows on water resources are substantial [8]. The aquatic ecosystem, human health, and other water users all suffer significant effects from the decline in the quality of water resources. Increased salinity, heavy metals, nutrients, organic debris, and sedimentation are a few of the frequent water quality issues [9].

To determine the amount of good water quality needed for residential use, the sustainable environmental or ecological flow of rivers, and the quality of water sources for various applications, it is crucial to monitor water quality and anthropogenic disturbances on water resources. Additionally, utilizing assessed factors, the water quality indices (WQIs) were created to depict the current water quality status of watercourses. Nevertheless, certain WQI models are not generic since they create uncertainty in the process of converting vast volumes of water quality data into a single index because they are typically created in accordance with site-specific rules for a specific region [10].

Numerous academics from around the world have used these indices to evaluate the state of surface water and groundwater bodies. This includes the Horton index, the National Sanitation Foundation WQI, the Scottish Research Development Department index, the water quality index of the Canadian Council of Ministers of the Environment (CCME–WQI), the Bascarón WQI, the fuzzy interface system, the Malaysian WQI and the West Java WQI [10–12], the comprehensive pollution index (CPI), organic pollution index (OPI), eutrophication index (EI), trace metal pollution index (TPI) [13], and the National Sanitation Foundation–Environmental Sanitation Technology Company of the State of São Paulo WQI) [14].

Previous studies used a modified pollution index to determine the water quality status of the Loskop Dam, South Africa, and reported nutrient enrichment and heavy metal pollution [15]. Son et al. [13] evaluated the water quality of the Cau River using a combination of indices, including the WQI, CPI, OPI, EI, and TPI. They found significant eutrophic conditions and organic pollution downstream. Tang et al. [16] used the comprehensive pollution index to assess the water quality of Qilu Lake. Matta et al. [17] used the CPI to assess the water quality of the Henwal River in India and found moderate and severe contamination at many sampling sites. Using a combination of indices (CPI, OPI, EI, and TPI), Mishra et al. [18] assessed the Surha Lake in India and discovered that it was eutrophic and moderately polluted. Lumb et al. [19] evaluated the water quality of the Mackenzie Great Bear Sub-basin using the CCME–WQI and found turbidity and trace metal contamination. In the Norte Chico zone of north-central Chile, Espejo et al. [20] also applied the water quality status of four watersheds (Huasco Basin, Elqui Basin, Choapa Basin, and Limarí Basin) using the CCME–WQI. They found that, overall, the water quality was fairly acceptable, with the Limarí Basin having particularly good quality.

Although these tools have been modified and applied in various reservoirs, the WQIs have been used to address these issues to better assess the general water quality for both surface water and groundwater; however, not much work has been conducted at the catchment level of the study area. For this study, a combination of both the CCME–WQI and CPI was used to understand the overall water quality at the study area. There are very limited studies where combined indices were used to assess coal mining impacts in water resources [21]. The main reasons for selecting the CCME–WQI in the study was because it is an easy application to use, and it offers flexibility in selecting the variables to analyze in the model [22] as well as its ability to identify the space and time of dynamics in water quality and provide a number to represent water quality in a specific area of interest and time. The CCME–WQI can also simplify a complex dataset to be easier to understand. Furthermore, the index can also communicate the water quality status to decision makers and the public at large.

The CPI is another essential technique for scientific reflection of the kinds and levels of pollution in water systems. The CPI was also used in Bangladesh to assess the Karnaphuli

River, where severe pollution was identified from different wastes generated from various industries situated at the banks of the river [23].

With the social and economic development of the town of Middelburg, South Africa, and the upper Olifants River catchment largely influenced by coal mining, McCarthy [24] confirmed that approximately 50 million liters per day of AMD were decanting from defunct mines. The water quality in the Loskop Dam situated just downstream of the town of Middelburg has over time shown contamination, with groundwater being impacted by mining taking place upstream [25]. Furthermore, Adler and Rascher [26] mentioned that the impacts of AMD found in the Witbank and Middelburg regions of South Africa included environmental, political, and socioeconomic impacts. The environmental impacts included pollution of surface water through heavy metals, affecting aquatic life [3,26,27]. Therefore, there was a need to assess the groundwater and surface water quality status of the Middelburg region and to determine their pollution sources.

According to Atangana [28] and Oberholster et al. [15], it is imperative to develop tools such as water monitoring indices that may be utilized to detect and mitigate pollution in these types of catchments. An effective way to enhance data interpretation and analysis is to assess the present status of water quality in the watershed using a variety of water quality indices (e.g., CCME–WQI and CPI) in combination with water quality guidelines. Using the pollution indices (CPI and CCME–WQI), in relation to the water quality guidelines can provide better scientific evidence on the level and extent of pollution in water resources, which will be essential for decision makers. Very little is known about the use of combined WQIs in catchments impacted by AMD in South Africa; it will therefore be important to improve monitoring data analysis and interpretation. The study therefore aimed at achieving the following objectives:

1. To use existing water quality monitoring data for both surface water and groundwater sampling points in the Middelburg region and compare it with the water quality resource objectives of the study area;
2. To use different water quality indices (CCME–WQI and CPI) in combination with the water quality guidelines to determine the impacts of coal mining activities on surface water and groundwater quality around the study area;
3. To evaluate the interrelationship trends of the surface water and groundwater quality data;
4. To provide possible and efficient mitigation measures for the protection of water resources from coal mining and other related land-use activities.

2. Materials and Methods

2.1. Description of Study Area

The case study is situated approximately three kilometers south of the town of Middelburg near the N4 National Road and the R35 Middelburg–Bethal Provincial Road within the Steve Tshwete Local Municipality of the Nkangala District in the Mpumalanga Province, South Africa. The mining industry around the area has been contributing effectively to the socioeconomic development of the country by providing employment opportunities. The study area falls within the summer rainfall regions of the Mpumalanga province, where a typical Highveld climate with warm summers are observed at the range of 12 °C to 29 °C and winters from −3 °C to 20 °C. The Mpumalanga Highveld has distinct wet and dry seasons, with about 91% of the area's mean annual rainfall experienced from October to April [29].

The topography of the study area is mostly characterized by subdued relief and consists of gently shallow sand slopes [30]. Most of the coal mines in and around the area are exploiting coal, which occurs in shallow depths, influencing the dominating use of opencast mining methods [31]. The area is underlain by sediments of the Ecca Group that forms part of the Karoo Supergroup. The Ecca Group consists mainly of dark grey shale that is carbonaceous in some instances, with interbedded whitish sandstone and greyish gritstone as well as occasional coal bands [32].

The area falls within the Olifants Water Management Area 2 (WMA 2). The water management area is characterized by extensive agricultural and mining activities, mostly associated with coal deposits dominating the northern part of the area [25]. The catchment in which the study area falls consists of perennial streams situated on both the eastern and western side, draining towards the major rivers, namely the Olifants and Klein-Olifants Rivers. Within the B11H quaternary catchment, the tributaries flow in a north–west direction to the confluence with the Olifants River; at B12D, the streams flow north–east to the Klein-Olifants River, which then drains to the major Olifants River. Figure 1 shows the regional and drainage locality of the study area.

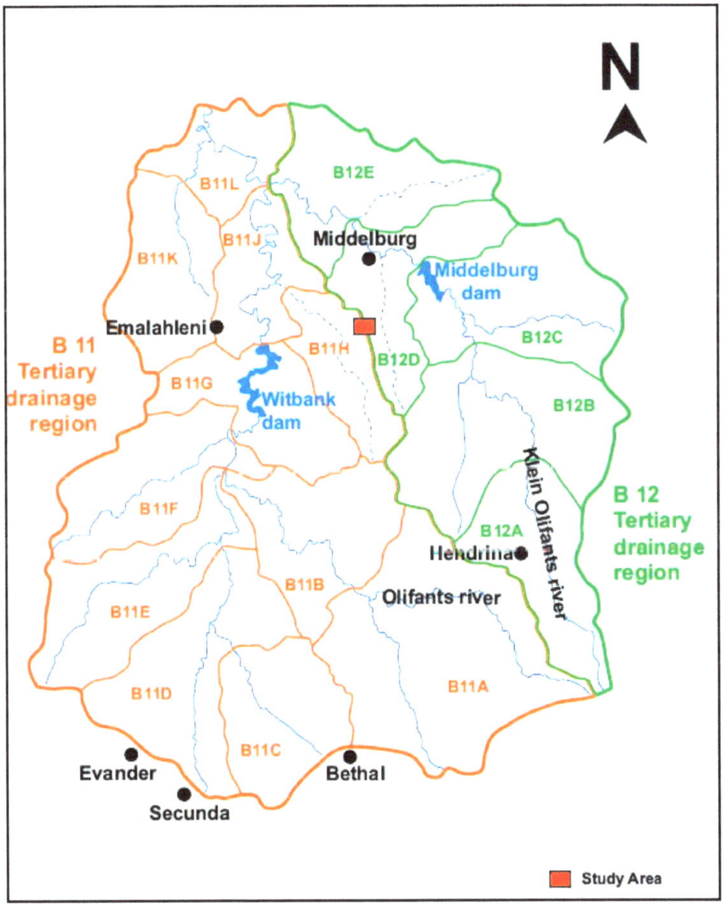

Figure 1. Regional and drainage regions of the study area, South Africa.

2.2. Field Sampling and Analysis

Due to increasing water pollution caused by mining activities, especially in the coalfields, authorities such as the Department of Water and Sanitation have enforced the importance of water quality management, which entails monitoring. The National Water Act 36 of 1998 has further introduced an integrated water resource management concept, compromising all water resources aspects such as water quality monitoring [33].

For the current study, historical data from the surface water and groundwater monitoring points, as shown in Figures 2 and 3, were utilized to achieve the objectives set. The data were obtained from an established monitoring program on a monthly basis for surface water and on a quarterly basis for groundwater for a period of five years from 2017 to 2021. The monitoring program was developed to provide the mining companies with the necessary information to assess the successes and failures of their water management strategies and comply with monitoring commitments stipulated in issued licenses and environmental permits. Using historical data provided answers and different perspectives on issues [34,35]. The water quality data were compared against the following standards to assess the level of contamination: resource quality objectives (RQO) [36], target water quality range (TWQR) [37,38], the CCME–WQI [39], and the guidelines of the Australian and New Zealand Environment and Conservation Council and the Agriculture and Resource Management Council of Australia and New Zealand (ANZECC/ARMCANZ) [40].

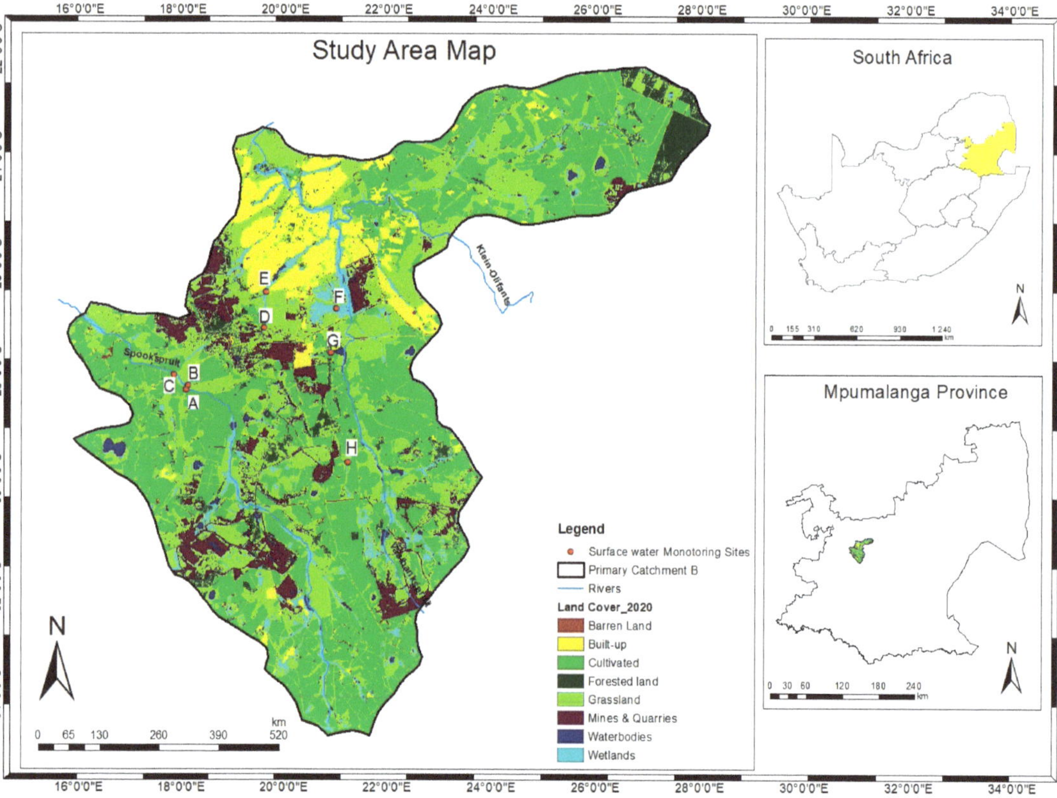

Figure 2. Surface water monitoring sites sampled from 2017 to 2021 (A–H sampling sites) [41].

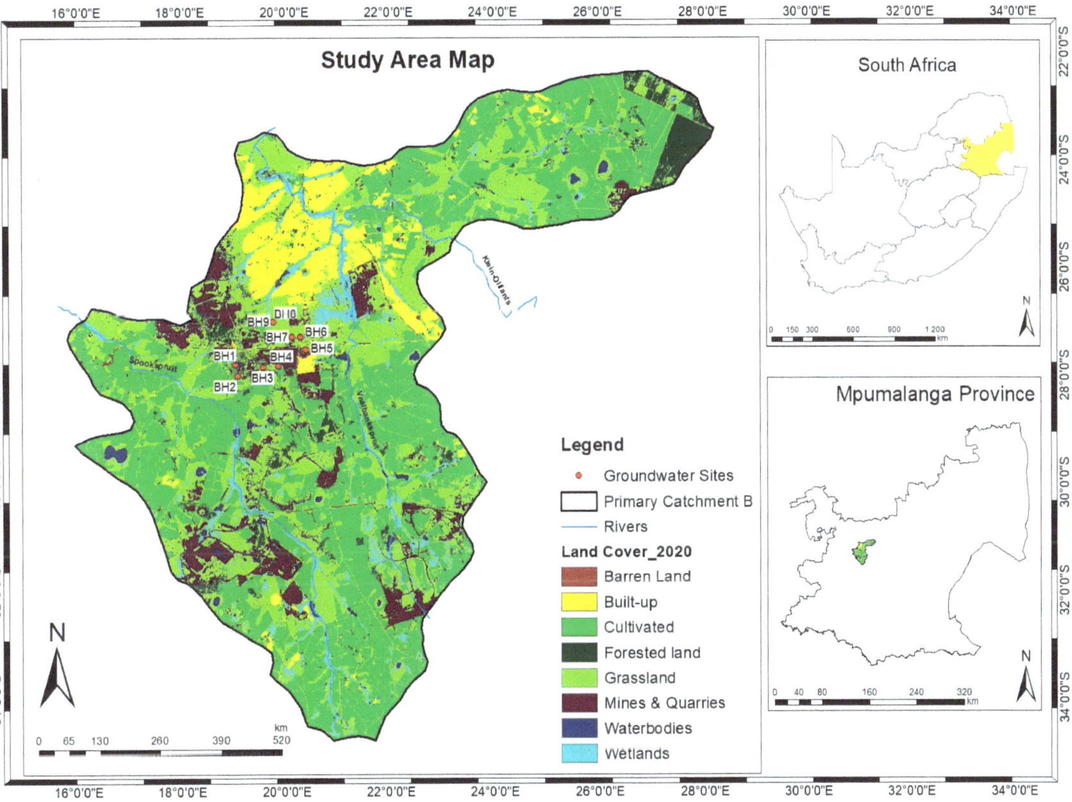

Figure 3. Groundwater sampling sites sampled from 2017 to 2021 (BH = borehole) [41].

2.3. Surface Water

The sampling procedure followed for the monitoring was the grab sampling method, which involved dipping a plastic bailer into the water to collect samples from the center of the streams, which were collected in clear, marked, sterilized one-liter plastic bottles. The bailer was rinsed with water from each sampling site to prevent cross-contamination. Monitoring of surface water was conducted on a monthly basis from 2017 to 2021 at eight sampling sites. The name of the sample, date, and time of collection were marked on each of the sampling bottles using a permanent marker. Water quality sampling was conducted during the day, and no on-site analysis was conducted. The samples were then transported in a cool insulated container under strict protocols within 24 h to an accredited laboratory for analysis, namely the Regen Waters Laboratory in Witbank, Mpumalanga, which is accredited for the chemical and microbiological analysis of water samples in terms of the South African National Accreditation System.

2.4. Groundwater

Contrary to surface water, groundwater monitoring was conducted during the day on a quarterly basis from 2017 to 2021 at eight boreholes using the grab sampling method. A plastic bailer tied with a rope was used, which involved dipping it into the borehole to collect samples from the borehole, which were collected into clear, marked, sterilized one-liter plastic bottles and labelled similarly to those of the surface water. The bailer was also rinsed with water from each sampling site to avoid the possibility of cross-contamination.

All water samples were kept in a cool insulated container under strict protocols to be tested within 24 h.

All collected samples were then transported to an accredited laboratory, Regen Waters Laboratory, in Witbank, Mpumalanga, for analysis in terms of the South African National Accreditation System. The analyses were conducted following the methods specified by the South African Bureau of Standards. For nutrient analysis such as chloride, nitrate, and ammonia, the laboratory used the procedures for the analysis of chloride, nitrate, and ammonia and methods such as spectroscopy and inductively coupled plasma–optimal emission spectrometry (ICP–OES). Furthermore, the inductively coupled plasma–mass spectrometry (ICP–MS) was used by Regen Waters Laboratory for metal analyses due to its sensitivity, as it detects very low concentrations.

2.5. Selection of Sampling Sites

The sampling sites were strategically selected to determine potential contamination of water courses. Surface water monitoring sites were selected from upstream and downstream of coal mining and related activities to monitor the water quality before and after the study. Groundwater boreholes around the area were assessed for any plume movement away from the mining and related activities. All sampling sites depicted a good representation of what the study aimed to achieve.

Figure 2 indicates the location and description of the surface water monitoring points, and Figure 3 shows the location of the boreholes sampled at the groundwater sampling sites.

2.6. Data Analysis

2.6.1. Water Quality Analysis

The Department of Water Affairs and Forestry [36] established the RQO of different catchments and the TWQR [37,38] portraying the maximum limits for concentrations of the variables allowed in surface water. The CCME [39] as well as the ANZECC/ARMCANZ [40] guidelines were also used to assess the level of water quality in the study area. The World Health Organization's [42] guidelines for drinking water quality was used to specifically assess the potential risks on groundwater in the receiving environment.

2.6.2. Water Pollution Indices

Understanding the source of pollution will help to recommend mitigation and management measures to reduce any further potential risks [43]. The available water quality dataset was used to provide empirical evidence in making future environmental decisions. The WQIs also represented the quality levels using the analyzed parameters. The CCME–WQI was used to understand the overall water quality for surface water resources. The benefit of using this index is its ability to identify water quality over space and time. It can also simplify complex water quality data to be easily understandable [22].

The following equations for the CCME–WQI were used:

$$CCME\text{–}WQI = 100 - \frac{\sqrt{F_1{}^2 + F_2{}^2 + F_3{}^2}}{1.732} \quad (1)$$

where

F1 (scope) represents the number of variables whose objectives were not met.

$$F_1 = \left(\frac{Number\ of\ failed\ variables}{Total\ number\ of\ variables} \right) \times 100 \quad (2)$$

F2 (frequency) represents the number of items by which the objectives were not met.

$$F_2 = \left(\frac{Number\ of\ failed\ tests}{Total\ number\ of\ tests} \right) \times 100 \quad (3)$$

F3 (amplitude) represents the amount by which the objectives were not met.

$$Amplitude = \left(\frac{nse}{0.01_{nse} + 0.01}\right) \quad (4)$$

Excursion:

$$Normalised\ sum\ of\ excursions(nse) = \frac{\sum_{i=1}^{n} excursion}{Number\ of\ tests} \quad (5)$$

$$Excursion = \left(\frac{Failed\ test\ value}{Guideline\ value}\right) - 1 \quad (6)$$

As indicated above, the understanding of water quality is important for managing the environment and possible mitigation of pollution. The CPI is another method that can be used to evaluate water quality [13]. The CPI technique is good for scientific reflection of the kinds and extent of pollution in water systems. The pollution index (PI) was used to undertake the groundwater assessment in this study to identify the coal mining impacts.

The CPI method is obtained from the following formula:

$$CPI = \frac{1}{n}\sum_{1=0} PI \quad (7)$$

where

PI represents pollution index of individual parameters.

N is the number of parameters.

$$PI = \left(\frac{C_i}{S_i}\right) \quad (8)$$

where

C_i is the measured concentration of ith paramete

S_i is the standard value of ith parameter.

The WQIs and values were classified into different categories as shown in Table 1.

Table 1. Water quality index ratings [44].

Water Quality Indices	CCME–WQI	Comprehensive Pollution Index
Excellent	91–100	<0.2
Good	71–90	0.21–0.40
Poor/fair	51–70	0.41–1.00
Very poor/marginal	26–50	1.01–2.0
Unsuitable/poor	0–25	>2.01

2.6.3. Statistical Analysis

All water quality analyses for surface water and groundwater samples were calculated and standardized using Microsoft Excel 2016 and are represented as mean and standard deviation. For this research, multivariate statistical techniques were also employed to characterize, assess, and verify the seasonality of temporal and regional fluctuations induced by human causes connected to surface water and freshwater quality. The water quality datasets of the study sites, which includes all the parameters, were analyzed using a multivariate approach such as principal component analysis (PCA) and regression analysis. Using the IBM SPSS program version 13, a constrained variant of the linear ordination approach of PCA, also known as a redundancy analysis (RDA), was carried out. Microsoft Excel 2016 was used for the regression analysis.

2.6.4. Principal Component Analysis

According to various researchers, the PCA is a crucial technique for establishing links between environmental quality and landscape [45]. The variation in a collection of response factors such as water quality variables explained by a set of explanatory variables such as land use of watersheds was summarized by the RDA.

It is further used to reduce data and to extract principal components for analyses of relationships between variables. PCA can reduce the number of correlated variables to a smaller set of orthogonal factors, making it easier to interpret a given multidimensional system by displaying the correlations among the original variables. PCA and derivative methods have been widely applied to various environmental streams, such as water, air, and soil, to identify sources of pollution through natural versus anthropogenic activities.

2.6.5. Multiple Linear Regression

Statistical methods such as the multiple linear regression model were used to measure the strength of the relationship between a measured dependent variable and independent variables of water quality parameters. The collected data from the surface and groundwater quality monitoring station were analyzed to assess the differences in values at the different sampling points.

Multiple linear regressions can be expressed using the following equation:

$$\gamma = \beta_0 + \beta_1 x_1 + \beta_2 \beta x_2 + \cdots \ldots\ldots\ldots\ldots\ldots + \beta_n x_n + \varepsilon \tag{9}$$

where

γ represents the dependent variable.
$x_1 \ldots, x_2$ represents several independent variables.
$\beta_0 \ldots \beta_n$ represents the regression coefficient.
ε represents the random error.

According to Bhatnagar and Devi [46], the coefficient of determination (R^2) measures the model's goodness of fit with the measured variables, which indicates the percentage of variance for the developed models, as shown in Equation (15).

Equation (10): The coefficient of determination:

$$R^2 = \frac{\sum_{i=1}^{N} (Y_i - \overline{y})^2}{\sum_{i=1}^{N} (y_i - \overline{y})^2} = \frac{SSR}{SST} = 1 - \frac{SSE}{SST} \tag{10}$$

where

SSR is the regression sum of squares.
SST is the observed variance.
SSE is sum of squares of the residuals.

R^2 can be adjusted to reflect the different independent variables used and is expressed as in Equation (10) above.

Equation (11): The independent variables:

$$Adjusted\ R^2 = 1 - \frac{N-1}{N-K-1}\frac{SSE}{SST} \tag{11}$$

Root mean square error is also used to evaluate the performance of the developed model, as shown in Equation (12).

Equation (12): The root mean square error:

$$Root\ Mean\ Square\ Error = \sqrt{\frac{1}{n}\sum_{i=1}^{N} (x_i - \hat{x}_i)^2} \tag{12}$$

The x_i is the actual value.

$\hat{x}_i$ is the predicted value, and n is the number of observations.

R^2, adjusted R^2, and root mean square error can be obtained directly from the output data results.

3. Results

3.1. Water Chemistry

Generally, surface water monitoring sites become contaminated rapidly due to exposure to various anthropogenic activities and natural causes. Looking at the average concentrations of the chemical variables from the analyzed water quality data, all the surface water monitoring points in the study showed some level of contamination that can be associated with coal mining and related activities.

Total dissolved solids (TDS) is a measure of organic and inorganic materials dissolved in water, such as salts, metals, ions, and different minerals. Anthropogenic activities are known to produce TDS in water. Agricultural runoff and mining impacts such as AMD can increase the presence of TDS in a water course. The increase in TDS at monitoring sites A and B exceeded the permissible limits in all different seasons, as shown in Figure 4. All other sites were below the selected guideline. A maximum average of 1373 mg/L was analyzed during the spring season and an average of 1217 mg/L recorded for TDS in summer.

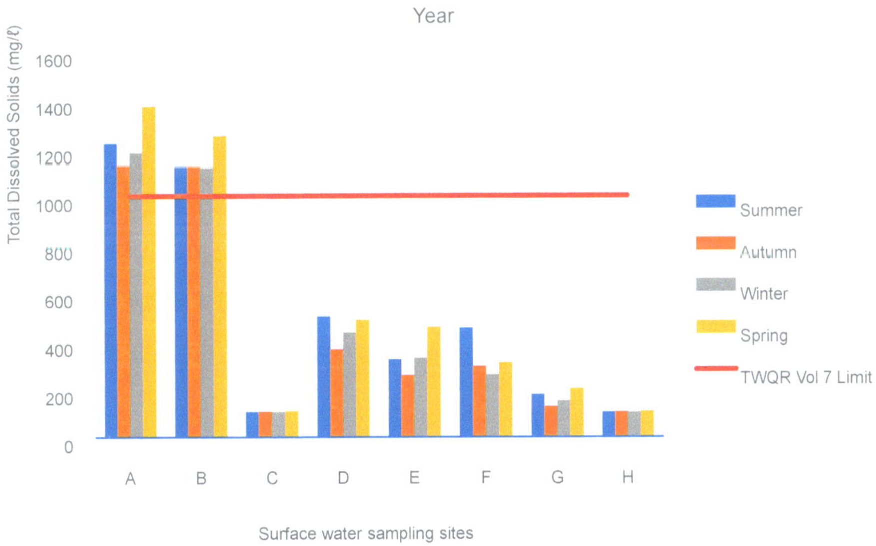

Figure 4. Average total dissolved solids (mg/L) from 2017 to 2021.

Sulfates represent another naturally occurring variable on the earth from different rock and soil types. Mining activities are likely to produce SO_4 in abundance, which is washed off to rivers and groundwater resources in the vicinity. The concentration of SO_4 at monitoring sites A and B was recorded as higher than the 500 mg/L limit of the RQO guideline [36] in all the seasons, ranging from 725 mg/L to 1174 mg/L for the entire five years of monitoring (see Figure 5 and Table 2).

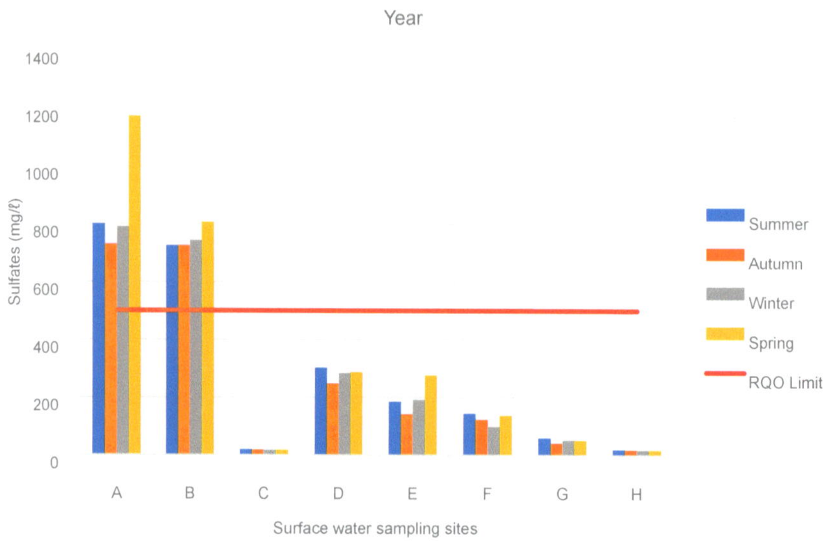

Figure 5. Average sulfates (mg/L) from 2017 to 2021.

Monitoring site B also recorded an elevated average concentration of electrical conductivity (EC) in the seasons above 30 mS/m when compared to the ANZECC/ARMCANZ [40] guidelines. Most of the surface water monitoring sites located away from the mining activities, such as C, G, and H, showed an improved water quality. Monitoring points D, E, and F had an average concentration exceeding the ANZECC/ARMCANZ [40] permissible limits in all the seasons for each point, as shown in Figure 6. Table 2 indicates the average concentrations of EC in all surface water monitoring sites.

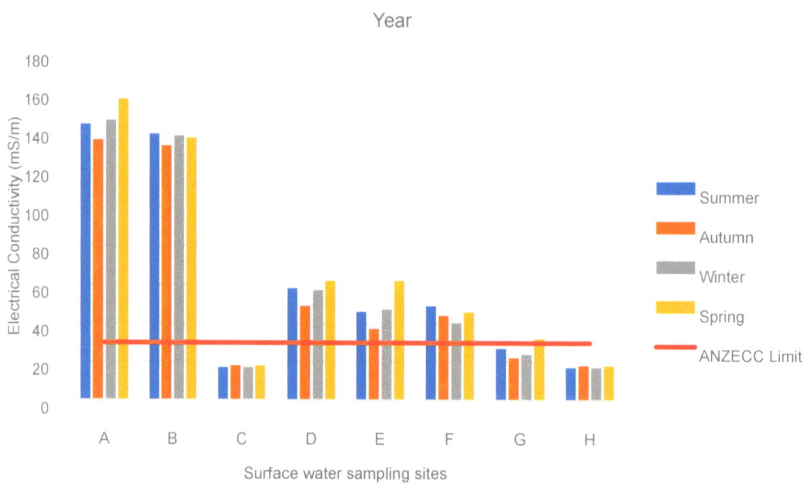

Figure 6. Average electrical conductivity (mS/m) from 2017 to 2021.

Sodium (Na), magnesium (Mg), and calcium (Ca) are largely found on the Earth's surface and absorbed by water as it moves across different soil and rock types. All sampling points fell within all the expected limits. Figures 7 and 8 provide graphical presentations of Ca and Mg, respectively.

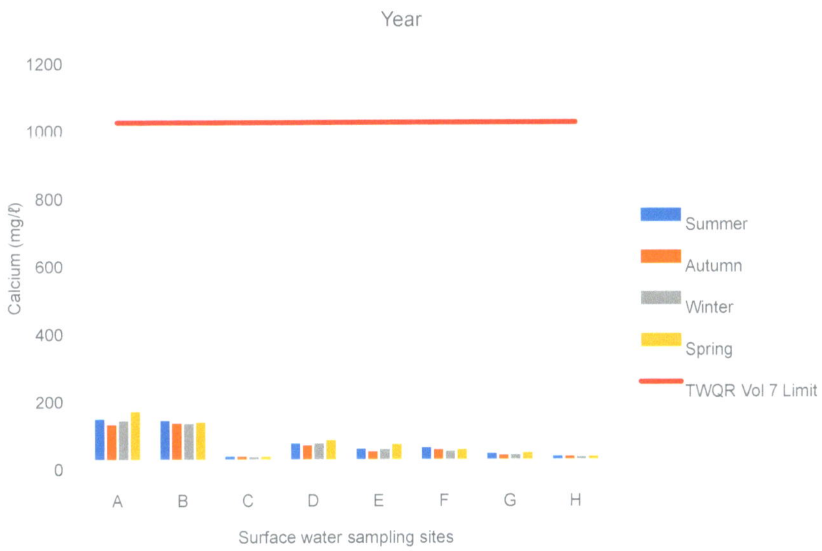

Figure 7. Average calcium (mg/L) from 2017 to 2021.

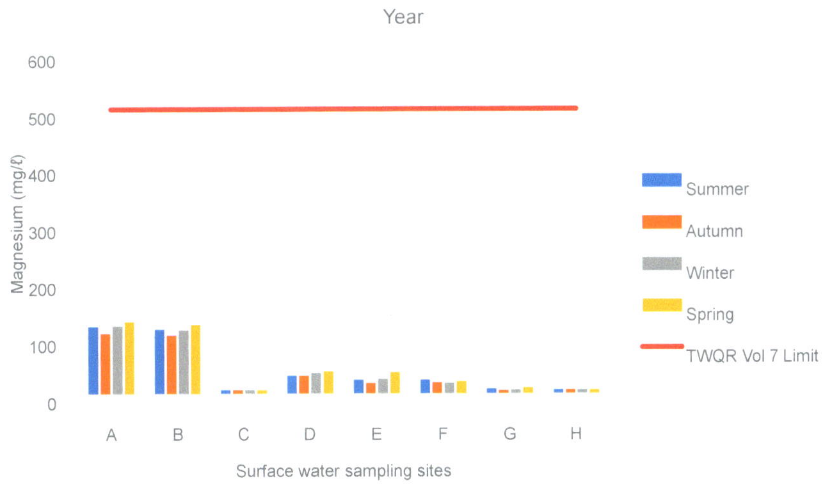

Figure 8. Average magnesium (mg/L) from 2017 to 2021.

pH determines the acidity and alkalinity in water. It is expressed as a negative log of the concentration of H+ ions. All the monitoring sites showed a pH level falling within the

8.0–9.0 range. Only monitoring site D indicated acidic conditions of six in all the seasons found below the expected limits (see Figure 9).

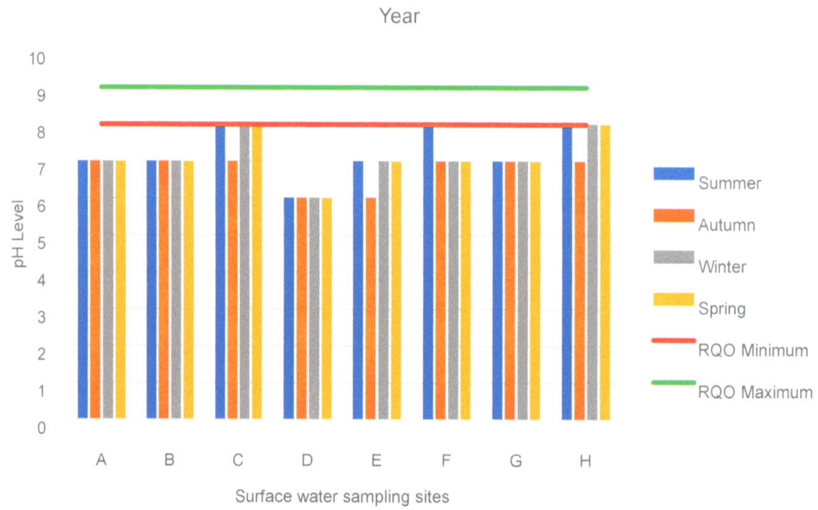

Figure 9. Average pH Level from 2017 to 2021.

Table 2. Average surface water quality results over four seasons for a period of five years (2017–2021) in Middelburg, South Africa.

Points	Season	TDS (mg/L)	SO_4 (mg/L)	Ca (mg/L)	Mg (mg/L)	Na (mg/L)	Fe (mg/L)	Mn (mg/L)	EC (mg/L)	pH	Al (mg/L)
		1000 ***	500 *	1000 ***	500 ***	2000 ***	0.3 ****	0.18 **	30 *****	8.0–9.0 *	5 **
A	Summer	1217	800	121	118	39	0.1	1	143	7	0.1
	Autumn	1123	730	105	106	38	0.04	0.4	135	7	0.03
	Winter	1179	790	116	119	36	0.03	0.1	145	7	0.03
	Spring	1373	1174	144	127	44	0.1	0.3	156	7	0.04
B	Summer	1160	725	116	113	37	0.1	1	138	7	0.1
	Autumn	1118	725	109	103	37	0.1	0.1	132	7	0.04
	Winter	1113	743	108	112	36	0.1	0.1	137	7	0.04
	Spring	1249	807	113	122	42	0.01	0.2	136	7	0.1
C	Summer	104	19	11	7	13	1	2	17	8	0.2
	Autumn	105	18	11	7	13	1	0.2	18	7	0.1
	Winter	102	17	9	7	14	0.2	0.03	17	8	0.1
	Spring	108	18	11	7	14	0.2	1	18	8	0.1
D	Summer	499	302	49	32	10	18	5	58	6	1
	Autumn	363	248	43	32	7	1	3	49	6	0.1
	Winter	433	283	49	37	11	3	3	57	6	0.3
	Spring	486	288	59	40	13	2	3	62	6	1
E	Summer	323	185	33	25	19	0.2	0.4	46	7	0.04
	Autumn	257	141	25	19	17	0.2	0.2	37	6	0.2
	Winter	327	191	32	26	20	0.0	0.2	47	7	0.02
	Spring	457	276	47	38	25	0.1	0.4	62	7	0.1

Table 2. *Cont.*

Points	Season	TDS (mg/L)	SO_4 (mg/L)	Ca (mg/L)	Mg (mg/L)	Na (mg/L)	Fe (mg/L)	Mn (mg/L)	EC (mg/L)	pH	Al (mg/L)
		1000 ***	500 *	1000 ***	500 ***	2000 ***	0.3 ****	0.18 **	30 *****	8.0–9.0 *	5 **
F	Summer	451	143	37	25	63	0.2	1	49	8	0.1
	Autumn	295	122	31	20	26	0.2	0.2	44	7	0.2
	Winter	258	98	26	19	26	0.1	0.1	40	7	0.1
	Spring	310	137	32	22	27	0.1	0.3	46	7	0.1
G	Summer	177	59	20	9	17	1	0.3	27	7	0.1
	Autumn	128	41	15	6	14	1	0.02	22	7	0.03
	Winter	150	51	16	7	18	0.1	0.01	24	7	0.03
	Spring	202	51	22	11	21	1	0.4	32	7	0.2
H	Summer	104	19	11	7	13	0.5	2	17	8	0.2
	Autumn	105	18	11	7	13	1	0.2	18	7	0.1
	Winter	102	17	9	7	14	0.2	0.0	17	8	0.1
	Spring	108	18	11	7	14	0.2	1	18	8	0.1

* RQO [36]; ** TWQR [37]; *** TWQR [38]; **** CCME [39]; ***** ANZECC/ARMCANZ [40].

Groundwater does not become contaminated as rapidly as surface water; however, mining activities such as blasting and seepages of dirty mine water can contribute to its pollution. The overall groundwater quality at the study area was poor to fair, with certain parameters exceeded their required limits from the WHO guidelines 42].

TDS concentrations were very low in most groundwater sampling sites in comparison to the WHO guidelines [42]. Monitoring site BH 2 was most contaminated, with TDS concentrations ranging from 1961 mg/L to 2456 mg/L, exceeding the 600 mg/L limit, as shown in Table 3 below. BH 7 and BH 9 also had some elevated average concentrations in autumn (1.046 mg/L) and summer (1048 mg/L) in the monitoring period of five years, as indicated in Figure 10.

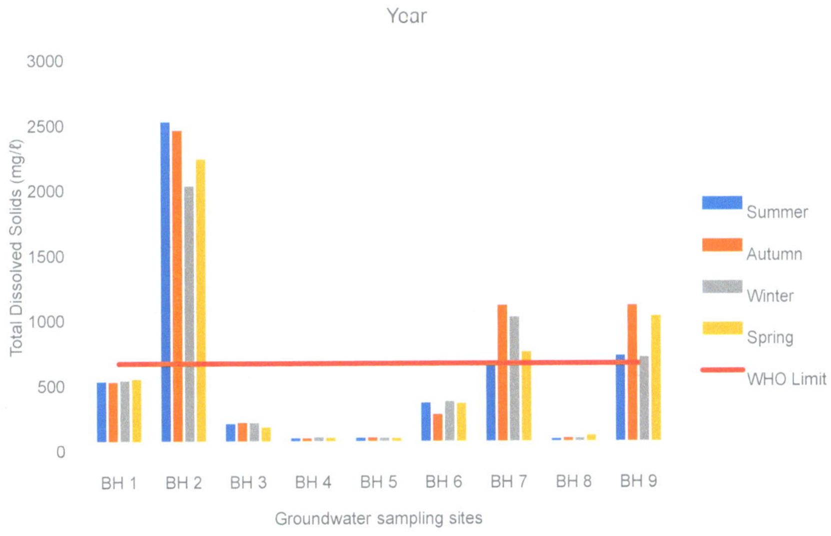

Figure 10. Average total dissolved solids (mg/L) from 2017 to 2021.

Borehole BH 1 had an increased SO_4 concentration (296 mg/L to 315 mg/L) higher than the 250 mg/L of the WHO (2017) benchmark [42]. BH 9 also indicated an exceedance in all the seasons: spring (438 mg/L), summer (674 mg/L), autumn (445 mg/L), and winter (650 mg/L) (Figure 11 and Table 3).

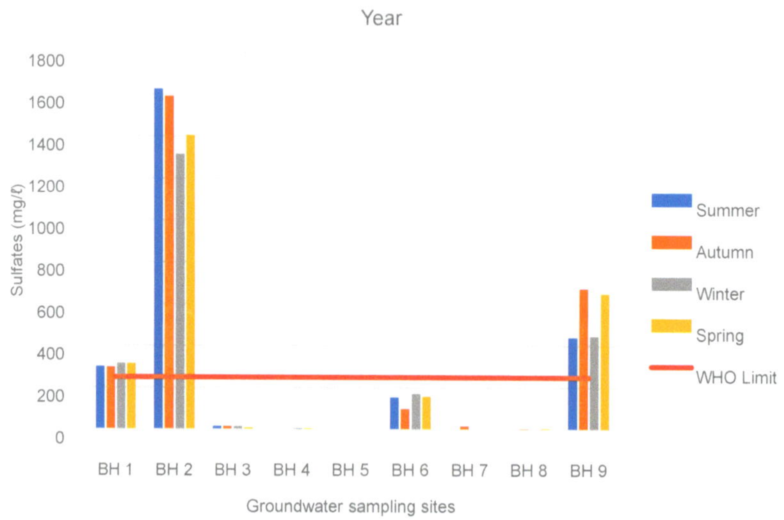

Figure 11. Average sulfates (mg/L) from 2017 to 2021.

The EC of all groundwater monitoring sites was within the WHO guidelines [42]. Monitoring sites BH 2 and BH 7 had increased average concentrations in all five years in all the seasons but still complied with the WHO guidelines [42] (Figure 12 and Table 3).

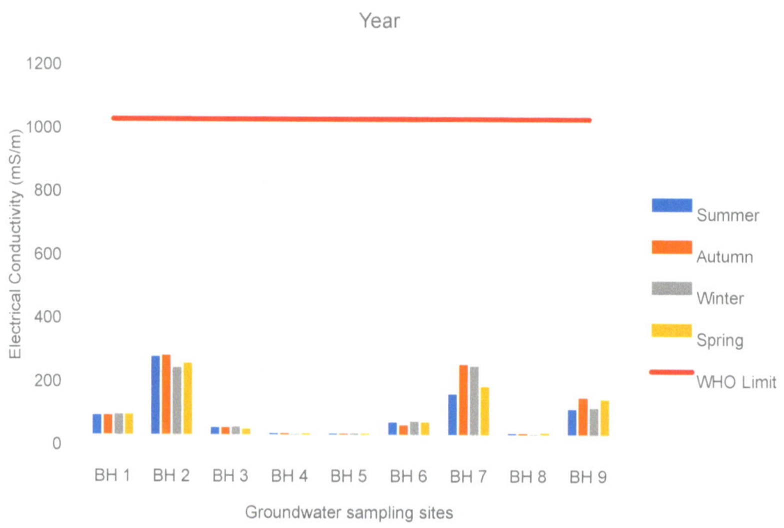

Figure 12. Average electrical conductivity (mS/m) from 2017 to 2021.

Monitoring site BH 2 had higher concentrations in comparison to the WHO [42] limits. This site showed a level of contamination that can be associated with the mining activities in the vicinity. Figure 13 and Table 3 show site BH 7 with a high average Ca concentration during the spring season in the five years of monitoring.

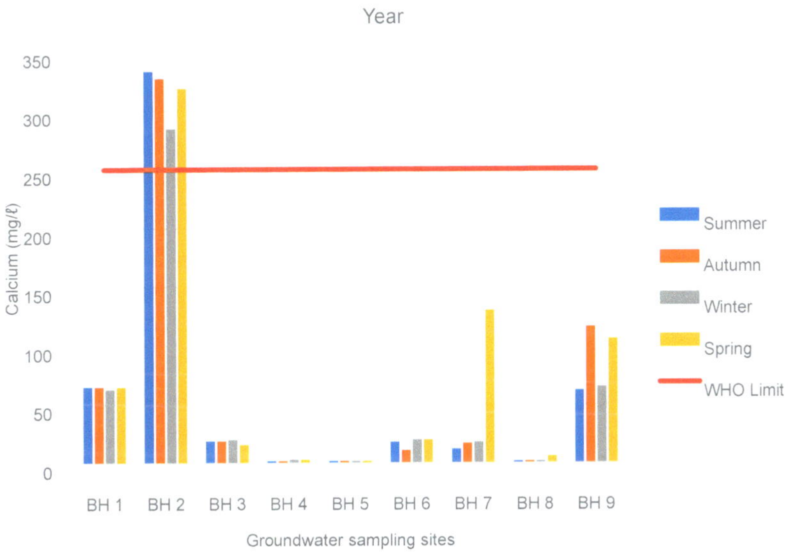

Figure 13. Average calcium (mg/L) from 2017 to 2021.

Table 3. Average groundwater quality results over four seasons for a period of five years (2017–2021) in Middelburg, South Africa.

Points	Season	TDS (mg/L)	SO_4 (mg/L)	Ca (mg/L)	Mg (mg/L)	Na (mg/L)	Fe (mg/L)	Mn (mg/L)	EC (mg/L)	pH	Al (mg/L)
		600	250	250	300	200	0.3	0.1	1000	6.5–8.5	0.9
BH 1	Summer	462	300	65	9	12	0.1	0.2	63	7	0.05
	Autumn	459	296	65	29	14	0.1	0.1	63	7	0.1
	Winter	468	315	63	33	13	0.1	0.1	66	8	0.0
	Spring	480	315	65	33	14	0.0	0.1	66	8	0.0
BH 2	Summer	2456	1625	333	167	72	3	10	248	5	2
	Autumn	2392	1589	327	158	75	9	9	252	6	2
	Winter	1961	1313	284	122	67	3	6	214	6	0.2
	Spring	2172	1405	319	139	76	3	7	228	6	3
BH 3	Summer	139	16	19	10	11	0.1	0.0	24	7	0.2
	Autumn	150	16	19	9	17	0.1	0.1	25	7	0.2
	Winter	147	16	20	10	12	0.1	0.0	26	7	0.2
	Spring	114	10	16	8	11	0.0	0.0	20	7	0.1
BH 4	Summer	27	2	2	1	5	1	0.3	6	5	0.02
	Autumn	28	3	2	1	5	0.4	0.2	6	6	0.02
	Winter	35	6	3	2	5	2	0.1	4	6	0.03
	Spring	33	6	3	2	5	0.1	0.1	6	5	0.01

Table 3. *Cont.*

Points	Season	TDS (mg/L)	SO_4 (mg/L)	Ca (mg/L)	Mg (mg/L)	Na (mg/L)	Fe (mg/L)	Mn (mg/L)	EC (mg/L)	pH	Al (mg/L)
		600	**250**	**250**	**300**	**200**	**0.3**	**0.1**	**1000**	**6.5–8.5**	**0.9**
BH 5	Summer	30	1	2	1	4	1	0.4	5	6	0.01
	Autumn	32	2	2	1	5	1	0.2	5	6	0.1
	Winter	30	2	2	1	5	0.1	0.2	6	6	0.03
	Spring	30	2	2	1	4	0.2	0.2	6	6	0.02
BH 6	Summer	298	154	18	21	30	7	0.4	41	6	0.1
	Autumn	209	98	11	13	26	2	0.2	32	6	0.1
	Winter	309	171	20	23	29	2	1	44	6	0.1
	Spring	297	158	20	22	30	1	0.3	42	6	0.1
BH 7	Summer	595	2	12	21	23	0.2	0.1	131	7	0.02
	Autumn	1046	17	17	15	28	2	0.1	224	8	0.1
	Winter	953	3	18	14	35	1	0.1	219	8	0.03
	Spring	691	3	130	13	24	1	0.1	155	8	0.1
BH 8	Summer	24	3	2	1	5	0.1	0.01	6	7	0.03
	Autumn	30	5	2	1	5	0.1	0.02	6	6	0.1
	Winter	29	4	2	1	5	0.1	0.02	4	6	0.1
	Spring	52	6	6	3	7	0.1	0.01	8	6	0.1
BH 9	Spring	658	438	62	59	14	6	2	82	6	0.1
	Summer	1048	674	116	90	21	22	4	119	7	0.04
	Autumn	646	445	65	65	14	3	2	86	6	0.03
	Winter	966	650	106	91	20	5	4	114	6	0.04

3.2. Canadian Council of Ministers of the Environment Water Quality Index

The results of the CCME–WQI were determined using the calculations and formulas shown in Section 3 above. Most of the surface monitoring sites in the study area indicted a fair to excellent water quality throughout all the seasons of the monitoring period of five years, as shown by the index. Monitoring sites A and B ranged between 59% and 72%, showing a fair quality, while monitoring site D had a very poor quality during the summer and winter seasons between 2017 and 2021 at 36% and 48%, respectively. A very good to excellent quality was determined in monitoring points E, F, G, and H, where in all the seasons, the WQI ranged between 73% and 100%.

The scope (F1) represented the number of variables whose objectives were not met, ranging from 0% to 66% in all the surface water monitoring points, as shown in Table 4. The frequency (F2) of monitoring sites A and B was high in all the monitoring seasons for the period of five years, which depicted a poor to fair water quality in comparison with the CCME index. The highest amplitude (F3) was recorded in monitoring site D during the summer and winter seasons between 2017 to 2021 at 90.5% and 72.5%, respectively, showing very poor analyzed water quality.

Table 4. CCME–WQI over four seasons for a period of five years (2017–2021) in Middelburg, South Africa.

Monitoring Points	Seasons	F1	F2	F3	CCME–WQI	WQI Ranking
A	Summer	40	30	40.7	62.8	Poor/fair water quality
	Autumn	40	28	31.7	66.4	Poor/fair water quality
	Winter	40	28.6	32	66.1	Poor/fair water quality
	Spring	50	29	40	59.3	Poor/fair water quality
B	Summer	40	28	38	64.2	Poor/fair water quality
	Autumn	30	24.6	29.3	71.9	Good water quality
	Winter	40	30.8	36.8	63.9	Poor/fair water quality
	Spring	40	29.3	33.3	65.5	Poor/fair water quality

Table 4. *Cont.*

Monitoring Points	Seasons	F1	F2	F3	CCME–WQI	WQI Ranking
C	Summer	20	9.3	46	70.5	Poor/fair water quality
	Autumn	20	6.66	14	85.4	Good water quality
	Winter	10	1.9	0.97	94	Excellent water quality
	Spring	10	0.67	2.6	94	Excellent water quality
D	Summer	60	20.6	90.5	36.2	Very poor water quality
	Autumn	50	13.3	42.6	61.3	Poor/fair water quality
	Winter	50	16.6	72.5	48.3	Very poor water quality
	Spring	40	16	65.3	54.8	Poor/fair water quality
E	Summer	30	11.3	16.8	79.1	Good water quality
	Autumn	30	10.6	10.7	80.6	Good water quality
	Winter	20	10.6	8.3	86.1	Good water quality
	Spring	40	13.3	19.1	73.3	Good water quality
F	Summer	30	10.7	23.2	77.2	Good water quality
	Autumn	30	10	10.8	80.7	Good water quality
	Winter	13	8.7	4.3	87.2	Good water quality
	Spring	30	11.3	15.9	79.3	Good water quality
G	Summer	30	13.3	32.1	73.5	Good water quality
	Autumn	10	2	12.3	90.8	Good water quality
	Winter	10	2	12.2	90.8	Good water quality
	Spring	0	0	0	100	Excellent water quality
H	Summer	20	7.3	46.7	70.4	Poor/fair water quality
	Autumn	20	6	17.9	84.1	Good water quality
	Winter	0	1.3	0.6	99.2	Excellent water quality
	Spring	20	4	15.9	85.0	Good water quality

Water quality indices	CCME–WQI
Excellent	91–100
Good	71–90
Poor/fair	51–70
Very poor/marginal	26–50
Unsuitable/poor	0–25

Monitoring borehole BH 1 showed a marginal or very poor quality in summer with a score of 40, improving during the other seasons to a good state ranging from 86 to 90. Borehole BH 2 also highlighted some level of pollution during all seasons, with all results falling within the very poor range over the whole monitoring period from 2017 to 2021. Furthermore, monitoring boreholes BH 3, BH 4, BH 5, and BH 8 showed good to excellent results, with scores between 80 and 100, with no indication of contamination that may be associated with the dominating mining activities at these points. However, BH 9 was contaminated and very poor during the summer and autumn seasons, with an improvement to a fair state in the winter and spring seasons.

The scope for F1 represented the number of variables whose objectives were not met, ranging from 0% to 60% in all the groundwater monitoring points, as shown in Table 5. The frequency (F2) of monitoring point BH 3 (0%) and BH 8 (0%) was low over all the monitoring seasons for the period of five years, which indicated an excellent water quality related to the guidelines. The highest amplitude (F3) was recorded in BH 1 during the summer season at 97.2%, indicating very poor analyzed water quality.

Table 5. CCME–WQI over four seasons for a period of five years (2017–2021) in Middelburg, South Africa.

Monitoring Points	Seasons	F1	F2	F3	CCME–WQI	WQI Ranking
BH 1	Summer	30	18	97.2	49.3	Very poor water quality
	Autumn	20	10	10.9	85.6	Good water quality
	Winter	10	10	10.6	89.8	Good water quality
	Spring	10	10	10.8	89.7	Good water quality
BH 2	Summer	60	48	88.2	32.4	Very poor water quality
	Autumn	60	44	87.3	33.7	Very poor water quality
	Winter	60	40	82.9	36.5	Very poor water quality
	Spring	60	44	84.7	34.8	Very poor water quality
BH 3	Summer	0	0	0	100	Excellent water quality
	Autumn	0	2	1.9	98.3	Excellent water quality
	Winter	0	0	0	100	Excellent water quality
	Spring	0	0	0	100	Excellent water quality
BH 4	Summer	20	8	23.3	81.6	Good water quality
	Autumn	20	8	10.6	86.1	Good water quality
	Winter	10	4	31.8	80.5	Good water quality
	Spring	10	2	0.6	94.1	Excellent water quality
BH 5	Summer	20	10	21.4	82.0	Good water quality
	Autumn	20	8	13.9	85.1	Good water quality
	Winter	10	4	2.3	93.6	Excellent water quality
	Spring	60	10	8.6	64.5	Poor/fair water quality
BH 6	Summer	20	22	70.9	55.5	Poor/fair water quality
	Autumn	20	14	35.6	75.0	Good water quality
	Winter	10	30	43.5	68.9	Poor/fair water quality
	Spring	60	18	19.6	62	Poor/fair water quality
BH 7	Summer	40	16	27.8	70.3	Poor/fair water quality
	Autumn	20	16	53.7	65.5	Poor/fair water quality
	Winter	20	14	48.1	68.8	Poor/fair water quality
	Spring	30	16	27.3	74.8	Good water quality
BH 8	Summer	0	0	0	100	Excellent water quality
	Autumn	0	0	0	100	Excellent water quality
	Winter	0	0	0	100	Excellent water quality
	Spring	0	0	0	100	Excellent water quality
BH 9	Summer	50	26	67.1	49.4	Very poor water quality
	Autumn	50	28	90.8	37.9	Very poor water quality
	Winter	40	24	68.0	52.3	Poor/fair water quality
	Spring	50	26	54.4	54.7	Poor/fair water quality

Water quality indices	CCME–WQI
Excellent	91–100
Good	71–90
Poor/fair	51–70
Very poor/marginal	26–50
Unsuitable/poor	0–25

3.3. Comprehensive Pollution Index

The surface water monitoring sites A, B, and C were very poor in the summer seasons, indicating some improvement during the rest of the seasons and falling within the permissible ranges of the CPI. All of monitoring site D's readings were greater than the suitable limit of 2.01, as shown in Table 6. However, all the findings at monitoring sites E, F, G, and H were compliant with the CPI limits, highlighting a fair to good state of water quality.

Table 6. Comprehensive pollution index and water quality index over four seasons for a period of five years (2017–2021) in Middelburg, South Africa.

Sampling Points	Comprehensive Pollution Index			
	Summer	Autumn	Winter	Spring
A	1.2 Very poor	1.0 Very poor	0.97 Poor	1.2 Very poor
B	1.1 Very poor	0.9 Fair	0.9 Fair	0.9 Fair
C	1.2 Very poor	0.5 Poor	0.3 Good	0.5 Poor
D	9.4 Very poor	2.2 Very poor	3.1 Very poor	2.8 Very poor
E	0.6 Fair	0.5 Fair	0.4 Good	0.7 Fair
F	0.8 Fair	0.5 Fair	0.4 Fair	0.5 Fair
G	0.8 Fair	0.4 Good	0.2 Excellent	0.8 Fair
H	1.2 Very poor	0.5 Poor	0.3 Good	0.5 Poor
BH 1	0.5 Fair	0.5 Fair	0.5 Fair	0.5 Fair
BH 2	8.2 Very poor	7.6 Very poor	5.4 Very poor	6.2 Very poor
BH 3	0.2 Excellent	0.3 Good	0.2 Excellent	0.2 Excellent
BH 4	0.5 Fair	0.3 Good	0.7 Fair	0.2 Excellent
BH 5	0.5 Fair	0.4 Good	0.3 Good	0.3 Good
BH 6	3.0 Very poor	0.9 Fair	1.3 Very poor	0.6 Fair
BH 7	0.7 Fair	1.6 Poor	1.3 Very poor	1.1 Very poor
BH 8	0.1 Excellent	0.1 Excellent	0.1 Excellent	0.1 Excellent
BH 9	3.5 Very poor	10.4 Very poor	2.6 Very poor	0.1 Excellent

Water quality indices	CPI
Excellent	<0.2
Good	0.21–0.40
Poor/fair	0.41–1.00
Very poor/marginal	1.01–2.0
Unsuitable/poor	>2.01

All of groundwater site BH 1's readings fell within the fair range over all seasons, while BH 2 was very polluted, ranging from 5 to 8. Boreholes BH 3 to BH 5 had good to excellent water quality. No impacts associated with coal mining could be identified in these boreholes using the comprehensive pollution index. BH 6 experienced different levels, with exceedances in the summer and winter seasons at 2.9 and 1.3, respectively. BH 9 was also contaminated, exceeding the permissible limits in terms of the CPI.

3.4. Comparison between the Two Indices

The CCME–WQI for surface water was found within the range of 36.2 to 100, showing different water qualities at different monitoring sites. Most sites that were located closer to coal mining activities in the study area showed poor to fair water quality scores, and those further away indicated an excellent water quality score. Boreholes BH 2 and BH 9 were contaminated, with low WQI scores between 32.4 to 54.7 in the different seasons. BH 8 had an excellent water quality at 100% in all the monitoring seasons. Similar to CCME–WQI, the CPI depicted some level of pollution in surface water monitoring points that exceeded the limits. Points in close proximity of the study area, like site A, had very poor CPI.

3.4.1. Multivariate Statistical Analysis Results

Principal Component Analysis

The analyses were conducted on the z-scores scale, ensuring that transformed data were normally distributed and that there were no non-normality issues hindering the interpretation of findings. The three components accounted for 78.6% of the total variance.

The PCA plotting diagrams (Figures 14–24) isolate the components elements that are pertinent in explaining the water quality variations at different sites. Panel-crossing this information with the descriptive analysis performed earlier, the sources of contamination at each site can be reasonably inferred, having on one side parameters whose roles are significant in the resulting quality and on the other their measured levels at each site.

In the surface water monitoring site A, three major components are shown: Component 1's group of elements included TDS (0.943), SO_4 (0.623), EC (0.926), Mg (0.847), Na (0.808), and Ca (0.845). Component 2 also indicated a remarkably high concentration of Mn and Al, respectively. Component 3 also showed an increased concentration of Fe (0.852). According to the PCA analysis and the locality of site A, the sources of contamination can be associated with the mining industry and the natural sources of the soil (Figure 14).

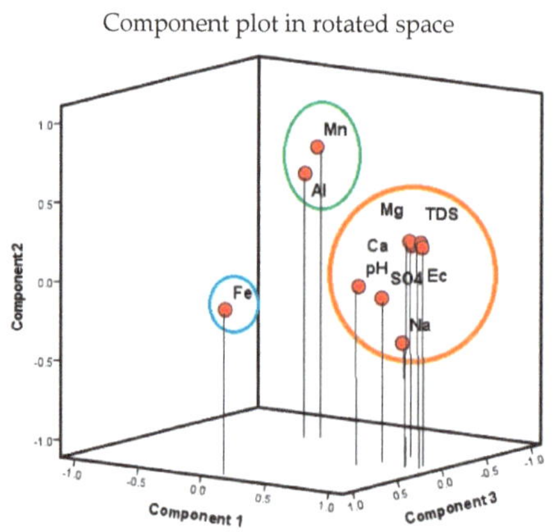

Figure 14. Principal component analysis—component plot for surface water monitoring at site A from 2017 to 2021.

The extracted components accounted for 86.6% of the total variance. Monitoring site B also had three components identified in the analyses in common with component 1, having a group of elements that included TDS (0.984), Mg (0.970), Na (0.678), SO_4 (0.976), EC (0.957), and Ca (0.969). These variables can be associated with anthropogenic activities occurring upstream of the sampling point. Component 2 had an increased level of Mn (0.734). Component 3 had Al at 0.853 and Fe at 0.765. Considering the location of sampling site B, the possibility of seepages from the notable mining activities could contribute to the elevations analyzed (Figure 15).

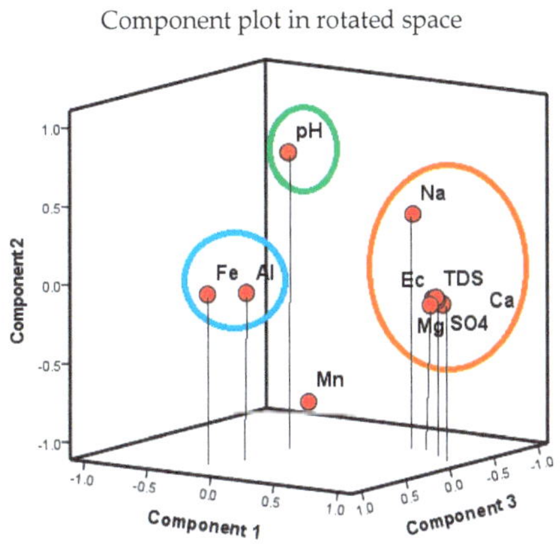

Figure 15. Principal component analysis—component plot for surface water monitoring at site B from 2017 to 2021.

The extracted components accounted for 33.27% of the total variance. At site C, four components were identified, with a group of parameters including TDS, Ca, Mg, EC, SO_4, and Na, showing a strong correlation with the PCA. Component 2 showed the elements Fe (0.846) and Al (0.778). Component 3 accounted for 11.77% of the total variance. The group of elements identified included Mn and Al. For component 4, the group had pH accounts that formed 11.5% of the total variance. These variables were already present in the soil; however, they were elevated through anthropogenic activities such as mining (Figure 16).

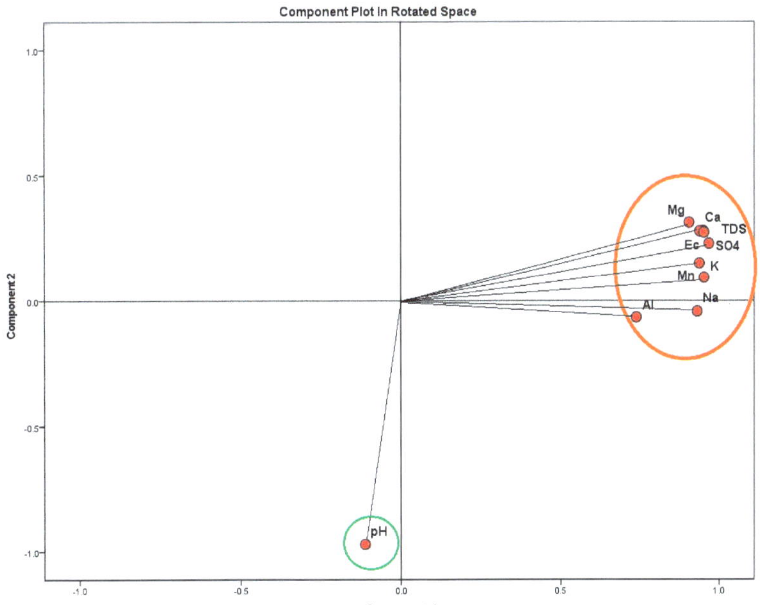

Figure 16. Principal component analysis—component plot for surface water monitoring at site D from 2017 to 2021.

There were two components that were extracted from site D, where component 1 had a group of elements—TDS, SO_4, Ca, Na, Mn, and EC—that showed a remarkably high concentration in the PCA. Component 2 explained a total variance of 10%. Mining activities are assumed to have contributed to the water quality status at this site (Figure 17).

The extracted components accounted for 78.5% of the total variance. The group of elements in the first component had a total variance of 60.8%, explained with a strong and high mean concentration of TDS, SO_4, Ca, Mg, Na, and EC in the PCA at site E. In component 2, the elements Fe, Mn, and Al explained a total variance of 17.7%. The source of these variables could be associated with anthropogenic activities, mainly coal mining activities, and some are mostly found naturally occurring on the surface.

In monitoring site F, component 1 had high concentrations of SO_4 (0.899), Ca (0.941), Mg (0.936), and EC (0.960). Component 2 had a group of elements that included TDS (0.835), Na (0.950), and Mn (0.686). Component 3 had concentrations primarily of Fe (0.905) and Al (0.903). The dominance of these variables could be associated with mining and industrial activities closer to the monitoring point.

The extracted components accounted for 80.3% of the total variance. In monitoring site G, the first component explained 49.5% of the total variance and a high value of EC (0.960), Ca (0.941), Mg (0.936), and SO_4 (0.899) The second component had 16.8% of the total variance, containing an elevated concentration of Na (0.950) and TDS (0.835). Fe and Al had increased levels at 0.905 and 0.903. This explained 14% of the total variance. This point indicates that anthropogenic activities like mining and industrialization could be impacting the resources in the area (Figure 17).

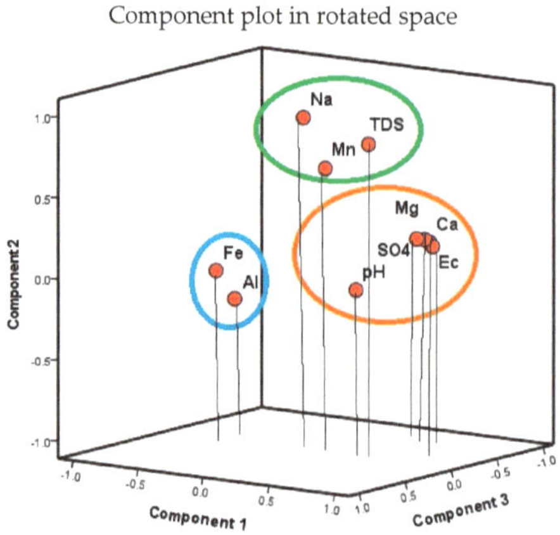

Figure 17. Principal component analysis—component plot for surface water monitoring at site G from 2017 to 2021.

The extracted components accounted for 80.9% of the total variance. The component of site H had high concentrations of TDS, EC, Na, Mg, and Ca, which accounted for a total variance of 54.6%. Component 2 had Fe (0.897) and Al (0.771) with a total variance of 16.3%. The third component had a total variance of 10%, with a high concentration of Mn and SO_4 at 0.901 and 0.501, respectively (Figure 18).

The extracted components accounted for 86.6% of the total variance. The groundwater monitoring site BH 1 showed that the first component consisted of increased concentrations of SO_4, EC, TDS, Ca, and Mg, accounting for a total variance of 58.9%. Component 2 had a group of elements that included Fe, Mn, and Al, with 16.4% of the total variance. Component 3 had a total variance of 11.3% with the level of pH (0.867) and the concentrations of Na (0.684) and Al (0.510) (Figure 19).

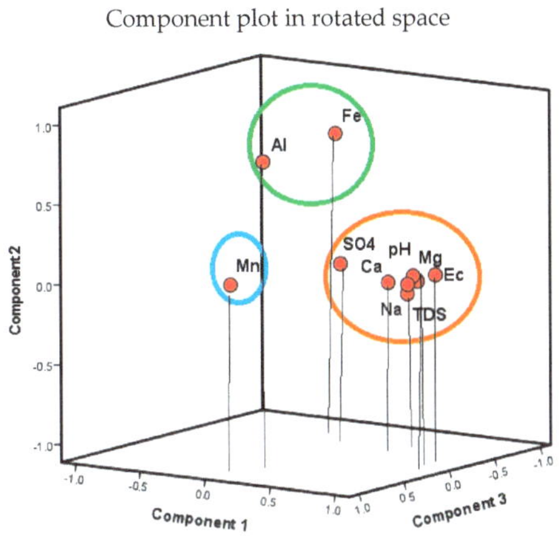

Figure 18. Principal component analysis—component plot for surface water monitoring at site H from 2017 to 2021.

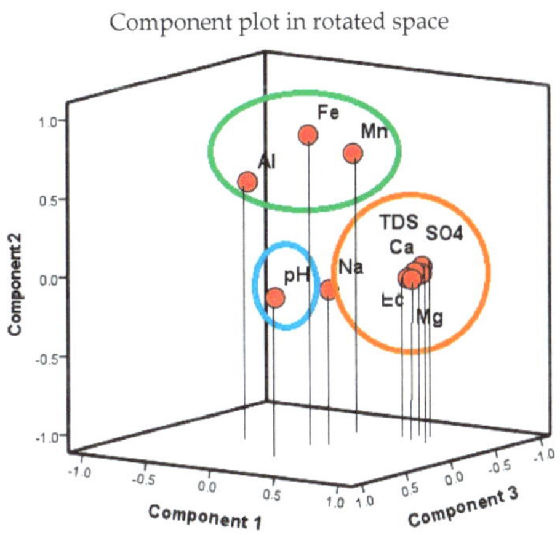

Figure 19. Principal component analysis—component plot for groundwater monitoring at site BH 1 from 2017 to 2021.

Borehole 2 only had two components from the PCA. Component 1 had a high concentration of EC (0.973), Ca (980), TDS (0.46), and SO_4 (942). The total variance explained was 65.6%. Component 2 had high metals with Mn (0.831), Al (0.733), and Fe (0.717), which accounted for 17.5% of the total variance (Figure 20).

The extracted components accounted for 80.1% of the total variance. Two components were determined in BH 3. The group of elements identified in this component included EC (0.977), TDS (0.968), Mg (0.949), and Ca (0.937), accounting for 60.8% of the total variance. Component 2 had elevated concentrations of metals such as Fe (0.933), Mn (0.841), and Al (0.673), with a total variance of 19.3%.

The extracted components accounted for 78% of the total variance. The analysis at BH 4 resulted in three components, where component 1 had an increased concentration of Na (0.923), TDS (0.899), and Mg (0.865), accounting for a total variance of 52.5%. Component 2 had a total variance of 13.5%, with high concentrations of Al, Fe, and SO_4. Component 3 had a dominance of Mn and Fe at 0.915 and 0.645, respectively, with 12% of the total variance.

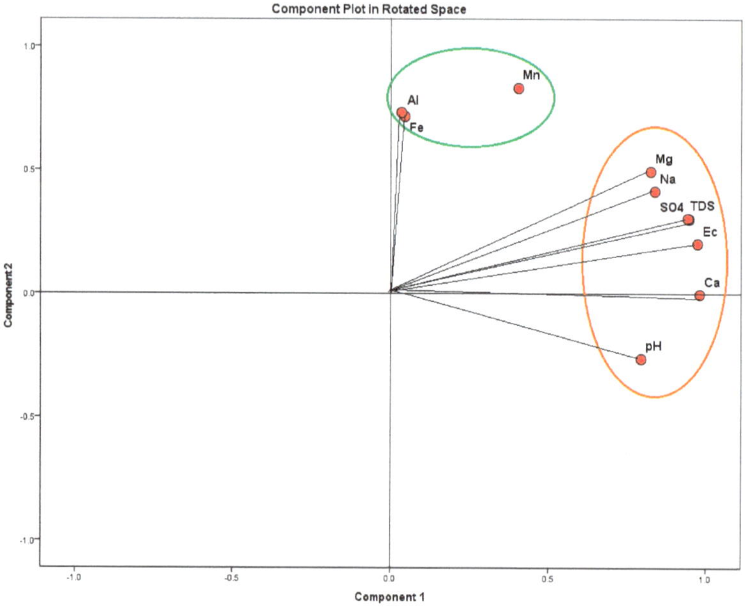

Figure 20. Principal component analysis—component plot for groundwater monitoring site BH 2 from 2017 to 2021.

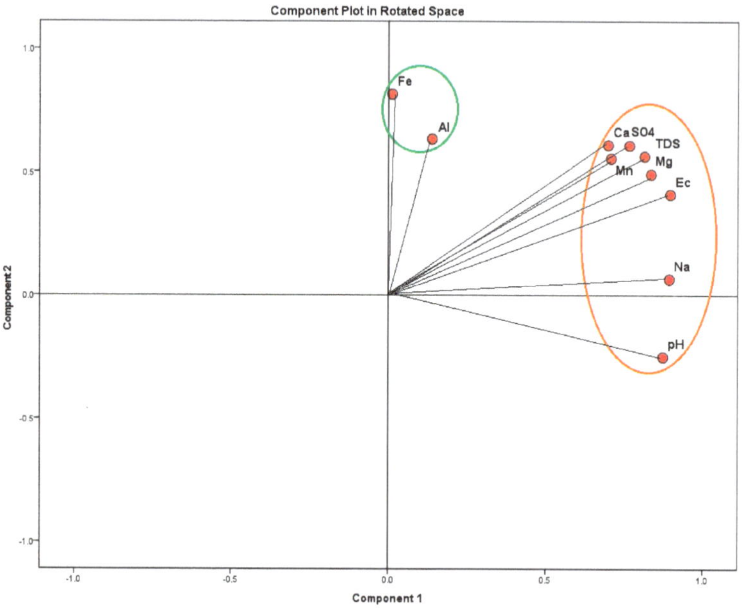

Figure 21. Principal component analysis—component plot for groundwater monitoring at site BH 6 from 2017 to 2021.

The extracted components accounted for 82.4% of the total variance. In BH 5, the group of elements that was found with high concentrations at the first components included TDS, Mg, pH, EC, and SO_4. Component 1 accounted for a total variance of 52.4%. The second component had a total variance of 17.1%, while component 3 had 12.9% of the total variance.

The extracted components accounted for 81.8% of the total variance. Two components were identified in this analysis at monitoring site BH 6. Component 1 had high concentrations of Na, EC, pH, Mg, and TDS, with 68% of the total variance explained. Component 2 had a total variance of 13.8%, with an increase in Fe, Al, Ca, and SO_4 (Figure 21).

The extracted components accounted for 66.6% of the total variance. Borehole 7 consisted of a group of elements that included TDS (0.945), EC (0.940), and Na (0.819), which explained a total variance of 39%. Component 2 had 16.6% of the total variance, with a dominance of pH, Ca, and Mg. The third component had high Al, Fe, and Mn, which accounted for 11% of the total variance (Figure 22).

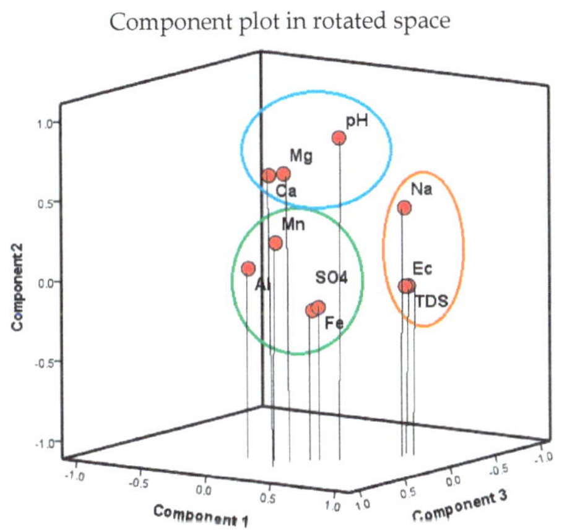

Figure 22. Principal component analysis—component plot for groundwater monitoring at site BH 7 from 2017 to 2021.

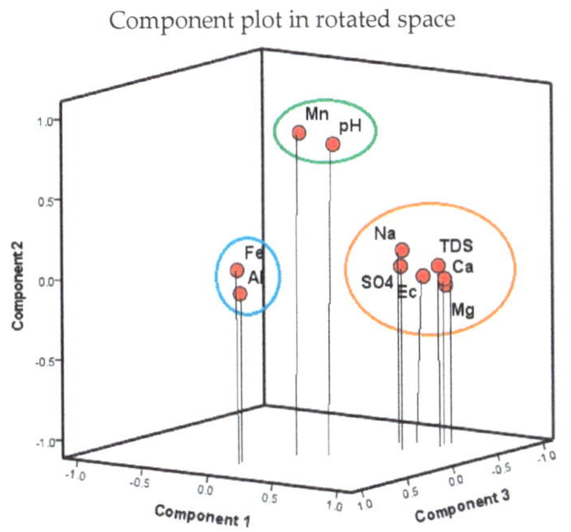

Figure 23. Principal component analysis—component plot for groundwater monitoring at site BH 8 from 2017 to 2021.

The extracted components accounted for 83.9% of the total variance. In BH 8, component 1 had a group of elements that contributed elevated TDS (0.989), Ca (0.983), EC (0.959), and Mg (0.976), with a total variance of 56%. Component 2 had an explained total variance of 16%. Component 3 only had Al (0.833) and Fe (0.760), with high concentrations at a total variance of 11.9%. The source of contamination at this point could be associated with natural sources such as the soils or the stratigraphical units (Figure 23).

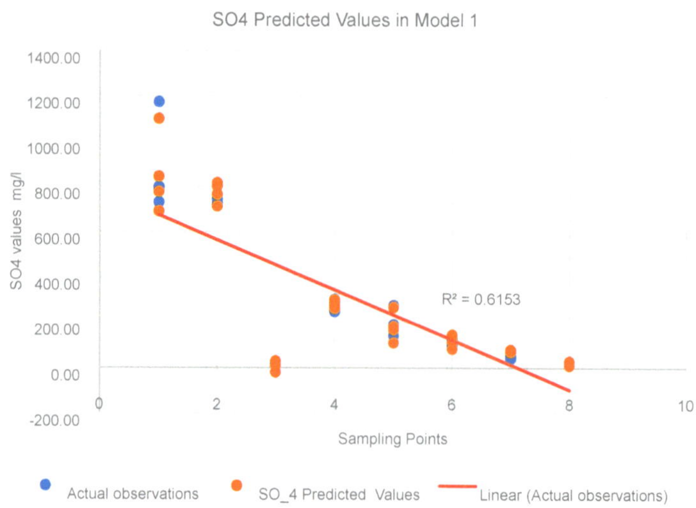

Figure 24. Predicted sulfate values in model 1 for both surface water and groundwater monitoring sites from 2017 to 2021.

Monitoring siteFihu BH 9 only had two components from the PCA. Component 1 had increased concentrations of TDS (0.982), SO_4 (0.982), Ca (0.984), Mn (0.982), Na (0.942), and Mg (0.945), explaining a total variance of 76%. The second component had high levels of pH (0.835) and Al (0.691), accounting for a total variance of 11.5%.

Linear Regression Analysis

The data were classified by the four seasons, where the average concentrations were calculated for each of the surface water and groundwater sampling points and each of the water chemical parameters. Four candidate regression models were fitted using the sulfate concentration as the dependent variable and the others as independent variables. The model was developed to assess the SO_4 ration against the dependent variables.

In model 1, the *p*-values of EC, Ca, and Fe were smaller than 0.05, indicating an effective model and presenting a good fit (Figure 24). The R^2 of 0.615 in the model shows that the dependent variables explained the variability of SO_4 as an independent variable, showing a strong correlation.

$$SO_4 = 53.12487 - 4.81874\ Ph - 6.65227\ Ec + 0.587389\ TDS + 11.77652\ Ca \\ -1.93564\ Mg - 2.77687\ Na - 166.58\beta_7\ Al + 13.40986\ Fe \\ -27.9553\ Mn \tag{13}$$

The R^2 value of the second model indicates that the water parameters pH, EC, TDS, Ca, Mg, Na, Al, Fe, and Mn explained 0.6153 of the variability of SO_4. The greater R^2 value indicated the strong relationship between the independent and dependent variables. The *p*-values of EC, Ca, Al, Fe, and Mn were below 0.05. The best fit of the model is presented below.

$$SO_4 = 2.206989808\ Ph - 6.42775334\ Ec + 0.621488677\ TDS \\ +11.81236424\ Ca + 2.483595419\ Mg + 3.021507737\ Na \\ +170.3619\ Al + 13.15460242\ Fe + 25.64761114\ Mn \tag{14}$$

Of the five variables of the model, only the last one was not significant. The R^2 value was at 0.6153, showing EC, Ca, Al, Fe, and Mn as a strong fit for the model. The *p*-values less than 0.05 also confirmed the good fit of the model. The presentation of the model is indicated in Equation (15) and Figure 25.

$$SO_4 = -22.2575 - 5.74723\ EC + 14.19187Ca - 199.464\ Al + 17.56244\ Fe - 18.3914\ Mn \quad (15)$$

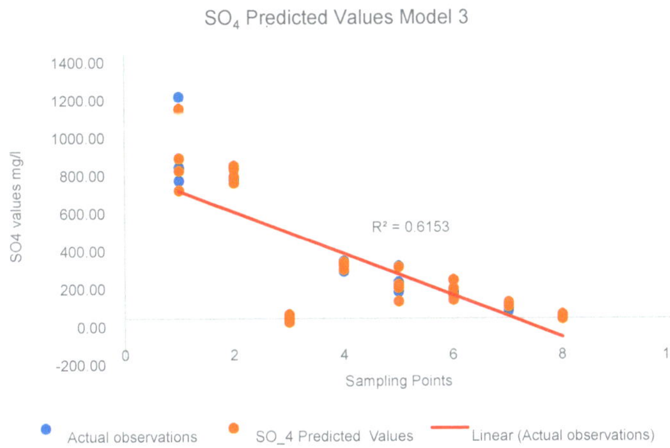

Figure 25. Predicted sulfate values in model 3 for both surface water and groundwater monitoring sites from 2017 to 2021.

In model 4 of the analysis, a strong correlation was determined, showing a good fit with an R^2 of 0.6153. (Figure 26). The R^2 in the model showed that dependent variables explained the 99% variability of the independent variable, indicating a strong correlation with the dependent variables. The *p*-values of EC, Ca, Al, Fe, and Mn were all below 0.05. The best fit of the model is presented below.

$$SO_4 = -7.39536\ EC + 15.98919Ca - 231.237\ Al + 20.5739\ Fe{-}23.9072\ Mn \quad (16)$$

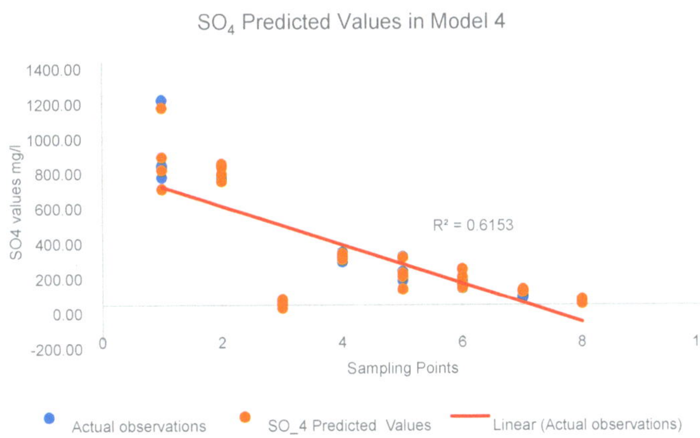

Figure 26. Predicted sulfate values in model 4 for both surface water and groundwater monitoring sites from 2017 to 2021.

4. Discussion

The current study was effective in determining if indeed anthropogenic activities such as coal mining do have an impact on water resources by using different WQIs. The use of the CCME–WQI and CPI has been utilized by developed countries with extensive mining activities, such as the United States, Spain, and China. In Spain, the continuous impacts of mining were determined in the Tinto River [5]. Using these indices, the China Coal Industry Association showed levels of poor water quality as a result of closed mines [6]. Oberholster et al. [15] indicated a higher concentration of Cl in the Loskop Dam greater than what the South African, Canadian, Australian, and New Zealand guidelines suggest.

In the present study, a full analysis of the water quality data was conducted to determine the extent to which the study area has been impacted by the coal mining activities in its vicinity. The water quality displayed high levels of the analyzed parameters, which are associated with mining and related activities, which are widespread in the proximity of the town of Middelburg. The analyses of metals and physicochemical parameters that were conducted using the WQIs indicated an in-depth understanding of the impact of coal mining and related activities. The study indicated that some water quality parameters from the used data collected from the year 2017 to 2021 indeed exceeded the limits from the RQO [36], TWQR [37,38], CMME [39], and ANZECC/ARMCANZ [40] guidelines.

The water quality status of monitoring sites A and B showed some pollution from mainly SO_4, total dissolved solids, and electrical conductivity. Groundwater monitoring boreholes BH 1, BH 2, BH 6, BH 7, and BH 9 were the most contaminated when compared with the WHO guidelines [42].

The use of the selected and combined indices (CCME–WQI and CPI) was a success in determining the water quality status of both the surface water and groundwater resources, confirming the presence of pollution from anthropogenic activities such as coal mining. The outcome of the current study can be related to the study by Hamlat et al. [47], who applied different WQIs in assessing the water quality in the Tafna catchment in Algeria. Their study aimed at determining the status and trends of the water quality in the Tafna catchment, although this was not limited to the impacts of coal mining.

The use of these indices has provided evidence that mining activities have an impact on water resources. The CCME–WQI of the surface monitoring points A and B ranged from 63 to 64 points, respectively, and were found within the poor to fair range in the summer season. The CPI for both points was between 1.1 and 1.2, which lies within the very poor range. These two points were notably in the vicinity of mining sites. The sampling points that were situated at a distance from the coal mining activities, such as sites E, F, G and H, had CCME–WQI scores ranging from 86 to 100 in the winter, indicating a good to excellent water quality. The CPI was between 0.2 and 0.4, found within the limits for good status. Boreholes BH 1, BH 2, and BH 6 experienced a poor CCME–WQI over all the seasons. The CPI also indicated poor qualities. The application of different WQIs was effective in determining the impacts of anthropogenic activities in the study area.

The CCME–WQI and CPI have provided reliable results of the degree of contamination by mining and related activities [14]. A study conducted by Atangana et al. [28] obtained similar results when applying CCME and CPI to assessment the surface water quality in the Vaalwaterspruit, a stream in Mpumalanga, South Africa. Good and fair results for the CCME–WQI and CPI best categorized the water quality, which deteriorated more in the downstream area due to nutrients (fluoride), trace metals (Al, Fe, and Mn), and particulate matter caused by mining activities in the area during the four-year (2017–2021) study period.

In the absence of similar studies within the catchment, the study conducted at the uMngeni River in South Africa, using the CCME–WQI, indicated that the index is critical in identifying influential parameters affecting water quality in water resources. The use of the CCME–WQI effectively simplified the complex dataset to be easier to understand. It is mostly used because it is an easy application to use, and it offers flexibility in selecting the variables to analyze in the model [22]. The CPI technique is essential for scientific reflection of the kinds and levels of pollution in water systems. The use of a combination of these indices can be applied in South Africa and the continent at large to predict water quality impacts from coal mining activities and manage it on catchment levels.

Application of the PCA statistical analysis using the SPSS software identified the dominance of parameters such as TDS, SO_4, Mg, EC, and Ca in most of the surface water points. These findings showed consistency in interpretation, similar to the analyses undertaken by Atangana et al. [28], indicating that some of the variables are sourced from anthropogenic activities such as mining. These chemical parameters are naturally occurring on the ground, with a strong correlation when exposed through anthropogenic activities. Point D also showed an increase in SO_4, TDS, Ca, Na, Mn, and EC, accounting for a total variance of 80.2%. In groundwater, BH 1, BH 2, and BH 3 showed a total variance ranging between 59% and 61%, indicating a strong correlation.

The analyses from the linear regression were used for the water quality variables that were found to have significant high levels in the models. The regression analyses carried out for the water quality parameters indicated a better and higher level of significance in the correlation coefficients. From the analyses, it was evident that the analyses parameters significantly correlated with the sulfates in all models, with all R^2 at 0.9.

5. Conclusions and Recommendations

The present study provided an overview of the current state of water quality and showed that anthropogenic activities such as coal mining and related activities impacted the water resources in the Middelburg area during the years 2017–2021. The study used historical water quality data for a period of five years, and the data were analyzed using WQIs such as CCME–WQI and CPI and multivariate statistics. Water quality results indicated elevated concentrations of the chemical parameters, contamination that can be associated with coal mining and related activities through the high concentration of TDS, SO_4, and EC in both surface and groundwater samples.

The two indices applied in the study were found to generate fair results in evaluating the water quality status of the surface and groundwater and could be considered by the water utility manager in assessing the water quality status of water resources impacted by acid mine drainage. The CCME–WQI index results showed surface water quality of fair to excellent quality throughout all seasons, with groundwater showing some level of contamination in some boreholes. The CPI result also indicated some improvement at some of the surface water points and also picked up pollution in some of the monitoring boreholes, mostly during the summer and winter seasons. Addressing the sampling area pollution point sources by the use of multivariate analysis, PCA identified contamination in both surface water and groundwater that could be associated with the mining industry, with some natural occurrences in some of the chemical parameters. The linear regression analysis also identified the dominance of SO_4 in most of the sampling points, which could also be linked to mining activities. In view of the above, the combination of the above indices suggested that there was contamination in water quality, showing increased concentrations of the analyzed chemical parameters. Using both indices can effectively predict water quality pollution by coal mines. The indices can be included in the water quality monitoring programs to supplement water quality data by simplifying large datasets.

It is recommended to enforce an extensive monitoring program on water resources and compare them against the guidelines of the Department of Water and Sanitation (previously Department of Water Affairs and Forestry), together with the guidelines for the indices. Mining companies must improve in the application of their water management systems and regularly monitor their effectiveness. It is also important that the government improve the implementation of the environmental legislations. Various treatment options are available to deal with the contamination arising from mining activities if managing their activities is not productive. A passive treatment system associated with constructed wetlands may pose real solutions for addressing pollution arising from abandoned mines and small-scale mines.

Author Contributions: Conceptualization, M.M.; methodology, M.M., E.A. and P.O.; software, M.M. and E.A.; validation, M.M., E.A. and P.O.; formal analysis, M.M.; investigation, M.M.; resources, M.M.; data curation, M.M.; writing—original draft preparation, M.M.; writing—review and editing, M.M., E.A. and P.O.; visualization, M.M.; supervision, P.O. and E.A.; project administration, M.M., P.O. and E.A.; funding acquisition, M.M. All authors have read and agreed to the published version of the manuscript.

Funding: We would like to thank Prof. Paul Oberholster and the Research Department at the University of the Free State for funding this study.

Data Availability Statement: For the research project, ethical consideration was required since using existing water quality monitoring data poses ethical concerns and duties that must be complied with; therefore, the data are unavailable due to privacy and ethical restrictions.

Acknowledgments: I would like to express my deepest appreciation to my supervisors, Paul Oberholster and Ernestine Atangana, for their tremendous mentorship. I could not have done this work alone. I would like to thank you for your encouragement; your advice throughout my research have been amazingly special. You were by far the best supervisors. I would like to acknowledge my former employer, Geovicon Environmental (Pty) Ltd., for granting me the permission to use their existing water quality data to carry out my research work. To my former colleagues at Geovicon, thank you so much for always believing in me and inspiring me to work harder. Above all, a special acknowledgement to my son, his mother, and my family. Thank you for the support, endless love, and encouragement. This would have never been possible without you. Your patience and understanding when I was always busy with my academic work gave me the strength I needed to pull through.

Conflicts of Interest: The authors declare no conflicts of interest.

Ethical Approval: The authors listed in the manuscripts have granted their agreement to be listed as co-authors, have reviewed and approved the work, and have given their approval for submission and publishing.

References

1. Eberhard, A. *The Future of South African Coal: Market, Investment, and Policy Challenges*; Working Paper 100; Program on Energy and Sustainable Development; Freeman Spogli Institute for International Studies; Stanford University: Stanford, CA, USA, 2011. Available online: https://fsi9-prod.s3.us-west-1.amazonaws.com/s3fs-public/WP_100_Eberhard_Future_of_South_African_Coal.pdf (accessed on 22 July 2024).
2. Makgae, M. The Status and Implications of the AMD Legacy Facing South Africa. In Proceedings of the International Mine Water Association Annual Conference, Bunbury, Australia, 29 September–4 October 2012; pp. 327–334. Available online: http://mwen.info/docs/imwa_2012/IMWA2012_Makgae_327.pdf (accessed on 22 July 2024).
3. Oelofse, S.H.H.; Hobbs, P.J.; Rascher, J.; Cobbing, J.E. The Pollution and Destruction Threat of Gold Mining Waste on the Witwatersrand—A West Rand Case Study. In Proceedings of the 10th International Symposium on Environmental Issues and Waste Management in Energy and Mineral Production (SWEMP 2007), Bangkok, Thailand, 11–13 December 2007; pp. 617–627. Available online: https://www.anthonyturton.com/assets/my_documents/my_files/983_SWEMP.pdf (accessed on 22 July 2024).
4. Romer, C.D. What ended the great depression? *J. Econ. Hist.* **1992**, *52*, 757–784. Available online: https://www.jstor.org/stable/2123226 (accessed on 22 July 2024). [CrossRef]
5. Grande, J.A.; Borrego, J.; Morales, J.A.; De la Torre, M.L. A description of how metal pollution occurs in the Tinto–Odiel rias (Huelva–Spain) through the application of cluster analysis. *Mar. Pollut. Bull.* **2003**, *46*, 475–480. [CrossRef]
6. Shang, Y.; Qi, P. Study on environmental negative effect and control of waste coal mine. *Coal Eng.* **2018**, *50*, 147–150. (In Chinese)
7. Ashton, P.J.; Harwick, D.; Breen, C.M. Changes in water availability and demand within South Africa's shared river basins as determinants of regional social ecological resilience. In *Exploring Sustainability Science: A South African Perspective*; Burns, M.J., Weaver, A., Clark, W., Kates, R., Eds.; African Sun Media: Stellenbosch, South Africa, 2008; pp. 279–310.
8. Oberholster, P.J.; Botha, A.-M.; Cloete, T.E. Biological and chemical evaluation of sewage water pollution in the Rietvlei nature reserve wetland area, South Africa. *Environ. Pollut.* **2008**, *156*, 184–192. [CrossRef]
9. Quibell, G.; Van Vliet, H.; Van der Merwe, W. Characterizing cause-and effect relationships in support of catchment water quality management. *Water SA* **1997**, *23*, 193–199. Available online: https://www.wrc.org.za/wp-content/uploads/mdocs/WaterSA_1997_03_1021.PDF (accessed on 22 July 2024).
10. Uddin, M.G.; Nash, S.; Olbert, A.I. A review of water quality index models and their use for assessing surface water quality. *Ecol. Indic.* **2021**, *122*, 107218. [CrossRef]
11. Kachroud, M.; Trolard, F.; Kefi, M.; Jebari, S.; Bourrié, G. Water quality indices: Challenges and application limits in the literature. *Water* **2019**, *11*, 361. [CrossRef]

12. Tirkey, P.; Bhattacharya, T.; Chakraborty, S. Water quality indices: Important tools for water quality assessment: A review. *Int. J. Adv. Chem.* **2013**, *1*, 15–28. Available online: https://airccse.com/ijac/papers/1115ijac02.pdf (accessed on 22 July 2024).
13. Son, C.T.; Giang, N.T.H.; Thao, T.P.; Nui, N.H.; Lam, N.T.; Cong, V.H. Assessment of Cau River water quality assessment using a combination of water quality and pollution indices. *J. Water Supply Res. Technol.—AQUA* **2020**, *69*, 160–172. [CrossRef]
14. Finotti, A.R.; Finkler, R.; Susin, N.; Schneider, V.E. Use of water quality index as a tool for urban water resources management. *Int. J. Sustain. Dev.* **2015**, *10*, 781–794. [CrossRef]
15. Oberholster, P.F.; Goldin, J.; Xu, Y.; Kanyerere, T.; Oberholster, P.J.; Botha, A.-M. Assessing the adverse effects of a mixture of AMD and sewage effluent on a sub-tropical dam situated in a nature conservation area using a modified pollution index. *Int. J. Environ. Res.* **2021**, *15*, 321–333. [CrossRef]
16. Tang, T.; Zhai, Y.; Huang, K. Water Quality Analysis and Recommendations through Comprehensive Pollution Index Method. *J. Manag. Sci. Eng.* **2011**, *5*, 95–100. [CrossRef]
17. Matta, G.; Kumar, A.; Uniyal, D.P.; Singh, P.; Kumar, A.; Dhingra, G.K.; Kumar, A.; Naik, P.K.; Shrivastva, N.G. Temporal assessment using WQI of River Henwal, a tributary of river Ganga in Himalayan region. *Int. J. Env. Rehab. Conserv.* **2017**, *VIII*, 187–204. Available online: https://essence-journal.com/wp-content/uploads/Volume_VIII/Issue_1/Temporal-assessment-using-WQI-of-River-Henwal-a-Tributary-of-River-Ganga-in-Himalayan-Region.pdf (accessed on 22 July 2024).
18. Mishra, S.; Sharma, M.P.; Kumar, A. Assessment of surface water quality in Surha Lake using pollution index, India. *J. Matr. Environ. Sci.* **2016**, *7*, 713–719. Available online: https://jmaterenvironsci.com/Document/vol7/vol7_N3/84-JMES-1858-2015-Mishra.pdf (accessed on 22 July 2024).
19. Lumb, A.; Halliwell, D.; Sharma, T. Application of CCME water quality index to monitor water quality: A case of the Mackenzie River Basin, Canada. *Environ. Monit. Assess.* **2006**, *113*, 411–429. [CrossRef] [PubMed]
20. Espejo, L.; Kretschmer, N.; Oyarzún, J.; Meza, F.; Núñez, J.; Maturana, H.; Soto, G.; Oyarzo, P.; Garrido, M.; Suckel, F.; et al. Application of water quality indices and analysis of the surface water quality monitoring network in semiarid North-Central Chile. *Environ. Monit. Assess.* **2012**, *184*, 5571–5588. [CrossRef] [PubMed]
21. Islam, M.M.; Karim, R.; Zheng, X.; Li, X. Heavy metal and metalloid pollution of soil, water and foods in Bangladesh: A critical review. *Int. J. Environ. Res. Public Health* **2018**, *15*, 2825. [CrossRef] [PubMed]
22. Uddin, M.G.; Moniruzzaman, M.; Khan, M. Evaluation of groundwater quality using CCME water quality index in the Rooppur nuclear power plant area, Ishwardi, Pabna, Bangladesh. *Am. J. Environ. Prot.* **2017**, *5*, 33–43. [CrossRef]
23. Karim, M.; Das, S.K.; Paul, S.C.; Islam, M.F.; Hossain, S. Water quality assessment of Karrnaphuli River, Bangladesh using multivariate analysis and pollution indices. *AJEE* **2018**, *7*, 1–11. [CrossRef]
24. McCarthy, T.S. The impact of acid mine drainage in South Africa. *S. Afr. J. Sci.* **2011**, *107*, 1–7. [CrossRef]
25. Department of Water Affairs and Forestry. *Olifants Water Management Area: Internal Strategic Perspective*; Report PWMA 04/0000/00/0304; Department of Water Affairs and Forestry: Pretoria, South Africa, 2004. Available online: https://www.dws.gov.za/Documents/Other/WMA/Olifants_ISP.pdf (accessed on 22 July 2024).
26. Adler, R.; Rascher, J. *A Strategy for the Management of Acid Mine Drainage from Gold Mines in Gauteng*; Report no. CSIR/NRE/PW/ER/2007/0053/C; Council for Scientific and Industrial Research: Pretoria, South Africa, 2007.
27. Pulles, W.; Banister, S.; Van Biljon, M. *The Development of Appropriate Procedures towards and after Closure of Underground Gold Mines from a Water Management Perspective*; Report 1215/1/05; Water Research Commission: Pretoria, South Africa, 2005; Available online: https://www.wrc.org.za/wp-content/uploads/mdocs/1215-1-051.pdf (accessed on 22 July 2024).
28. Atangana, E. An indices-based water quality model to evaluate surface water quality: A case study in Vaalwaterspruit, Mpumalanga, South Africa. *J. Afr. Earth Sci.* **2023**, *205*, 105001. [CrossRef]
29. Midgley, D.C.; Pitman, W.V.; Middleton, B.J. *Surface Water Resources of South Africa*; WRC Report No 298/1.1/94; Water Research Commission: Pretoria, South Africa, 1990; Volume 1.
30. Cairncross, B.; McCarthy, T.S. A geological investigation of Klippan in Mpumalanga province, South Africa. *S. Afr. J. Geol.* **2008**, *111*, 421–428. [CrossRef]
31. Wilson, M.G.C.; Anhaeusser, C.R. *The Mineral Resources of South Africa*, 6th ed.; Handbook 16; Council for Geoscience: Pretoria, South Africa, 1998.
32. Visser, J.N.J. The permo-carboniferous Dwyka Formation of Southern Africa: Deposition by a predominantly subpolar marine ice sheet. *Palaeogeogr. Palaeoclimatol. Palaeoecol.* **1989**, *70*, 377–391. [CrossRef]
33. Department of Water and Sanitation, South Africa. *National Water Act, 1998 (Act No. 36 of 1998): Classes and 25 Resource Quality Objectives of Water Resources for the Olifants Catchment*; Government Gazette No. 466; Department of Water and Sanitation, South Africa: Pretoria, South Africa, 2016. Available online: https://www.gov.za/sites/default/files/gcis_document/201604/39943gon466.pdf (accessed on 22 July 2024).
34. Heaton, J. Secondary analysis of qualitative data: An overview. *Hist. Soc. Res.* **2008**, *33*, 33–45. Available online: https://www.jstor.org/stable/20762299 (accessed on 22 July 2024).
35. Smith, E. *Pitfalls and Promises: The Use of Secondary Data Analysis in Educational Research*; McGraw-Hill Education: New York, NY, USA, 2008.
36. Department of Water Affairs and Forestry. *Best Practice Guideline G3: Water Monitoring Systems*; Department of Water Affairs and Forestry: Pretoria, South Africa, 2006. Available online: https://www.mineralscouncil.org.za/work/environment/environmental-resources/send/26-environmental-resources/358-g3-water-monitoring-systems (accessed on 22 July 2024).

37. Department of Water Affairs and Forestry. *South African Water Quality Guidelines. Volume 5: Agricultural use: Livestock Watering*, 2nd ed.; Department of Water Affairs and Forestry: Pretoria, South Africa, 1996. Available online: https://www.dws.gov.za/iwqs/wq_guide/edited/Pol_saWQguideFRESH_vol5_Livestockwatering.pdf (accessed on 1 June 2023).
38. Department of Water Affairs and Forestry. *South African Water Quality Guidelines. Volume 7: Aquatic Ecosystems*, 2nd ed.; Department of Water Affairs and Forestry: Pretoria, South Africa, 1996. Available online: https://www.dws.gov.za/iwqs/wq_guide/edited/Pol_saWQguideFRESH_vol7_Aquaticecosystems.pdf (accessed on 1 June 2023).
39. Canadian Council of Ministers of the Environment. Canadian water quality guidelines for the protection of aquatic life. In *Canadian Environmental Quality Guidelines*; Canadian Council of Ministers of the Environment: Winnipeg, MB, Canada, 2012; Available online: https://ccme.ca/en/resources/water-aquatic-life (accessed on 15 September 2023).
40. Australian and New Zealand Environment and Conservation Council and Agriculture and Resource Management Council of Australia and New Zealand. National water quality management strategy: Paper No. 4. In *Australian and New Zealand Guidelines for Fresh and Marine Water Quality. Vol. 1: The Guidelines*; Australian Water Association: Canberra, Australia, 2000. Available online: https://www.waterquality.gov.au/sites/default/files/documents/anzecc-armcanz-2000-guidelines-vol1.pdf (accessed on 22 July 2024).
41. QGIS Geographic Information System. Open Source Geospatial Foundation Project. 2021. Available online: http://qgis.org (accessed on 22 July 2024).
42. World Health Organization. *Guidelines for Drinking-Water Quality: Fourth Edition Incorporating the First Addendum*, 4th ed.; World Health Organization: Geneva, Switzerland, 2017; License: CC BY-NC-SA 3.0 IGO. Available online: https://iris.who.int/bitstream/handle/10665/254637/9789241549950-eng.pdf?sequence=1 (accessed on 18 September 2023).
43. Carroll, S.P.; Dawes, L.A.; Goonetilleke, A.; Hargreaves, M. *Water Quality Profile of an Urbanising Catchment—Ningi Creek Catchment*; Technical Report; School of Urban Development, Queensland University of Technology: Brisbane, Australia, 2006; (Unpublished). Available online: https://eprints.qut.edu.au/4198/1/4198.pdf (accessed on 22 July 2024).
44. Ramakrishnaiah, C.R.; Sadashivaiah, C.; Ranganna, G. Assessment of water quality index for the groundwater in Tumkur Taluk, Karnataka State, India. *J. Chem.* **2009**, *6*, 757424. [CrossRef]
45. Ding, J.; Jiang, Y.; Liu, Q.; Hou, Z.; Liao, J.; Fu, L.; Peng, Q. Influences of the land use pattern on water quality in low-order streams of the Dongjiang River basin, China: A multi-scale analysis. *Sci. Total Environ.* **2016**, *551–552*, 205–216. Available online: https://daneshyari.com/article/preview/6323131.pdf (accessed on 22 July 2024). [CrossRef]
46. Bhatnagar, A.; Devi, P. Applications of correlation and regression analysis in assessing lentic water quality: A case study at Brahmsarovar Kurukshetra, India. *Int. J. Environ. Sci.* **2012**, *3*, 813–820. [CrossRef]
47. Hamlat, A.; Guidoum, A.; Koulala, I. Status and trends of water quality in the Tafna catchment: A comparative study using water quality indices. *J. Wat. Reuse Desal.* **2017**, *7*, 228–245. [CrossRef]

Disclaimer/Publisher's Note: The statements, opinions and data contained in all publications are solely those of the individual author(s) and contributor(s) and not of MDPI and/or the editor(s). MDPI and/or the editor(s) disclaim responsibility for any injury to people or property resulting from any ideas, methods, instructions or products referred to in the content.

hydrology

Article

An Analysis of the Spatiotemporal Variability of Key Water Quality Parameters in China

Kexin Li [1], Qichun Yang [1,2,*] and Xia Li [3]

1 Thrust of Earth, Ocean and Atmospheric Sciences, The Hong Kong University of Science and Technology (Guangzhou), Guangzhou 511453, China; kli373@connect.hkust-gz.edu.cn
2 Center for Ocean Research in Hong Kong and Macau, Department of Ocean Science, Hong Kong University of Science and Technology, Hong Kong 999077, China
3 Research and Development Center for Watershed Environmental Eco-Engineering, Beijing Normal University, Zhuhai 519087, China; lixiabnu@bnu.edu.cn
* Correspondence: qichunyang@hkust-gz.edu.cn; Tel.: +86-13543061953

Abstract: Intensifying anthropogenic disturbances have caused water pollution in China in recent decades. China has a vast territory with diverse climate conditions, land use types, and human activities, leading to significant water quality variability. However, few studies have investigated nationwide spatiotemporal patterns of key water quality parameters. In this study, we analyze monthly water quality observations from 3647 gauge stations to understand how water quality changes over time and space in China. We group the stations by water resource regions and adopt Python and SPSS to analyze the spatiotemporal variability and intercorrelations of eight water quality parameters. Results indicate that the concentrations of biochemical oxygen demand of 5 days (BOD5), chemical oxygen demand (COD), dissolved oxygen (DO), ammonia nitrogen (NH3-N), total nitrogen (TN), and total phosphorus (TP) show similar spatial patterns, with higher concentrations in the northern parts than the southern regions of China. The concentrations of COD and TP are higher in the rainy season than in the dry season, while DO, NH3-N, and TN show the opposite seasonal patterns. Strong positive correlations were found between BOD and COD, NH3-N and TP. The annual cumulative distribution figures demonstrate that all parameters showed slightly lower concentrations in 2022 and 2023 than in 2021, except for DO and TN. The TN/TP ratios across different water resource regions in China are significantly higher than 16, indicating that phosphorus is the limiting factor of eutrophication. This investigation provides a comprehensive understanding of the spatiotemporal variability of water quality parameters across China. The results of this study are highly valuable for investigating mechanisms regulating water quality across large spatial scales, thus providing valuable implications for improving water quality and mitigating water pollution.

Keywords: water pollution; nutrient export; eutrophication; heavy metal; spatiotemporal variability; water quality management

Citation: Li, K.; Yang, Q.; Li, X. An Analysis of the Spatiotemporal Variability of Key Water Quality Parameters in China. *Hydrology* **2024**, *11*, 135. https://doi.org/10.3390/hydrology11090135

Academic Editors: Meysam Vadiati and Dingjiang Chen

Received: 29 June 2024
Revised: 27 July 2024
Accepted: 21 August 2024
Published: 26 August 2024

Copyright: © 2024 by the authors. Licensee MDPI, Basel, Switzerland. This article is an open access article distributed under the terms and conditions of the Creative Commons Attribution (CC BY) license (https://creativecommons.org/licenses/by/4.0/).

1. Introduction

Water pollution has been a critical environmental problem that is increasingly concerning due to intensifying anthropogenic disturbances [1,2]. According to China's 2023 Surface Water Environmental Quality Report (https://www.mee.gov.cn/hjzl/shj/, accessed on 5 June 2024) and China's Surface Water Environmental Quality Standard (GB3838-2002) (https://www.mee.gov.cn/ywgz/fgbz/bz/, accessed on 13 May 2024), 10.6% of streams across the country cannot reach the Class III water quality condition, 1.5% of streams are categorized as Class V, while 0.7% of the streams have water quality conditions that are worse than Class V.

Intensifying human activities are the primary reason for water pollution. Anthropogenic disturbances, such as deforestation and the expansion of agricultural and urban areas, contribute to water quality degradation. Pollutants produced by human activities mentioned above are often insufficiently treated, and then transported to rivers and lakes through runoff, resulting in water pollution [3]. According to the 2022 Ecological Environment Statistical Annual Report of China (https://www.mee.gov.cn/hjzl/sthjzk/, accessed on 5 June 2024), agricultural and domestic sewage discharge were identified as two primary pollutant sources for chemical oxygen demand (COD), ammonia nitrogen (NH3-N), total nitrogen (TN), and total phosphorus (TP) in the surface waters of China.

Excessive nutrients transported to rivers cause a series of environmental problems. Sewage water and agricultural runoff containing large amounts of nutrients are the primary reasons for the boom of phytoplankton [4,5], which leads to decreases in the concentration of dissolved oxygen (DO) and increases in the biochemical oxygen demand of 5 days (BOD5) and COD. During algae bloom, toxic substances produced by algae and the decrease in DO can kill many organisms, accelerating eutrophication and the deterioration of water quality. Degraded water quality affects drinking water supply, aquaculture, and recreation [6]. For example, in 2014, excessive nitrogen and phosphorus caused an algae bloom in Lake Erie in the United States, resulting in a shutdown of water supplies to 500,000 residents over the past several decades [7]. In China, the cyanobacteria pollution incident in Lake Taihu in 2007 caused drinking water pollution, resulting in a serious water supply shortage for millions of residents in Wuxi City for nearly a week [8].

In addition to nutrients, heavy metals are another group of pollutants causing environmental problems. Watersheds with excessive heavy metals, such as Hg [9], Cd [10], and Cr [11], are not suitable for drinking water supply or water withdrawal for crop irrigation. Agricultural and industrial sewage have been identified as the primary sources of heavy metal pollution in rivers, especially in the Yellow River Basin and Northeast China [12]. Although lots of resources have been invested to control water pollution, it remains a problem across many parts of China.

China has a vast territory with various climate zones and population densities, resulting in substantial variability of water quality parameters over time and space. These parameters demonstrate strong seasonal variations due to changes in climate conditions and human activities over time [13]. Water quality also varies with the types of land use. For example, the concentrations of nutrients show a negative correlation with forest coverage and a positive correlation with the area of agricultural and urban land [14]. Water quality demonstrated close correlations with the drainage area properties, including reservoir densities, agricultural land areas, and rainfall, across the upper, middle, and downstream regions of the Yangtze River Basin [15]. Climate change also increases the risks of water pollution. For example, decreases in runoff due to drought can result in increases in nutrient concentrations, favoring the growth of algae and plankton [16].

Over the past four decades, since the reform and opening up, water bodies in China have experienced increasingly intensive anthropogenic disturbances. At the same time, the implementation of ecological protection policies has improved aquatic environment conditions at the local scale, mitigating the adverse impacts of the disturbances on water quality [17]. How these counteracting activities have jointly affected water quality remains unclear. As a result, an analysis of the spatiotemporal variability of water quality will provide valuable implications for balancing economic development and environmental protection [17,18].

Analyzing the spatial and temporal dynamics of water quality parameters [19,20] and their relationships with watershed properties [14] is needed to support water quality management [21]. Statistical analyses were adopted to analyze the characteristics of water quality [18,22], and to infer the driving forces of aquatic nutrients [17], as well as the relationships between human activities and water pollution [23]. For example, the Mann–Kendall trend analysis [24] and time-series decomposition methods were applied to investigate long-term trends and the seasonal variability of water quality [25]. Machine learning models were recently applied to estimate the spatial and temporal changes in nutrient concentrations [26–28]. Temporal trends of different land use types were also explored, to identify critical pollution sources.

However, previous investigations are limited by the lack of long-term nationwide monitoring data [17]. Existing analyses have mainly focused on one or a few watersheds [29–32], while only a few studies have paid attention to large spatial scales [33]. At the same time, the number of gauge stations and water quality parameters used in these studies is often small [32], and the water quality data used in some studies are relatively short [34]. For example, based on water quality observations in 2014 from 763 gauge stations, Wang et al. identified water pollution hotspots in North China [29]. Dou et al. analyzed water quality changes in the Huaihe River based on 78 gauge stations and identified the important impacts of river discharge and temperature on water quality [30]. Li et al. explored the spatiotemporal variability of TN, TP, COD, and NH3-N in Honghu Lake and highlighted the contribution of anthropogenic activities to water pollution [31]. Wu et al. investigated the water quality of the Pearl River Basin based on observations from 16 gauge stations, and suggested that regionalized management activities could help improve water quality [32]. Using data collected from 145 gauge stations, Zhou et al. [35] analyzed the impacts of population and economic development on water quality across China.

Long-term monitoring at a large spatial scale can provide a more comprehensive analysis of water quality conditions and the long-term impact of human activities on water quality [36,37]. In addition, existing studies in China mainly use comprehensive water quality indicators, such as the water quality index (WQI) and water pollution index (WPI), to evaluate the overall water quality of one or a few watersheds [38–41]. However, the spatiotemporal variability of individual water quality parameters has not been sufficiently investigated, limiting the applications of findings from these investigations to support water quality management. In addition, systematic water quality monitoring in China at the national scale only started in recent decades, and thus just accumulated much shorter periods of observations compared with other countries, such as the United States [36]. A national-scale analysis of water quality in China is needed to improve our understanding of how human activities affect water quality and watershed biogeochemistry.

In recognition of the necessity and existing challenges, this study provides a comprehensive analysis of the spatiotemporal variability of key water quality parameters across China, filling a gap in the current research on national-scale water pollution in China. We compiled a 3-year (2021–2023) monthly water quality dataset for 3647 gauge stations across China and analyzed the spatial distribution of water quality parameter concentrations across nine water resource regions. The seasonal and annual variation patterns of water quality parameters, the spatiotemporal variability of parameters, and the correlation between water quality parameters are investigated to understand the overall water pollution conditions in China and to infer the underlying mechanisms controlling water quality. The results of this study are expected to provide valuable information to policymakers to formulate pollution mitigation strategies.

The objectives of the study are to (1) provide an overall evaluation of water quality conditions in China, (2) understand how key water parameters vary over time and across China, and (3) explore the underlying mechanisms responsible for the variability of water quality. To achieve these objectives, we analyze the statistical characteristics of water quality parameters by water resource regions and months (Sections 3.1–3.3). To understand complex interactions among selected water quality parameters, we also quantify the intercorrelations of the parameter, as well as ratios between TN and TP (Section 3.4). Factors affecting water quality variability are discussed in Sections 4.1–4.4.

2. Materials and Methods

2.1. Study Area

The study area of this investigation is mainland China (excluding the Hong Kong, Macao, and Taiwan regions), which is characterized by a variety of land use types and cultivation patterns, high heterogeneity of urban development, and different levels and types of water pollution (Figure 1). Dominant land use types include forest areas in the northeastern, southwestern, and southeastern areas of China, cropland in northern

China, and desert in northwestern China. In northern China, wheat and corn are the primary crop types, while rice is mainly cultivated in southern China. The majority of the population is located in the eastern part of China. The climate types include a tropical monsoon climate and a subtropical monsoon climate located in southern China, as well as a temperate monsoon climate in northern China. The climate type in western China is a typical temperate continental climate, where Northwest parts of the country are characterized by an inland dry climate. Precipitation ranges from ca. 50 mm/year in the desert areas to over 2000 mm/year in the coastal areas of southern China. The average annual temperature ranges from ca. 5 to 25.1 °C.

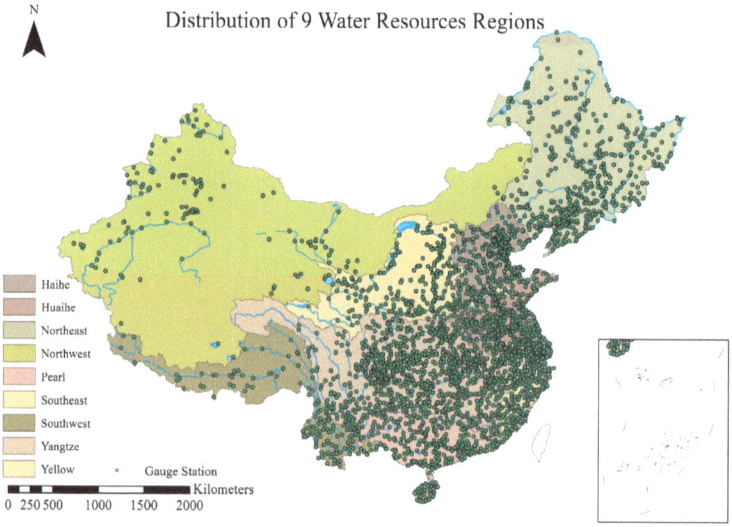

Figure 1. Locations of the water quality gauge stations and 9 water resource regions.

According to the classification for primary water resource areas, mainland China includes nine major regions, including the Haihe River Basin, the Huaihe River Basin, the Northeast Area, the Northwest Area, the Pearl River Basin, the Southeast Area, the Southwest Area, the Yangtze River Basin, and the Yellow River Basin (Figure 1). Water resource regions, such as the Northeast Area, the Northwest Area, the Pearl River Basin, the Southeast Area, and the Southwest Area, are all composed of many smaller watersheds. In this study, we compiled water quality data from 3647 gauge stations across China to investigate water quality conditions nationwide. Both point and non-point source pollution have contributed to the degraded water quality in China.

2.2. Data Collection

The water quality data of this study are from the automatic water quality monitoring platform of the China General Environmental Monitoring Station network (https://www.cnemc.cn, accessed on 28 June 2024). The dataset includes monthly concentrations of water quality parameters from 3647 gauge stations for the period 2021–2023. Water quality parameters were selected based on pollution types, such as the depletion of oxygen, eutrophication, and biochemical pollution. In this study, we primarily focus on nutrients (e.g., TN, TP, and NH3-N) in watersheds, parameters indicating the oxygen levels in water (e.g., BOD5, COD, and DO), and concentrations of heavy metals (e.g., Cr6 and Pb).

According to the environmental quality standards for surface water (GB3838-2002) in Table 1, the concentration of water quality parameters of gauge stations is categorized into five classes (Table 1). For water bodies designated as sources of drinking water, the water quality must meet at least Class III standards. Classes IV and V represent mild and moderate pollution levels, respectively. Concentrations surpassing the Class V standard indicate severe pollution, which means measures must be taken to mitigate pollution. For heavy metal elements, higher standards are required for water sources that meet drinking water standards [42].

Table 1. Environmental quality standards of China for surface waters.

Parameters [1]	Classes				
	I	II	III	IV	V
BOD≤	3	3	4	6	10
COD≤	15	15	20	30	40
Cr6≤	0.01	0.05	0.05	0.05	0.1
DO≥	7.5	6	5	3	2
NH3-N≤	0.15	0.5	1.0	1.5	2.0
Pb≤	0.01	0.01	0.05	0.05	0.1
TN≤	0.2	0.5	1.0	1.5	2.0
TP≤	0.02	0.1	0.2	0.3	0.4

[1] The unit of water quality parameters is mg/L.

2.3. Data Analysis

In this research, we first calculated the averages of the selected parameters to demonstrate the magnitudes of the pollution levels. The spatiotemporal analysis was conducted using Box and Whisker Plots, which demonstrate the median, upper and lower quantiles, as well as the upper and lower extremes of the parameters, enabling the comparisons of water quality by month or water resource regions. Pearson correlation was adopted to analyze the intercorrelations among the selected water quality parameters [43]. We also used the cumulative distribution function to compare water quality conditions in different years [44].

The spatial distributions of water quality parameters were illustrated using ArcGIS based on the average concentrations of each gauge station (https://www.esri.com/, accessed on 26 July 2024). The cumulative probability distribution of water quality parameters was analyzed using Python (https://www.python.org/, accessed on 26 July 2024). We also used Python to plot the ratio of total nitrogen to total phosphorus across nine water resource regions. For the correlation analysis, we used IBM SPSS Statistics 23 to process the data and visualize the results (https://www.ibm.com/spss, accessed on 25 July 2024).

3. Results

3.1. Spatial Pattern of Water Quality Parameters

The distributions of water quality concentrations across the whole of China provide details of the spatial variability of water quality (Figure 2). As depicted in Figure 2a, a few gauge stations located along the downstream region of the Haihe River and the Yellow River have recorded BOD concentrations exceeding 4 mg/L, exceeding the Class III water quality standards for BOD. In Northeast China, COD concentrations above 20 mg/L are recorded across many gauge stations, with a few stations even showing COD concentrations over 30 mg/L (Class IV water quality standard). DO concentrations are notably low at the confluence of major tributaries and lakes within the Yangtze River Basin, as well as in the estuaries of the Yangtze and Pearl Rivers, and in the southeastern coastal regions (Figure 2).

A few gauge stations in the Yellow River Basins show Cr6 concentrations higher than 0.01 mg/L (Figure 2c), especially in Shaanxi Province and Gansu Province, where the concentrations of Cr6 are particularly high, even more than 0.1 mg/L, exceeding the Class V water quality standard. Gauge stations located in Liaoning and Anhui Provinces have recorded Pb concentrations close to 0.01 mg/L, approaching the water quality standards of Class I. High concentrations of NH3-N and TP often appear simultaneously, as shown in Figure 2e,h, in the Southeast of Northeast China, the Huang Huai Hai region, upstream of the Yangtze River Basin (east of Yunnan Province), and the coastal areas of South China.

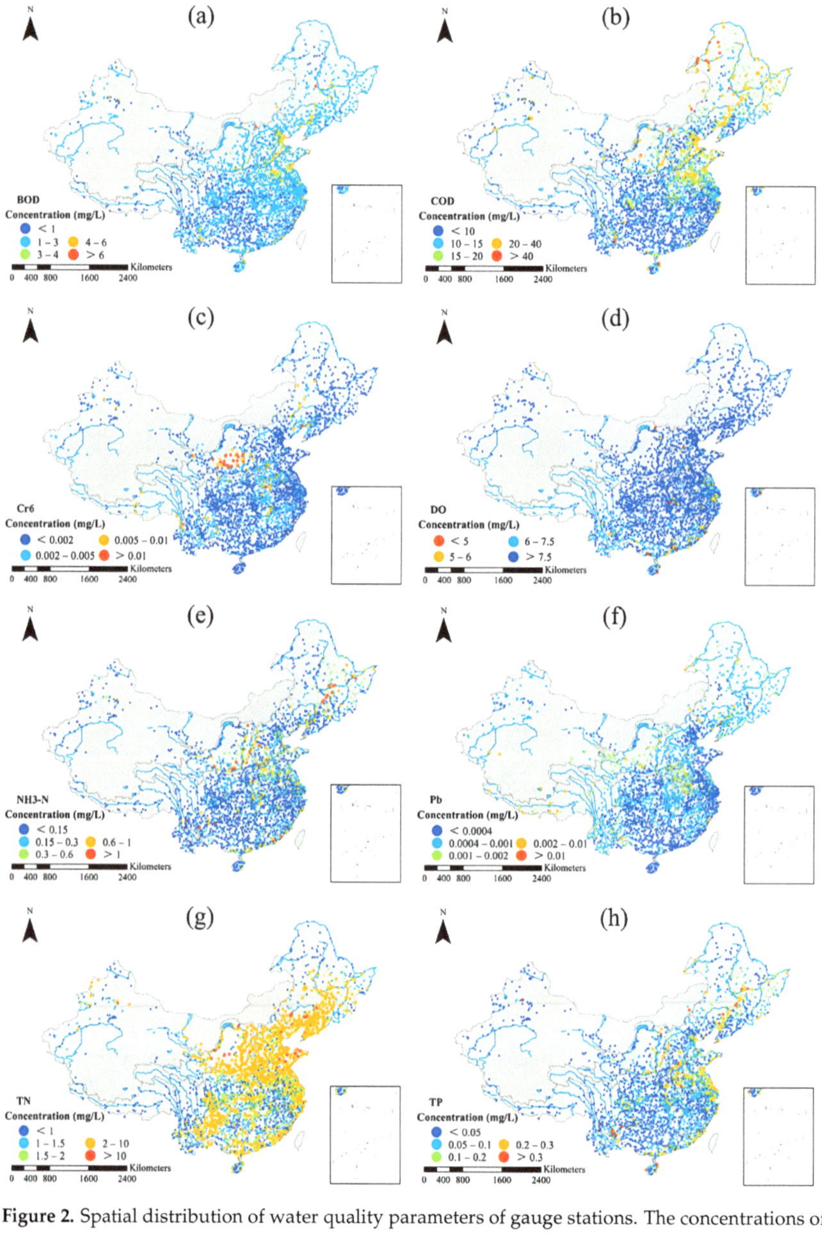

Figure 2. Spatial distribution of water quality parameters of gauge stations. The concentrations of BOD, COD, Cr6, DO, NH3-N, Pb, TN, and TP are shown in subplots (**a**–**h**), respectively (the value of each gauge station is calculated based on the average of monthly observations from 2021 to 2023, and then classified according to the natural break method and China Surface Water Environmental Quality Standard).

The distribution of water quality parameters across different water resource regions shows significant variability (Figure 3). Specifically, BOD and COD concentrations are higher in the Huaihe River Basin, the Haihe River Basin, and Northeast regions, while being lower in the Pearl River Basin, Southwest, and Southeast China. In the latter three southern regions, levels of BOD and COD are categorized as Class I water quality conditions according to the China Environmental Quality standard of surface water. However, the concentrations of DO tend to be lower in southern regions. As can be seen in Figure 3d, the median values of DO concentrations in the Pearl River Basin and Southeast region are below 7.5 mg/L.

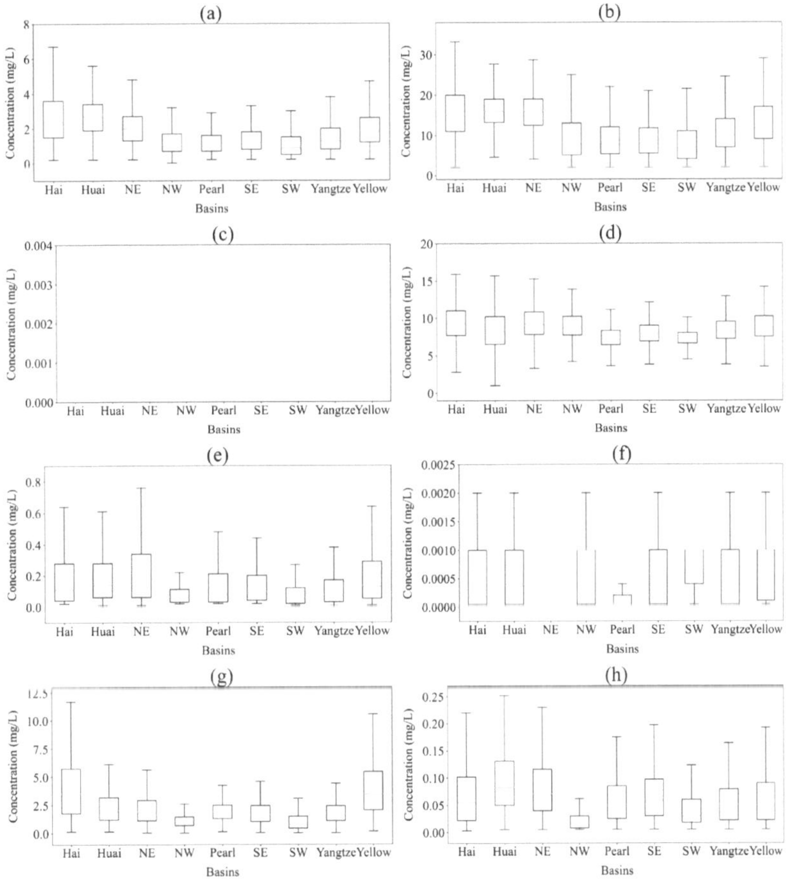

Figure 3. Water quality conditions across 9 water resource regions in 2021–2023. The concentrations of BOD, COD, Cr6, DO, NH3-N, Pb, TN, and TP are shown in subplots (**a**–**h**), respectively. Abbreviations on the x-axis refer to the water resource regions: the Haihe River Basin, the Huaihe River Basin, the Northeast Area, the Northwest Area, the Pearl River Basin, the Southeast Area, the Southwest Area, the Yangtze River Basin, and the Yellow River Basin.

For heavy metal cations, the concentrations of Cr6 and Pb in the nine water resource areas are usually lower than 0.01 mg/L, meeting the Class I water quality standard. Large amounts of gauge stations observed Cr6 concentrations of 0.002 mg/L. The concentrations of nutrients (TN and TP) are higher in the Huaihe River Basin, the Haihe River Basin, and Northeast China, while being lower in the Southwest and Southeast China. From Figure 3e, it can be seen that the NH3-N concentrations of most gauge stations are below 0.5 mg/L, meeting the Class II water quality standard. Figure 3g demonstrates that the TP concentration is generally consistent with NH3-N spatially. Except for the Huaihe River Basin and Northeast China, the majority of gauge stations have TP concentrations below 0.1 mg/L, meeting the standard of Class II. The concentrations of TN are generally higher than TP, especially in the Haihe River Basin and the Yellow River Basin. In these two regions, the median values of TN concentrations are greater than 2 mg/L, exceeding the Class V standard of TN according to the national water quality criteria (Table 1).

3.2. Temporal Variability of Water Quality Parameters

The water quality parameters also demonstrate evident temporal changes (Figure 4). The concentrations of BOD fluctuate throughout the year and are slightly higher in the first half of the year than in the latter half. The COD concentrations are higher in winter and spring, but lower in summer and autumn. The BOD and COD concentrations of most water bodies in China are below 4 and 20 over different seasons, meeting the Class III water quality standards. Figure 4d demonstrates that the level of DO declines at the beginning of the year and then rises from late summer to the end of the year, and it is slightly higher in the first half of the year than in the latter half. For most water bodies, the concentrations of DO are higher than 6 mg/L over the whole year, meeting Class II standards, and most water bodies meet Class I water quality standards in winter and spring.

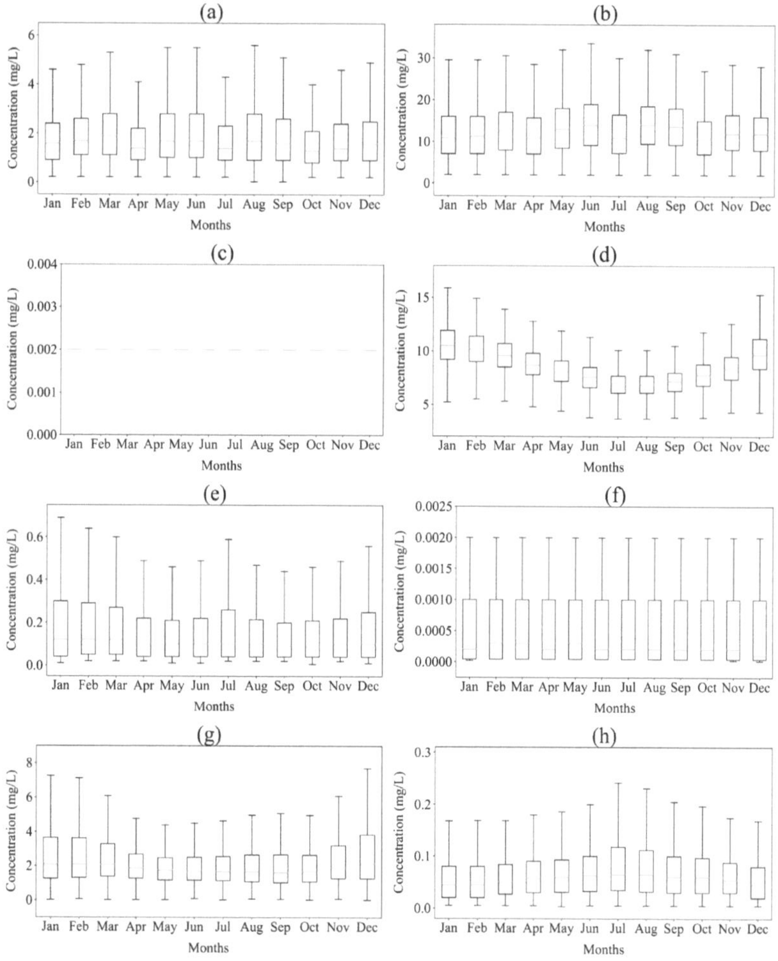

Figure 4. Seasonal variation in water quality parameters across 9 water resource regions from 2021 to 2023. The concentrations of BOD, COD, Cr6, DO, NH3-N, Pb, TN, and TP are shown in subplots (**a–h**), respectively.

Most gauge stations have Cr6 concentrations of 0.002 mg/L. Pb concentrations are higher in spring and lower in winter (Figure 4f). Similar seasonal patterns can be seen in NH3-N and TN, with higher concentrations in the dry season (winter) compared to the wet season (summer). For NH3-N, most water bodies have concentrations lower than 0.5 mg/L in the whole year, reaching the level of the Class II water quality standard. However, for TN concentrations, the median reached 2 mg/L almost every month, which is the water quality standard of Class V. Figure 4h illustrates an opposite seasonal variation pattern of TP compared to NH3-N and TN, with increasing concentrations at the beginning of the year and decreasing concentrations after July. The concentrations of TP in summer are higher than in other seasons.

We further analyze how the water quality parameters have changed over the three years (Figures 5 and 6). Compared to 2022 and 2023, slightly higher concentrations of BOD, COD, Cr6, NH3-N, and TP in 2021 accounted for a higher proportion, indicating that these water parameters might have improved over time. Especially for NH3-N, as depicted in Figure 5e, the concentrations decreased in the last two years. The average annual concentrations of DO were the lowest in 2022. No significant changes can be seen in the concentrations of TN (Figure 5g), which indicates that the issue of nitrogen pollution has not improved over time. More evident differences are observed in the plot of the high concentrations (top 10% for each parameter) (Figure 6). Except for Pb and TN, all the other six parameters show reductions in high concentrations, suggesting pollution was less severe in 2002 and 2003, relative to 2021.

3.3. Spatiotemporal Variability of Water Quality Parameters

Similarities can be found in the magnitude and spatiotemporal patterns of some of the selected water quality parameters (Figures 7 and 8). Figure 7 demonstrates similar magnitudes of parameters BOD, COD, and DO across nine water resource regions. The concentrations of BOD and COD in the Huaihe River Basin, the Haihe River Basin, and Northeast China exhibit significantly higher values compared to other regions, particularly in summer. The monthly median concentrations of COD in Northeast China, the Huaihe River Basin, and the Haihe River Basin consistently exceed 15 mg/L, which is the value of Class I and Class II water quality standards. In all water resource regions, DO has lower concentrations in the wet season, showing an opposite seasonal change from COD. A sharp decrease in DO concentrations is observed in the Huaihe River Basin during summer. Except for the Southwest region and the Pearl River Basin, the average concentrations of DO in other water resource regions exceed 7.5 mg/L in different months of the year.

The concentrations of Cr6 at a few gauge stations in the Yellow River Basin show values higher than 0.01 mg/L in June, surpassing the Class I water quality standard and significantly higher than other water resource regions (Figure 7), indicating potential pollution caused by this element. It is necessary to conduct a comprehensive investigation of the sources of pollution that are responsible for the abnormally high Cr6 levels in this region. The concentrations and seasonal variabilities of Pb in the Pearl River Basin are markedly lower than the other regions.

In the Huaihe River Basin and the Haihe River Basin, NH3-N concentrations decrease in spring, followed by a sudden increase in July and August. In other regions, the concentrations of NH3-N are lower during the wet season than in the dry season. Figure 8 shows that the concentrations of TN in the Haihe River Basin, the Huaihe River Basin, the Yellow River Basin, and Northeast China are higher across all months compared to other regions. In addition, the seasonal variation in TN in the northern water resource regions also exhibits a greater fluctuation than that in the southern regions. Specifically, TN concentrations in northern China tend to decrease during spring and summer, and then increase during autumn and winter, whereas the Southwest water resource region shows an opposite temporal trend. Similarly, TP concentration is generally higher in the northern water resource regions than those in the south, showing increases during spring and summer, followed by a decrease in autumn and winter. Notably, the concentrations of TP in July and August are significantly higher compared to other months.

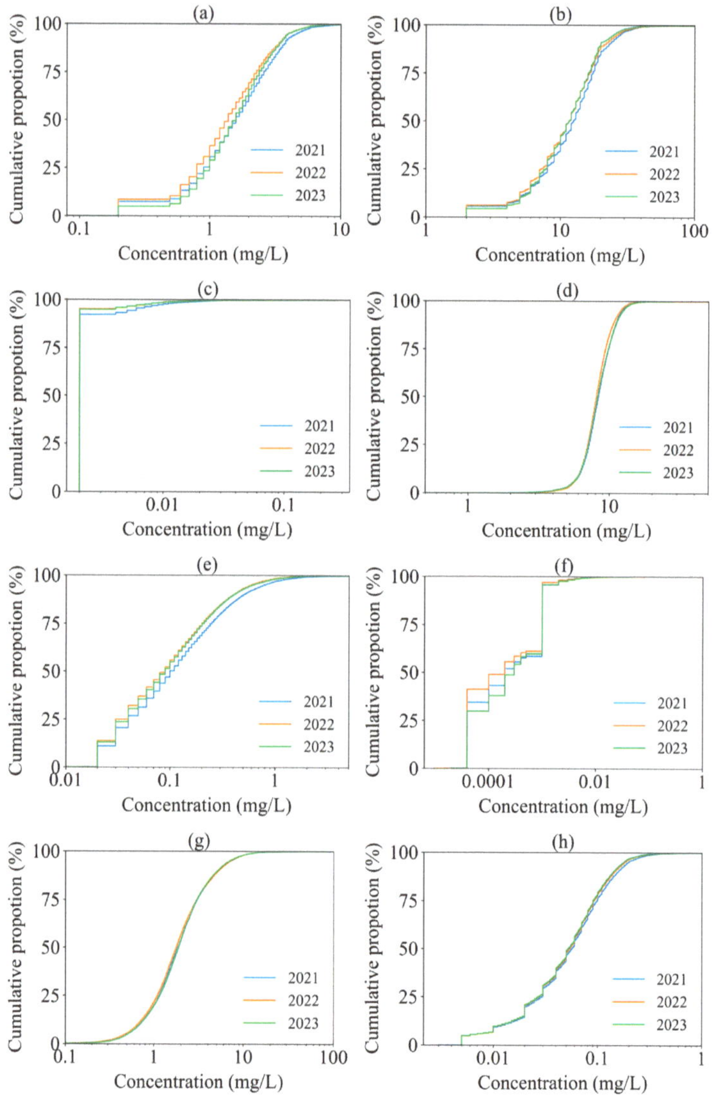

Figure 5. Annual variations in concentrations of water quality parameters from 2021 to 2023. The concentrations of BOD, COD, Cr6, DO, NH3-N, Pb, TN, and TP are shown in subplots (**a–h**), respectively.

3.4. Intercorrelations among Parameters

Some parameters demonstrate strong intercorrelations (Figure 9). A strong positive correlation is observed between BOD and COD across all water resource regions. TP shows strong positive correlations with BOD and COD, while NH3-N displays a correlation with these two parameters as well. The concentrations of TN show weaker correlations with BOD and COD, especially in the Huaihe River Basin, the Haihe River Basin, and Northeast China. However, a strong positive correlation can be seen between TN, BOD, and COD concentrations in Southwest and Southeast China. DO exhibits weak negative correlations with nutrients (TN, TP, and NH3-N). In the northern water resource regions, DO and TN show a weak positive correlation, while a weak negative correlation can be seen in the southern regions. In the Yellow River Basin, Cr6 shows a weak positive correlation with COD and a strong positive correlation with TN. NH3-N shows a strong positive correlation with TP. Weak positive correlations between NH3-N and TN, as well as TN and TP, are observed in the northern region. Conversely, in the southern water resource regions like the Yangtze River Basin, the Pearl River Basin, and Southeast and Southwest China, strong positive correlations between these nutrient parameters are identified. The correlations between heavy metal and other water quality parameters are not significant.

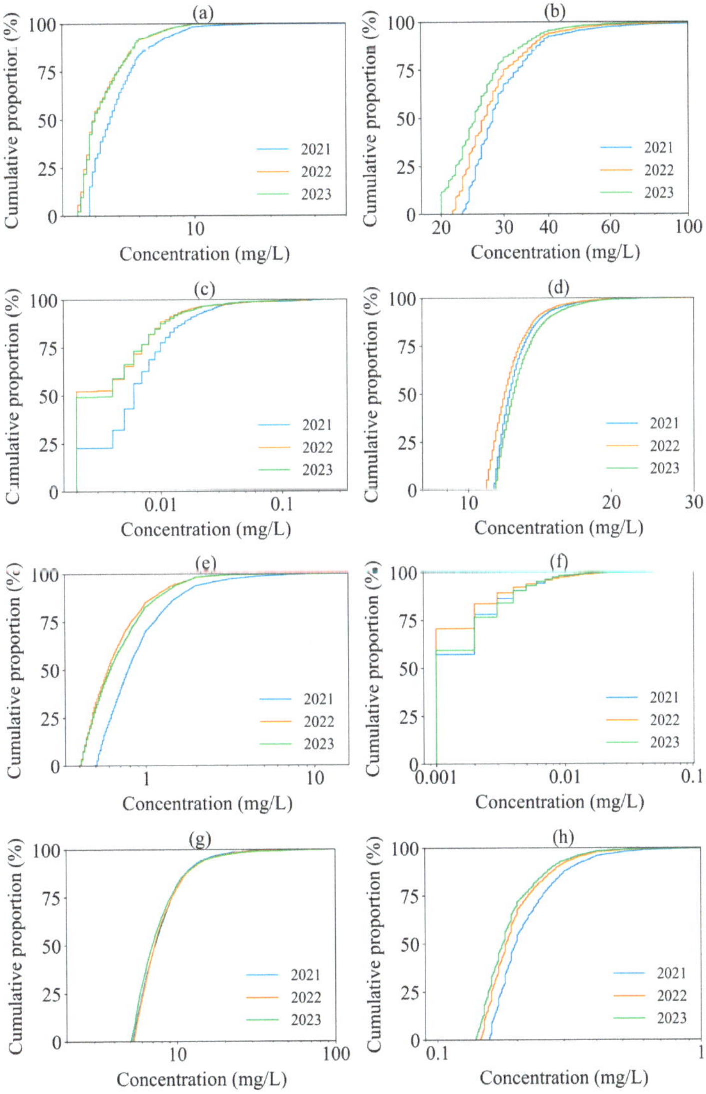

Figure 6. Annual variations in top 10% highest water quality parameter concentration records from 2021 to 2023. The concentrations of BOD, COD, Cr6, DO, NH3-N, Pb, TN, and TP are shown in subplots (**a–h**)respectively.

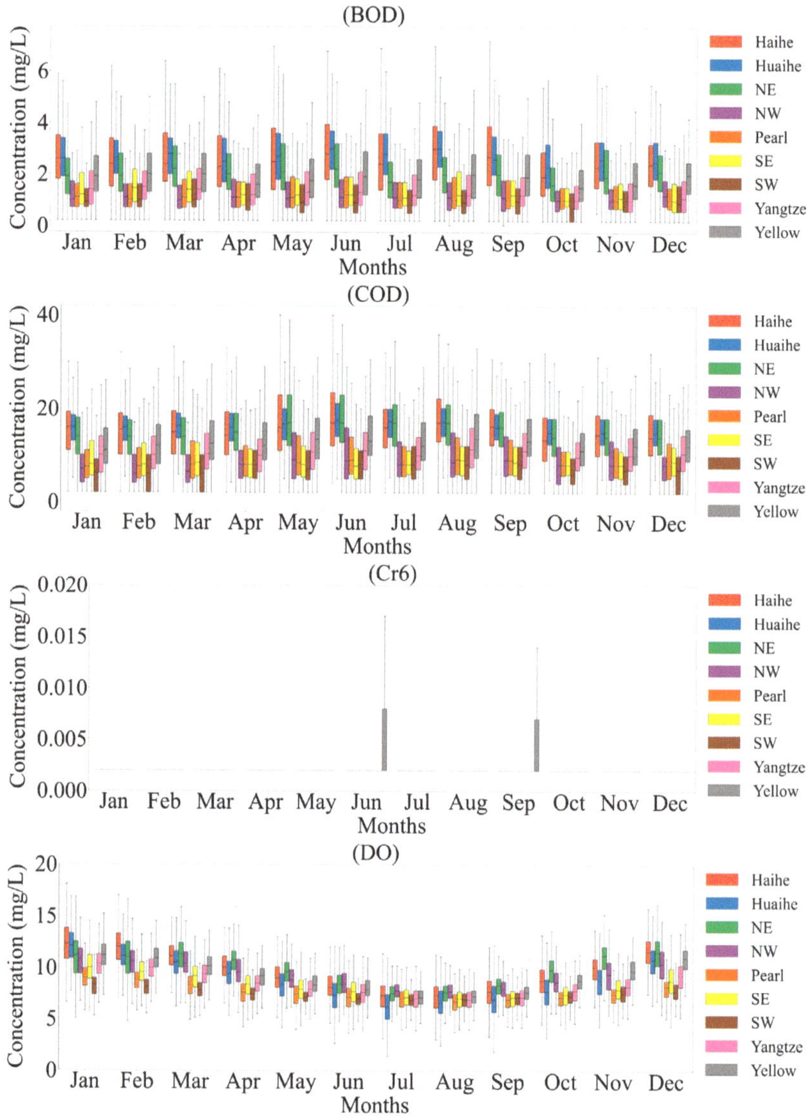

Figure 7. Spatiotemporal variability of BOD, COD, Cr6, and DO across 9 water resource regions in 2021–2023.

The ratios of TN and TP demonstrate temporal and spatial variations across different water resource regions throughout the year (Figure 10). The TN/TP ratio is significantly higher than 16, which is considered the most suitable molar ratio for plankton growth. Higher ratios are observed in the Yellow River Basin, the Huaihe River Basin, and Northwest China, whereas relatively low ratios are observed in the Yangtze River Basin, the Pearl River Basin, and Southwest and Southeast China. For the Yellow River Basin and the Huaihe River Basin, high concentrations of TN and TP are observed simultaneously. A relatively high TN/TP ratio is noted in Northwest China, possibly due to the low TP concentration in this region. The TN/TP ratio decreases in spring and summer, and then increases in autumn and winter. In the water resource regions with higher TN/TP ratios, the ratio exhibits a significant change within a year.

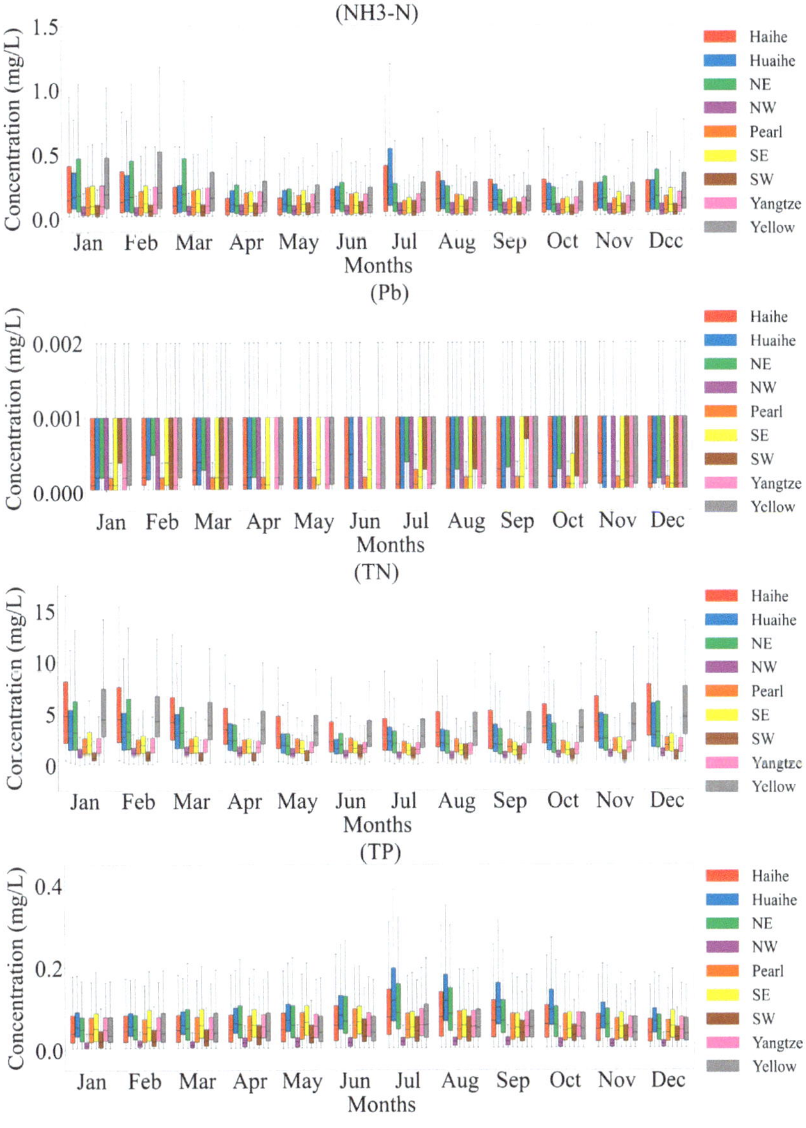

Figure 8. Same as Figure 7, but for NH3-N, Pb, TN, and TP.

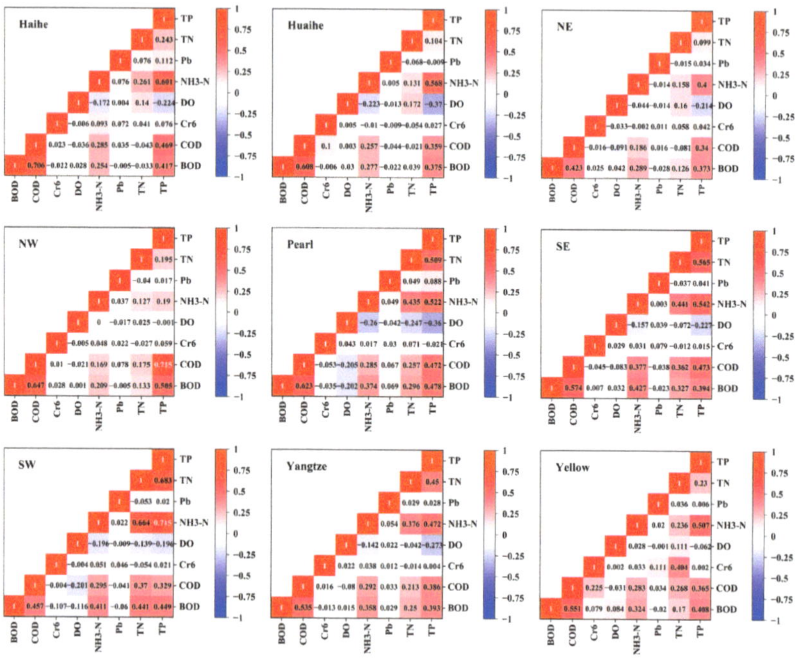

Figure 9. Inter-correlations of the water quality parameters in 9 water resource regions.

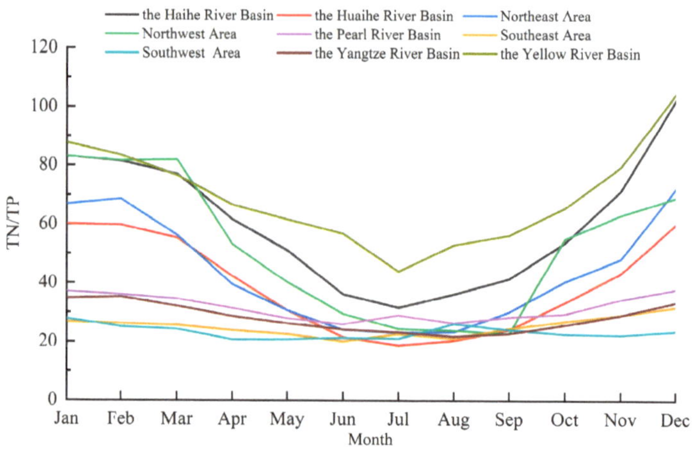

Figure 10. The ratio of TN/TP across 9 water resource regions.

4. Discussion

4.1. Overall Water Quality Condition in China

The spatial distribution of water quality parameters reveals that northern China exhibits higher concentrations for most of the selected water quality parameters than those in southern China, indicating poorer water quality conditions in the Huaihe River Basin, the Haihe River Basin, the Yellow River Basin, and Northeast China. This result aligns with previous investigations [45], which utilized the water quality index to assess water quality conditions across various water resource regions.

Water quality parameters also exhibit seasonal and annual fluctuations. The concentrations of COD and TP are higher in summer than in winter, but DO, NH3-N, and TN display opposite seasonal patterns. BOD, COD, Cr6, NH3-N, and TP show lower concentrations in 2022 and 2023 compared to 2021, suggesting an overall improvement in China's water quality over time. This temporal trend of water quality change found in this study is in line with previous studies [2,41]. Conversely, TN concentrations have not shown a clear trend over the same time period, and the average concentration of TN across all water resource regions is worse than the Grade IV water quality standard, except for Northwest and Southwest China, indicating that nitrogen pollution will remain a challenge for water quality management in China.

4.2. Factors Affecting the Variability of Water Quality Parameters

The variability of water quality parameters is significantly influenced by numerous factors. Higher temperatures can influence the physical and chemical conditions of watersheds, such as pH, salinity, solubility, viscosity, and diffusion rates, which in turn affect biochemical processes, like nitrification and denitrification [46]. This might be the reason for the strong seasonality of many parameters. Precipitation in northern China is much lower than that in the south, and the dilution effects of water may be responsible for the higher concentrations in northern areas and low concentrations in southern regions for the selected parameters [35].

In addition, the dilution effect of precipitation may also play a role in seasonal changes in concentrations of water quality parameters. In northern China, during spring and early summer, precipitation does not increase significantly, and warming temperatures lead to enhanced evapotranspiration and higher concentrations of the selected parameters [47]. Watersheds characterized by steep slopes usually experience rapid streamflow rates and soil erosion, resulting in shorter water retention times and a weaker nutrient retention capacity [17]. Gauge stations on the Yunnan-Guizhou Plateau have observed higher concentrations of Cr6 and Pb, which are attributable to parent materials and the low soil pH in those areas [48].

Human activities have increasingly impacted water quality over recent decades. Plains and lowland areas usually have a high population density. Intensified land use and urbanization can cause an increased discharge of agricultural, domestic, and industrial sewage, thereby contributing to both point and non-point source pollution [35,49]. For example, higher concentrations of nutrients were observed in Jilin and Liaoning Provinces, as well as downstream from the Yellow River Basin. This widespread phenomenon of nutrient pollution may be explained by the dominant land use types (agricultural land) in these areas. Concentrations of nutrients are more affected by human activities compared to climatic and geographical factors [17]. Gauge stations around Taihu Lake have higher TN concentrations, which can be explained by intensive urbanization, high population density, and active human activities indicated by the high GDP in these areas [50].

Reservoir operation also affects water quality. Li et al. [45] found that the construction of the Three Gorges Dam has decreased the flow velocity in the Yangtze River, reducing river runoff and diminishing its capacity for self-purification. However, Yang et al. [51] insisted that enhanced water retention due to reservoir damming can reduce sediment transport. Additionally, frequent mining activities are a possible explanation for the high levels of heavy metals in Yunnan Province [52], emphasizing the complex interaction between human disturbances and natural processes in affecting water quality.

4.3. Intercorrelation among Water Quality Parameters

Eutrophication is typically characterized by high concentrations of BOD, COD, and nutrients, as well as low levels of DO. However, in this study, the correlations between DO and other water quality parameters are weak, with only a slightly negative correlation between DO and concentrations of NH3-N or TP. In the Pearl River Basin, the correlation coefficient between DO and TP is -0.36, consistent with the negative correlation reported in previous studies for the same water resource region [32,37]. The correlation between DO and other parameters appears weaker in northern watersheds compared to the south, which may be related to differences in temperature among these regions. In southern China, higher temperatures lead to less oxygen dissolved in the water.

The correlation between TN and other water quality parameters also exhibits spatial variations. In the Hai River Basin, the Huai River Basin, and Northeast China, the concentrations of TN show low correlations with BOD and COD; high ratios of TN/TP also appear in these water resource regions. This may be due to higher anthropogenic N input in these regions, leading to high variabilities in water quality parameters [17]. The concentration of riverine COD is influenced by the activity of aquatic organisms, atmospheric deposition, and the influx from terrestrial vegetation and soil organic matter. Temperature fluctuations can alter water's physical and chemical properties [53], further affecting biochemical processes, and leading to the significant variability of COD. Further research on water quality analysis needs to quantify the contribution of anthropogenic and natural factors to regional water quality to unravel the underlying mechanisms responsible for the variability of water quality parameters.

4.4. Processes Leading to High Concentrations of Water Quality Parameters

The rapid development of industry, urban expansion, and changes in agricultural activities have greatly increased the discharge of pollutants, leading to widespread water pollution [54]. In Northeast China, many gauge stations have observed COD concentrations higher than 20 mg/L. This region, as an important base for grain production, has extensive agricultural land where excessive fertilizer use leads to high concentrations of BOD, COD, and nutrients, contributing to non-point source pollution [55]. The higher concentration of Cr6 in the Loess Plateau is attributable to the higher local soil background values and the recharge of groundwater with a high concentration of Cr6 [56]. In addition, the wastewater and soil erosion in the Loess Plateau also contribute to phosphorus transport [57]. Population densities and agricultural yields are relatively high in the middle and downstream of the Yellow River Basin [58,59]. Shanxi Province, located in the middle reaches of the Yellow River Basin, is an important base for coal production in China. The higher TN concentration in Shanxi Province can be explained by anthropogenic discharges, such as domestic sewage, livestock manure, nitrogen fertilizer, industrial sewage, etc. [60].

In southern China, eutrophication in the middle and downstream of the Yangtze River is closely linked to high population density and GDP. Point source pollution is the main component of water pollution in Yunnan Province, including mining activities [8], inadequate sewage treatment, and the expansion of built-up land [61]. The southern coastal areas of China are facing eutrophication challenges, driven by the development of big coastal cities and estuarine sewage discharges. Therefore, it is necessary to monitor and manage the water quality of estuaries to mitigate coastal eutrophication caused by human activities.

4.5. Implications for Pollution Control and Future Work

Our research highlights that northern water resource regions usually have worse water quality than southern regions of China. Water pollution is relatively light in Northwest and Southwest China, corresponding to the smaller population and lower impact of human activities in those regions. This spatial distribution has much to do with China's economic development strategy. Therefore, water treatment facilities should keep up with the economic development and minimize the human impacts on water quality. Since TN concentrations exceed the Class V standard (2 mg/L) for most measuring stations across the country, local management agencies should pay more attention to the attenuation of nutrient transport to rivers.

We note that a few water quality parameters showed lower concentrations in 2022 and 2023 compared to 2021, indicating that the water quality has improved compared to the results of the previous water quality survey [43]. However, urbanization is still the main reason for pollution emissions and will continue to exert pressure on water quality [2,59]. Even in watersheds where water quality has significantly improved, the issue of excessively high pollutant concentrations still occurs frequently [8]. Therefore, further mitigation of water pollution will still be a major challenge in the coming decades.

This investigation provides an overview of China's water quality and lays the foundation for future investigations. Further investigations on the spatiotemporal analysis of water quality and underlying mechanisms in each water resource area can be performed based on this study. At the same time, in addition to analyzing each water quality parameter, comprehensive evaluation methods and the water quality index can be used, as well as regional economic development, to make a comprehensive evaluation of water quality conditions [22]. In addition, this study can also provide a reference for the analysis of water quality with process-based water quality models.

Meanwhile, a few limitations should be noticed to better understand the results of this work. First, because of the low data availability, we primarily focused on water quality data in 2021–2023, which is a short time period for understanding the long-term trends in water quality. More efforts in data compilation are needed to extend the data to early years to show how water quality has changed over a long time period. Second, more quantitative analyses of how river basin properties have contributed to the variability of water quality are needed to derive a solid understanding of mechanisms regulating water quality. Third, linking water quantity and water quality will be needed to understand the control of the water cycle on water quality.

5. Conclusions

This study investigated the national-scale monthly water quality observation data from 2021 to 2023 and analyzed the spatiotemporal variability of key water quality parameters in China. Similarities can be found through the spatial distribution of BOD5, COD, DO, NH3-N, TN, and TP concentrations, with a lower magnitude in the south than in the north. The concentrations of heavy metal elements can mostly meet the Class I water quality standards. Nitrogen pollution should receive more attention, since it is becoming a national problem, especially during the dry season. When comparing water conditions in 2021 to the next two years, all parameters showed better conditions, except for DO and TN. The TN/TP ratios of all water resource regions were above 16, indicating that phosphorus was the limiting factor of eutrophication. The findings of this study can provide valuable information to water quality management agencies and support the formulation of water pollution control strategies. Reducing nitrogen export, particularly in dry seasons and in the northern parts of China, will be needed to improve water quality in China.

Author Contributions: Conceptualization, K.L. and Q.Y.; methodology, K.L. and Q.Y.; formal analysis, K.L.; investigation, K.L. and Q.Y.; resources, Q.Y.; data curation, Q.Y.; writing—original draft preparation, K.L.; writing—review and editing, K.L., Q.Y. and X.L; visualization, K.L.; supervision, Q.Y.; project administration, Q.Y. and X.L.; funding acquisition, Q.Y. and X.L. All authors have read and agreed to the published version of the manuscript.

Funding: This work was supported by the Hongkong-Macau Center of Ocean Research (CORE) 2023 program (CORE is a joint research center for ocean research between Laoshan Laboratory and HKUST), the Guangzhou Technology Bureau and the Hong Kong University of Science and Technology (Guangzhou) 2023 joint program (2024A03J0611), and the Chinese Academy of Science Earth System simulator program (elpt_2023_000430). The work described in this paper was substantially supported by a grant from the Research Grants Council of the Hong Kong Special Administrative Region, China (Project Reference Number: AoE/P-601/23-N). Xia Li was supported by the GuangDong Basic and Applied Basic Research Foundation (Grant Numbers: 2021A1515110830 and the National Natural Science Foundation of China (42206170).

Data Availability Statement: Data will be available upon request.

Conflicts of Interest: The authors declare no conflicts of interest.

References

1. Vörösmarty, C.J.; Sahagian, D. Anthropogenic Disturbance of the Terrestrial Water Cycle. *Bioscience* **2000**, *50*, 753–765. [CrossRef]
2. Ma, T.; Zhao, N.; Ni, Y.; Yi, J.; Wilson, J.P.; He, L.; Du, Y.; Pei, T.; Zhou, C.; Song, C.; et al. China's Improving Inland Surface Water Quality since 2003. *Sci. Adv.* **2020**, *6*, eaau3798. [CrossRef] [PubMed]
3. Pesce, S.F.; Wunderlin, D.A. Use of Water Quality Indices to Verify the Impact of Córdoba City (Argentina) on Suquía River. *Water Res.* **2000**, *34*, 2915–2926. [CrossRef]
4. Moore, K.; Jennings, E.; Allott, N.; May, L.; Järvinen, M.; Arvola, L.; Tamm, T.; Järvet, A.; Nõges, T.; Pierson, D.; et al. Modelling the Effects of Climate Change on the Supply of Inorganic Nitrogen. In *The Impact of Climate Change on European Lakes*; Springer: Dordrecht, The Netherlands, 2010. [CrossRef]
5. Niu, L.; van Gelder, P.; Luo, X.; Cai, H.; Zhang, T.; Yang, Q. Implications of Nutrient Enrichment and Related Environmental Impacts in the Pearl River Estuary, China: Characterizing the Seasonal Influence of Riverine Input. *Water* **2020**, *12*, 3245. [CrossRef]
6. Carpenter, S.R.; Caraco, N.F.; Correll, D.L.; Howarth, R.W.; Sharpley, A.N.; Smith, V.H. Nonpoint Pollution of Surface Waters with Phosphorus and Nitrogen. *Ecol. Appl.* **1998**, *8*, 559–568. [CrossRef]
7. Paerl, H.W.; Gardner, W.S.; McCarthy, M.J.; Peierls, B.L.; Wilhelm, S.W. Algal Blooms: Noteworthy Nitrogen. *Science* **2014**, *346*, 175. [CrossRef]
8. Qin, B.; Paerl, H.W.; Brookes, J.D.; Liu, J.; Jeppesen, E.; Zhu, G.; Zhang, Y.; Xu, H.; Shi, K.; Deng, J. Why Lake Taihu Continues to Be Plagued with Cyanobacterial Blooms through 10 years (2007–2017) Efforts. *Sci. Bull.* **2019**, *64*, 354–356. [CrossRef]
9. Harada, M. Minamata Disease: Methylmercury Poisoning in Japan Caused by Environmental Pollution. *Crit. Rev. Toxicol.* **1995**, *25*, 1–24. [CrossRef]
10. Kasuya, M.; Teranishi, H.; Aoshima, K.; Katoh, T.; Horiguchi, H.; Morikawa, Y.; Nishijo, M.; Iwata, K. Water Pollution by Cadmium and the Onset of Itai-Itai Disease. *Water Sci. Technol.* **1992**, *25*, 149–156. [CrossRef]
11. Zhang, Z.; Wang, J.J.; Ali, A.; DeLaune, R.D. Heavy Metal Distribution and Water Quality Characterization of Water Bodies in Louisiana's Lake Pontchartrain Basin, USA. *Environ. Monit. Assess.* **2016**, *188*, 628. [CrossRef]
12. Qiaoqiao, Z.H.O.U.; Bo, R.E.N.; Youzhi, L.I.; Yandong, N.I.U.; Xiaohui, D.I.N.G.; Xin, Y.A.O.; Hualin, B.I.A.N. Trends and Sources of Dissolved Heavy Metal Pollution in Water of Rivers and Lakes in China. *Environ. Chem.* **2020**, *8*, 2044–2054. [CrossRef]
13. Duan, H.; Loiselle, S.A.; Zhu, L.; Feng, L.; Zhang, Y.; Ma, R. Distribution and Incidence of Algal Blooms in Lake Taihu. *Aquat. Sci.* **2015**, *77*, 9–16. [CrossRef]
14. Zhang, W.; Chen, D.; Li, H. Spatio-Temporal Dynamics of Water Quality and Their Linkages with the Watershed Landscape in Highly Disturbed Headwater Watersheds in China. *Environ. Sci. Pollut. Res.* **2018**, *25*, 35287–35300. [CrossRef] [PubMed]
15. Deng, C.; Liu, L.; Li, H.; Peng, D.; Wu, Y.; Xia, H.; Zhang, Z.; Zhu, Q. A Data-Driven Framework for Spatiotemporal Characteristics, Complexity Dynamics, and Environmental Risk Evaluation of River Water Quality. *Sci. Total Environ.* **2021**, *785*, 147134. [CrossRef]
16. Paerl, H.W. Assessing and Managing Nutrient-Enhanced Eutrophication in Estuarine and Coastal Waters: Interactive Effects of Human and Climatic Perturbations. *Ecol. Eng.* **2006**, *26*, 40–54. [CrossRef]
17. Zhang, H.; Cao, X.; Huo, S.; Ma, C.; Li, W.; Liu, Y.; Tong, Y.; Wu, F. Changes in China's River Water Quality since 1980: Management Implications from Sustainable Development. *NPJ Clean Water* **2023**, *6*, 45. [CrossRef]
18. Fan, X.; Cui, B.; Zhao, H.; Zhang, Z.; Zhang, H. Assessment of River Water Quality in Pearl River Delta Using Multivariate Statistical Techniques. *Procedia Environ. Sci.* **2010**, *2*, 1220–1234. [CrossRef]
19. Wunderlin, D.A.; María Del Pilar, D.; María Valeria, A.; Fabiana, P.S.; Cecilia, H.A.; de Los Ángeles, B.M. Pattern Recognition Techniques for the Evaluation of Spatial and Temporal Variations in Water Quality. A Case Study:: Suquía River Basin (Córdoba–Argentina). *Water Res.* **2001**, *35*, 2881–2894. [CrossRef]
20. Li, J.; Xia, C. Drivers of Spatial and Temporal Dynamics in Water Turbidity of China Yangtze River Basin. *Water* **2023**, *15*, 1264. [CrossRef]
21. Xia, J.; Zhang, Y.Y.; Zhan, C.; Ye, A. Water Quality Management in China: The Case of the Huai River Basin. *Int. J. Water Resour. Dev.* **2011**, *27*, 167–180. [CrossRef]
22. Kisi, O.; Ay, M. Comparison of Mann–Kendall and Innovative Trend Method for Water Quality Parameters of the Kizilirmak River, Turkey. *J. Hydrol.* **2014**, *513*, 362–375. [CrossRef]
23. Duan, W.; He, B.; Chen, Y.; Zou, S.; Wang, Y.; Nover, D.; Chen, W.; Yang, G. Identification of Long-Term Trends and Seasonality in High-Frequency Water Quality Data from the Yangtze River Basin, China. *PLoS ONE* **2018**, *13*, e0188889. [CrossRef]
24. Huo, S.; Ma, C.; Li, W.; He, Z.; Zhang, H.; Yu, L.; Liu, Y.; Cao, X.; Wu, F. Spatiotemporal Differences in Riverine Nitrogen and Phosphorus Fluxes and Associated Drivers across China from 1980 to 2018. *Chemosphere* **2023**, *310*, 136827. [CrossRef] [PubMed]
25. Ghosh, H.; Tusher, M.A.; Rahat, I.S.; Khasim, S.; Mohanty, S.N. Water Quality Assessment Through Predictive Machine Learning. In *Lecture Notes in Networks and Systems*; Springer: Singapore, 2023; Volume 699 LNNS. [CrossRef]
26. Dong, L.; Gong, C.; Huai, H.; Wu, E.; Lu, Z.; Hu, Y.; Li, L.; Yang, Z. Retrieval of Water Quality Parameters in Dianshan Lake Based on Sentinel-2 MSI Imagery and Machine Learning: Algorithm Evaluation and Spatiotemporal Change Research. *Remote Sens.* **2023**, *15*, 5001. [CrossRef]
27. Liu, X.; Li, G.; Liu, Z.; Guo, W.; Gao, N. Water Pollution Characteristics and Assessment of Lower Reaches in Haihe River Basin. *Procedia Environ. Sci.* **2010**, *2*, 199–206. [CrossRef]
28. Wang, E.; Li, Q.; Hu, H.; Peng, F.; Zhang, P.; Li, J. Spatial Characteristics and Influencing Factors of River Pollution in China. *Water Environ. Res.* **2019**, *91*, 351–363. [CrossRef]

29. Bu, H.; Meng, W.; Zhang, Y. Spatial and Seasonal Characteristics of River Water Chemistry in the Taizi River in Northeast China. *Environ. Monit. Assess.* **2014**, *186*, 3619–3632. [CrossRef]
30. Dou, M.; Zhang, Y.; Li, G. Temporal and Spatial Characteristics of the Water Pollutant Concentration in Huaihe River Basin from 2003 to 2012, China. *Env. Monit Assess* **2016**, *188*, 522. [CrossRef]
31. Li, K.; Wang, L.; Li, Z.; Xie, Y.; Wang, X.; Fang, Q. Exploring the Spatial-Seasonal Dynamics of Water Quality, Submerged Aquatic Plants and Their Influencing Factors in Different Areas of a Lake. *Water* **2017**, *9*, 707. [CrossRef]
32. Wu, J.; Cheng, S.P.; He, L.Y.; Wang, Y.C.; Yue, Y.; Zeng, H.; Xu, N. Assessing Water Quality in the Pearl River for the Last Decade Based on Clustering: Characteristic, Evolution and Policy Implications. *Water Res.* **2023**, *244*, 120492. [CrossRef]
33. Sun, J.; Chen, Y.; Zhang, Z.; Wang, P.; Song, X.; Wei, X.; Feng, B. The Spatio-Temporal Variations of Surface Water Quality in China during the "Eleventh Five-Year Plan". *Environ. Monit. Assess.* **2015**, *187*, 64. [CrossRef] [PubMed]
34. Li, S.; Gu, S.; Liu, W.; Han, H.; Zhang, Q. Water Quality in Relation to Land Use and Land Cover in the Upper Han River Basin, China. *Catena* **2008**, *75*, 216–222. [CrossRef]
35. Zhou, Y.; Ma, J.; Zhang, Y.; Qin, B.; Jeppesen, E.; Shi, K.; Brookes, J.D.; Spencer, R.G.M.; Zhu, G.; Gao, G. Improving Water Quality in China: Environmental Investment Pays Dividends. *Water Res.* **2017**, *118*, 152–159. [CrossRef]
36. Stets, E.G.; Sprague, L.A.; Oelsner, G.P.; Johnson, H.M.; Murphy, J.C.; Ryberg, K.; Vecchia, A.V.; Zuellig, R.E.; Falcone, J.A.; Riskin, M.L. Landscape Drivers of Dynamic Change in Water Quality of U.S. Rivers. *Environ. Sci. Technol.* **2020**, *54*, 4336–4343. [CrossRef] [PubMed]
37. Risal, A.; Parajuli, P.B.; Dash, P.; Ouyang, Y.; Linhoss, A. Sensitivity of Hydrology and Water Quality to Variation in Land Use and Land Cover Data. *Agric. Water Manag.* **2020**, *241*, 106366. [CrossRef]
38. Poonam, T.; Tanushree, B.; Sukalyan, C. Water Quality Indices-Important Tools for Water Quality Assessment: A Review. *Int. J. Adv. Chem.* **2013**, *1*, 15–28.
39. Sutadian, A.D.; Muttil, N.; Yilmaz, A.G.; Perera, B.J.C. Development of River Water Quality Indices—A Review. *Environ. Monit. Assess.* **2016**, *188*, 58. [CrossRef]
40. Tunc Dede, O.; Telci, I.T.; Aral, M.M. The Use of Water Quality Index Models for the Evaluation of Surface Water Quality: A Case Study for Kirmir Basin, Ankara, Turkey. *Water Qual. Expo. Health* **2013**, *5*, 41–56. [CrossRef]
41. Huang, J.; Zhang, Y.; Bing, H.; Peng, J.; Dong, F.; Gao, J.; Arhonditsis, G.B. Characterizing the River Water Quality in China: Recent Progress and on-Going Challenges. *Water Res.* **2021**, *201*, 117309. [CrossRef]
42. Hu, G.; Bakhtavar, E.; Hewage, K.; Mohseni, M.; Sadiq, R. Heavy Metals Risk Assessment in Drinking Water: An Integrated Probabilistic-Fuzzy Approach. *J. Environ. Manag.* **2019**, *250*, 109514. [CrossRef]
43. Pearson, K. Note on Regression and Inheritance in the Case of Two Parents. *Proc. R. Soc. Lond.* **1895**, *58*, 240–242. [CrossRef]
44. Hogg, R.V.; McKean, J.W.; Craig, A.T. *Introduction to Mathematical Statistics*, 7th ed.; Pearson: Boston, MA, USA, 2013.
45. Li, J.; Meng, X.; Zhang, Y.; Li, J.; Xia, L.; Zheng, H. Analysis of the Temporal and Spatial Distribution of Water Quality in China's Major River Basins, and Trends between 2005 and 2010. *Front. Earth Sci.* **2015**, *9*, 463–472. [CrossRef]
46. Xia, X.; Zhang, S.; Li, S.; Zhang, L.; Wang, G.; Zhang, L.; Wang, J.; Li, Z. The Cycle of Nitrogen in River Systems: Sources, Transformation, and Flux. *Environ.Sci. Process. Impacts* **2018**, *20*, 863–891. [CrossRef] [PubMed]
47. Piao, S.; Ciais, P.; Huang, Y.; Shen, Z.; Peng, S.; Li, J.; Zhou, L.; Liu, H.; Ma, Y.; Ding, Y. The Impacts of Climate Change on Water Resources and Agriculture in China. *Nature* **2010**, *467*, 43–51. [CrossRef] [PubMed]
48. Xu, L.; Guan, J.; Ba, Y.; Chen, W.; Huang, J.; Cheng, Y.; Zhang, Y.; Qu, Q.; Zhao, M. Spatial Distribution Pattern and Driving Mechanism of Heavy Metal Elements in Soils of Middle-Alpine Hilly Region, Yunnan Province. *Geol. China* **2024**, *51*, 304–326. [CrossRef]
49. Foley, J.A.; DeFries, R.; Asner, G.P.; Barford, C.; Bonan, G.; Carpenter, S.R.; Stuart Chapin, F.; Coe, M.T.; Daily, G.C.; Gibbs, H.K.; et al. Global Consequences of Land Use. Global consequences of land use. *Science* **2005**, *309*, 570–574. [CrossRef] [PubMed]
50. Wang, J.; Fu, Z.; Qiao, H.; Liu, F. Assessment of Eutrophication and Water Quality in the Estuarine Area of Lake Wuli, Lake Taihu, China. *Sci. Total Environ.* **2019**, *650*, 1392–1402. [CrossRef]
51. Yang, S.L.; Milliman, J.D.; Xu, K.H.; Deng, B.; Zhang, X.Y.; Luo, X.X. Downstream Sedimentary and Geomorphic Impacts of the Three Gorges Dam on the Yangtze River. *Earth Sci. Rev.* **2014**, *138*, 469–486. [CrossRef]
52. Wang, L.; Jiao, Y.; Ming, Q.; He, L.; Zhou, H. Evaluation of Heavy Metal Pollution in Bijiang Basin in Yunnan Province. *Res. Environ. Sci.* **2009**, *22*, 595–600.
53. Yang, C.; Yang, P.; Geng, J.; Yin, H.; Chen, K. Sediment Internal Nutrient Loading in the Most Polluted Area of a Shallow Eutrophic Lake (Lake Chaohu, China) and Its Contribution to Lake Eutrophication. *Environ. Pollut.* **2020**, *262*, 114292. [CrossRef]
54. Hu, Y.; Cheng, H. Water Pollution during China's Industrial Transition. *Environ. Dev.* **2013**, *8*, 57–73. [CrossRef]
55. Lou, H.Z.; Wu, X.J.; Hao, F.H.; Yang, S.; Zhang, X. Research Progress of Non-Point Source Pollution in China over the Past 30 Years and Discussion of the Future Direction Development. *Acta Sci. Circumstantiae* **2020**, *40*, 1535–1949.
56. He, X.; Li, P. Surface Water Pollution in the Middle Chinese Loess Plateau with Special Focus on Hexavalent Chromium (Cr^{6+}): Occurrence, Sources and Health Risks. *Expo. Health* **2020**, *12*, 385–401. [CrossRef]
57. Hu, Y.; Zhang, K.; Li, Y.; Sun, Y.; Li, H.; Yang, G. Human Activities Increase the Nitrogen in Surface Water on the Eastern Loess Plateau. *Geofluids* **2021**, *2021*, 9957731. [CrossRef]
58. Ringler, C.; Cai, X.; Wang, J.; Ahmed, A.; Xue, Y.; Xu, Z.; Yang, E.; Jianshi, Z.; Zhu, T.; Cheng, L.; et al. Yellow River Basin: Living with Scarcity. *Water Int.* **2010**, *35*, 681–701. [CrossRef]

59. Li, Q.; Zhou, J.; Chen, B.; Huang, B.; Zeng, X.; Zhan, J.; Pan, X. Toxic Metal Contamination and Distribution in Soils and Plants of a Typical Metallurgical Industrial Area in Southwest of China. *Environ. Earth Sci.* **2014**, *72*, 2101–2109. [CrossRef]
60. Liu, W.; Li, S.; Bu, H.; Zhang, Q.; Liu, G. Eutrophication in the Yunnan Plateau Lakes: The Influence of Lake Morphology, Watershed Land Use, and Socioeconomic Factors. *Environ. Sci. Pollut. Res.* **2012**, *19*, 858–870. [CrossRef]
61. Pehme, K.M.; Burlakovs, J.; Kriipsalu, M.; Pilecka, J.; Grinfelde, I.; Tamm, T.; Yahya, J.; Hogland, W. Urban hydrology research fundamentals for waste management practices. *Res. Rural Dev.* **2019**, *1*, 160–167. [CrossRef]

Disclaimer/Publisher's Note: The statements, opinions and data contained in all publications are solely those of the individual author(s) and contributor(s) and not of MDPI and/or the editor(s). MDPI and/or the editor(s) disclaim responsibility for any injury to people or property resulting from any ideas, methods, instructions or products referred to in the content.

hydrology

MDPI

Article

Hydro Geochemical Characteristics and Mineralization Process of Groundwater in the Phosphatic Basin of Gafsa, Southwestern Tunisia

Nada Nasri [1,2,3,*], Fouad Souissi [1,2], Takoua Ben Attia [4], Amina Ismailia [1,2], Olfa Smida [1,2], Dhouha Tangour [3], Eduardo Alberto López Maldonado [5,*] and Radhia Souissi [1,2]

1 Department of Geology, Faculty of Sciences of Tunis, University of Tunis El Manar, Tunis 2092, Tunisia; foued.souissi@fst.utm.tn (F.S.); amina.ismailia@etudiant-fst.utm.tn (A.I.); smidaolfa@gmail.com (O.S.)

2 Laboratoire des Matériaux Utiles, Institut National de Recherche et d'Analyse Physico-Chimique (INRAP), Technopark of Sidi Thabet, Ariana 2020, Tunisia

3 Laboratoire de Chimie Organique, Centre International des Technologies de l'Environnement de Tunis (CITET), Boulevard de l'Environnement, La Charguia, Tunis 1080, Tunisia; unite.dev@citet.nat.tn

4 Laboratory of Research "LR 23/ES/10": Population Pathology, Environmental Aggressors, Alternative Therapy, Faculty of Medicine of Tunis, University of Tunis El Manar, Tunis 2092, Tunisia; takoua.benattia@fst.utm.tn

5 Faculty of Chemical Sciences and Engineering, Autonomous University of Baja California, Tijuana 22424, Mexico

* Correspondence: nada.nasri66@gmail.com (N.N.); elopez92@uabc.edu.mx (E.A.L.M.)

Abstract: The present study examines the water quality in the Quaternary Mio-Plio-Quaternary aquifer of the mining basin of Gafsa using a hydrochemical approach and multivariate statistical methods, to assess groundwater mineralization processes. Results from the analysis of groundwater quality collected during the winter (January 2020) and summer (June 2021) seasons reveal a pronounced stability in geochemical parameters, emphasizing a noteworthy consistency in water composition between the two seasons, with the dominance of the Na-Ca-Mg-SO4-Cl facies, in addition to the fact that all year round these concentrations are beyond their respective WHO limits. Despite the intensive extractive and transformation phosphate industry, the prolonged interaction of water with geological formations is the primary factor controlling their high mineralization. This results from the dissolution of carbonates (calcite, dolomite), gypsum, and halite. The results of the PCA represent two correlation classes. Class 1 comprises major elements sulfate, chloride, sodium, magnesium, and calcium strongly correlated with electrical conductivity (EC) and total dissolved solids (TDS). This correlation is indicative of the water mineralization process. Class 2 includes major elements nitrate and potassium weakly correlated with (TDS) and (EC) As regards heavy metals, their concentrations fall consistently below their respective potability standards established by the WHO across all water sampling points. Meanwhile, fluoride (F-) concentrations exhibited values ranging from (1.6 mg·L^{-1} to 2.9 mg·L^{-1}) in the winter of January 2020 and (1 to 2.9 mg·L^{-1}) in the summer of June 2021, surpassing its WHO limit (1.5 mg·L^{-1}) in almost all water samples. These findings allow us to conclude that the high mineralization of these waters is acquired due to the dissolution of carbonates (calcite, dolomite), gypsum, and halite due to their prolonged interaction with the geological formations. The deterioration of groundwater quality in the Gafsa mining basin associated with phosphate extraction and processing activities appears to be primarily due to the intensive exploitation of deep-water resources.

Keywords: phosphate basin; groundwater; hydrochemical facies; PCA analysis; water quality

Citation: Nasri, N.; Souissi, F.; Ben Attia, T.; Ismailia, A.; Smida, O.; Tangour, D.; Lopez Maldonado, E.A.; Souissi, R. Hydro Geochemical Characteristics and Mineralization Process of Groundwater in the Phosphatic Basin of Gafsa, Southwestern Tunisia. *Hydrology* **2024**, *11*, 142. https://doi.org/10.3390/hydrology11090142

Academic Editor: Meysam Vadiati

Received: 4 August 2024
Revised: 30 August 2024
Accepted: 4 September 2024
Published: 6 September 2024

Copyright: © 2024 by the authors. Licensee MDPI, Basel, Switzerland. This article is an open access article distributed under the terms and conditions of the Creative Commons Attribution (CC BY) license (https://creativecommons.org/licenses/by/4.0/).

1. Introduction

Groundwater is the second largest freshwater resource on the planet and meets more than one-third of global drinking water demand [1]. However, aquifers are at risk of contamination owing to extensive pumping and agricultural and industrial activities. Intensive pesticide and fertilizer applications, wastewater discharge, industrial effluences, and excessive groundwater abstraction are some of the activities that lead to groundwater contamination. These activities have resulted in the deterioration of water resources in various regions worldwide [2]. In most arid and semiarid regions worldwide, the availability of sufficient freshwater has become a limiting factor for development [3,4]. Water scarcity has always been a dominant problem in North Africa and the Middle East. It has been accentuated by rain shortages that have caused a decrease in groundwater resources [5,6]. Overexploitation and changes in land usage have provoked not only the reduction of available water but also the deterioration of water quality [7–9]. The degree and severity of soil salinization escalated rapidly during the latter part of the 20th century [8]. The Sahara oases of Northern Africa are no exception [8–10]. Generally, groundwater chemistry is primarily controlled by natural processes such as aquifer lithology, residence time, and water rock interactions in unsaturated and saturated zones [1,9]. However, anthropogenic processes, such as agricultural and industrial activities and urban development, have several harmful consequences, such as the overexploitation of groundwater resources [10], the return flow of irrigation water [11], and saltwater intrusion [8,12,13], which may contribute to the degradation of water quality. Many hydrogeological approaches have been used to understand these interactions and predict groundwater dynamics [3,13–15]. On the other hand, fluoride is found ubiquitously in natural waters, with varying concentrations across different sources. In groundwater, fluoride concentrations can vary widely depending on the geological characteristics of the rocks [16]. Flouride-rich waters are encountered in three major geological terrains: sedimentary basins, crystalline basement areas, and volcanic regions [16]. In groundwater, the fluoride content depends on various factors, including the porosity of rocks and soils through which water circulates, speed of circulation, temperature during the interaction between water and rocks, composition of salts encountered by water before reaching aquifers, and chemical capacity of water to solubilize fluoride [8,17]. Mining operations generally significantly impact the amount and quality of water resources in a neighborhood. During exploitation, dewatering leads to the drawdown of the water table, which results in the disappearance of specific springs or a decrease in their flow. After mine closure, rebound degrades groundwater quality [4,8,18]. Numerous studies have examined the groundwater chemistry in the Gafsa mining basin in Southwest Tunisia [19,20]. All these studies conducted on the hydrochemistry of water are concentrated mainly at specific sites in the basin. This means that these studies have yet to focus on the hydrochemistry of water in the entire basin, especially around industrial sites. However, this study focuses on water hydrochemistry in mining as the whole basin, especially those close to industrial sites (Moulares, Redeyef, M'etlaoui, and M'dhilla).

Chemical analyses conducted as part of the Gafsa North aquifer system investigation by [21] indicated that the Mio-plio-Quaternary waters are of the sodium sulfate type. The salinization of these waters is primarily attributed to the dissolution of gypsum and, to a lesser extent, halite. The geological condition of the region can explain this. According to [22], the influence of geological features, notably the Gafsa Fault, has emerged as a pivotal factor affecting the quality of groundwater, specifically its salinity levels in the vicinity.

The results from [23] study in the Northern Gafsa Basin, specifically in Gafsa city, revealed concentrations of Ca^{2+}, Mg^{2+}, Na^{+}, K^{+}, Cl^{-}, $SO_4{}^{2-}$, NO^{3-}, and HCO^{3-} that surpassed the standards set by the (WHO). The applied analysis of significant elements has substantiated an anthropogenic origin, predominantly linked to agricultural activities, for these major elements. These elements contributed significantly to the pronounced mineralization observed in the study area.

A study conducted by [24] in the same area revealed that groundwater in Northern Gafsa exceeds the standards set by the (WHO) for sulfate, calcium, and chloride concentrations, with dominant calcic sulfated facies. These results indicate that the chemical quality of water is primarily influenced by the geological nature of the aquifer, with these elements playing a significant role in the observed high salinity levels.

None of the potassium and nitrate values from either campaign exceeded the WHO standards, a finding consistent with the observations of [19,25,26]. This suggests that these elements may not significantly contribute to winter mineralization in the mining basin.

Refs. [27,28], investigated the adverse impacts of a phosphate industrial zone on groundwater in the Jorf Lasfar region of Morocco. Their study revealed a significant increase in the measured electrical conductivity of groundwater. The calcium and potassium concentrations met the compatibility standards set by the WHO. However, the chloride, sodium, nitrate, and magnesium concentrations exceeded the water quality standards. The increased conductivity was primarily attributed to chloride and sodium ions.

The study conducted by [29] in the phosphate region of central Jordan, in the Al Hisa province, revealed a significant concentration of chloride, magnesium, and sulfide, surpassing the standards set by the WHO. This investigation highlights the saline nature of the waters in the Al-Hisa zone, with water salinity attributed to the dissolution of soluble elements in the effluent waters. In addition, the concentrations of potassium and nitrate in the same phosphate region were below the standards established by the WHO. These observations are consistent with the findings of the present study.

This study aimed to identify and interpret the factors responsible for mineralization and the degradation of water quality in the mining basin of Gafsa. Therefore, it is essential to assess the processes and phenomena controlling the reactions of aquifer systems that may be exposed to environmental or climate changes. Hydrochemical techniques (Piper, Schoeller–Berkalov representations) were used to determine the dominant facies of groundwater in the study area, better understand the relationship between these elements, and specify the factors affecting mineralization. Statistical calculations (NPCA) using Python language were performed to better understand the link between the quantitative variability of cations and anions.

2. Materials and Methods

2.1. Presentation of the Study Area

2.1.1. Geology

The mining basin of Gafsa, situated in the southwest of Tunisia, encompasses most of the country's phosphate mining activities [8,30]. To its north lies the Gafsa Range, while to the east, it is bordered by the Guettar and Jebel Chamsi regions. To the south, it is delineated by Chott El Gharsa, and to the west lies Algeria [8,30]. Geologically, the outcrops of this region consist mainly of Quaternary to Mio-Pliocene formations, including the sands of the Beglia Formation and the sandy/silty clays and conglomerates of the Segui Formation in the regions of Redeyef and Metlaoui [8,30].

The Palaeogene (mainly Eocene) and Early/Late Cretaceous formations constitute the geological outcrops in the Metlaoui, Gafsa, and Sehib Ranges. According to [8,31,32] the geological units of the Metlaoui Group, which hosts the phosphate series of the Gafsa-Metlaoui basin, are attributed to the Eocene epoch. This primarily includes three formations: The Thelja Formation, characterized by dolomitic deposits alternating with gypsum and marl. The Chouabine Formation comprises four phosphate units (A, B, C, and D). The Kef-Eddour Formation consists of marl–phosphate alternations framed by two limestone bars.

The lithostratigraphic column (Figure 1), spanning the Cretaceous to the Quaternary, exhibits two primary stratigraphic sequences. The first sequence comprises marine Cretaceous and Paleogene deposits from formations including Sidi Khalif, Melloussi, Boudinar, Bouhedma, Sidi Aïch, Orbabta, Zebbeg, Aleg, Abiod, El Haria, Metlaoui, and Jeb's. In contrast, the second sequence consists of continental Neogene and Quaternary deposits from the Sehib, Beglia, and Segui formations. These two sequences are divided by a significant sedimentary hiatus extending from the Middle Eocene to the Early Miocene [33].

All sampling points are located within the Mio-Plio-Quaternary.

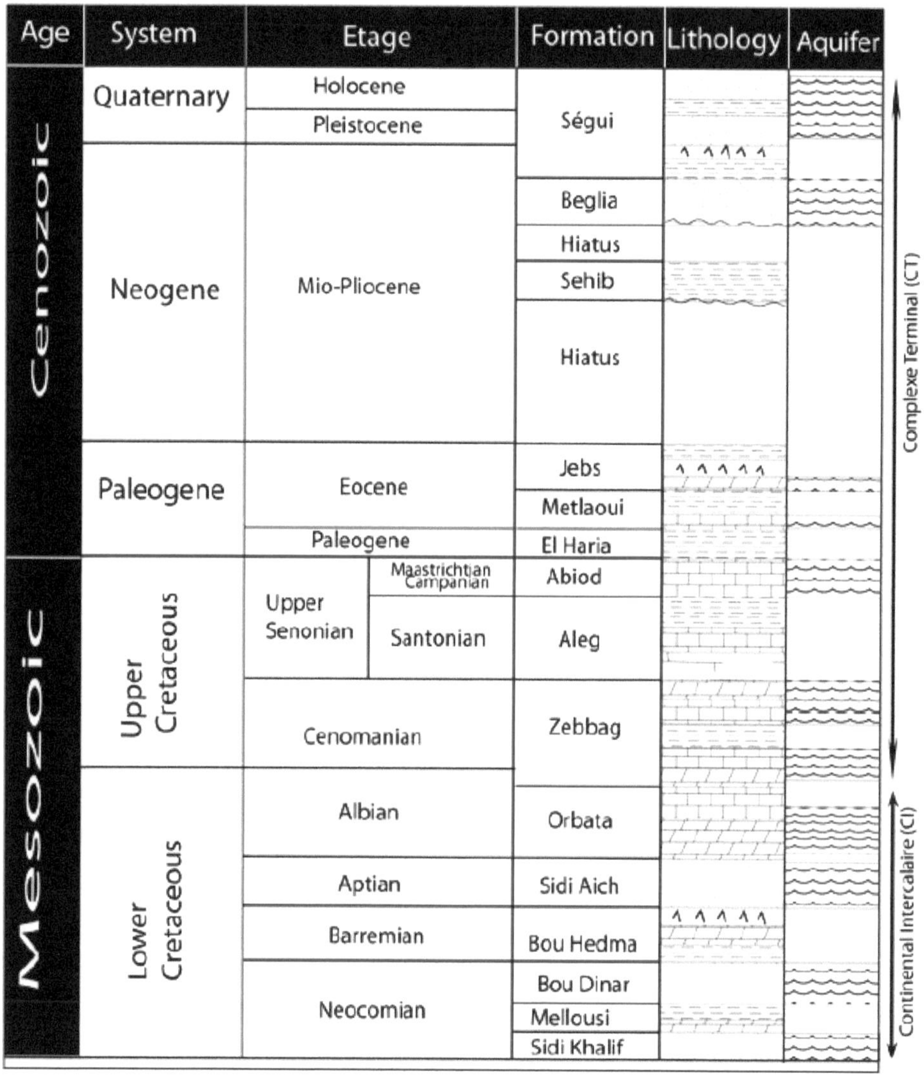

Figure 1. Lithostratigraphic column of the study area [8].

2.1.2. Phosphate Industry and Environment

The phosphate industry is considered one of Tunisia's most polluting (air, soil, freshwater, seawater, sediments, soil, etc.), activities [8,34]. The Phosphate Company of Gafsa (CPG) operates in eight open-pit mines (Kef Schfaier, Kef Eddour Central, Kef Eddour West, Metloui, Jallabia, M'dhilla, Redeyef and Moulares) and nine processing plants (laundries) (Figure 2). Most of the phosphate production (around 90%) is processed by the (CPG) and the Groupe Chimique Tunisien (GCT) in the Gafsa basin (M'dhilla) and other phosphoric acid-producing plants operating in the south of Tunisia (Gabes, Skhira, Sfax). These plants produce vast amounts of highly toxic effluents [35,36]; loaded with fluorides, radionuclides, metallic trace elements (MTE), sulfites, and phosphoric acid [36–38], these effluents are discharged into the hydrographic network (Wadi Bayech, EL Kbir, Wadi El Melah, Wadi Moulares, Wadi Magroun, and Wadi Tebedditt).

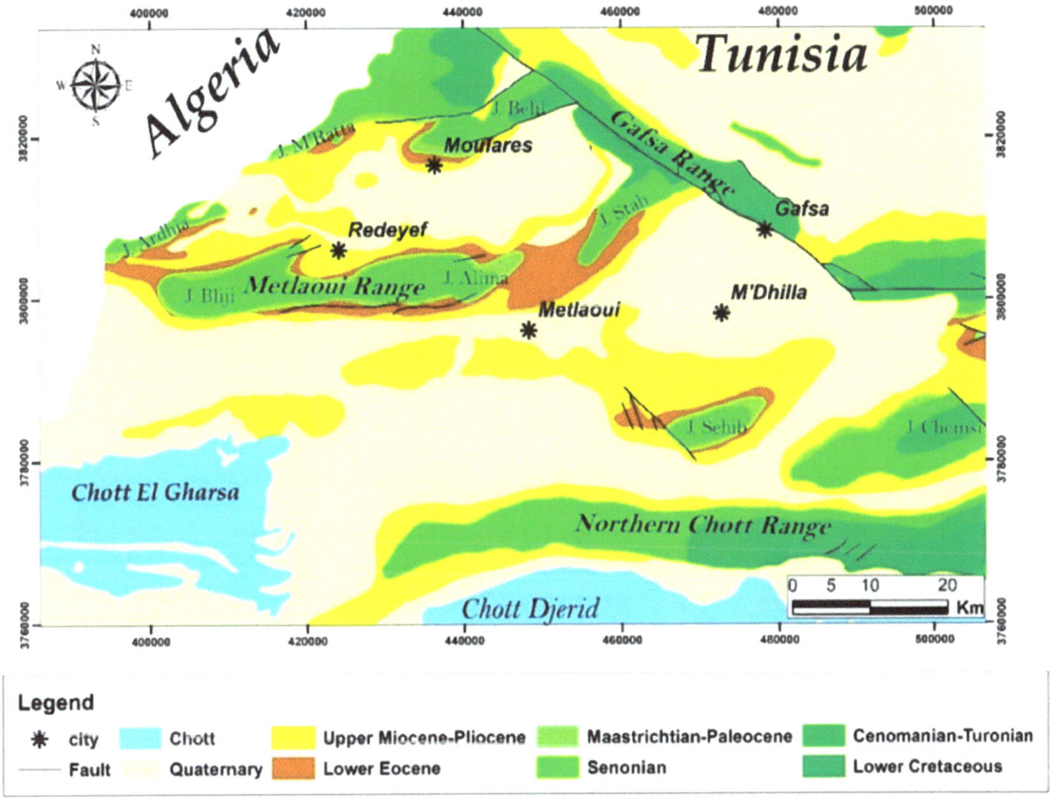

Figure 2. Geologic map of the Gafsa mining basin [19], showing the location of the phosphate laundries and the water sampling points.

2.1.3. Sampling and Methods

Groundwater samples were collected during wet (January 2020) and dry (June 2021) seasons from seven boreholes (80–250 m deep) exploiting the so-called Complexe Terminal (CT) Mio-Plio-Quaternary aquifer system (Figure 1). Borehole positions (34° 40–34° 20 N latitude and 8° 00–8° 40 E longitude), were located using a GPS.

Water samples were conditioned in polyethylene containers and kept in a cooler for 24 h. In the laboratory, samples were vacuum filtered through 0.45 µm Millipore filters MilliporeSigma, Darmstadt, Germany, a Merck company and stored at 4° C before being analyzed.

Temperature (T) and hydrogen potential (pH) were measured in situ using an ISO-SCAN PH meter. The electrical conductivity (EC) and total dissolved salt (TDS) were measured using a conductivity meter (WTW Model INOLAB Cond7310P).

Water analysis was performed at the laboratory of the Centre International des Technologie de l'Environnement (CITET), using Dionex chromatography (DX ICS-1100) for the analysis of anions (SO_4^{2-}, Cl^-, F^-, NO_3^-) and an inductively coupled plasma atomic emission spectrometer (ICP-AES; Optima 7300 DV) for the analysis of cations (Ca^{2+}, Mg^{2+}, K^+, Na^{2+}) and trace elements (F^-, Fe, Mn, Al, Zn, Cr, Pb, Cd). The total alkalinity (HCO_3^-) was determined by titration with 0.01 or 0.1 M HCl.

The accuracy of complete chemical analysis is also checked and ionically balanced by calculating the cation–anion equilibrium in terms of equivalents per milligram ($mg \cdot L^{-1}$). The ionic balance for all samples is within the $\pm 5\%$ interval.

Multivariate statistical techniques (PCA) and hydrochemical representations (Piper and Schoeller–Berkaloff) are used for results processing.

3. Results

Water samples collected from boreholes during both winters (January 2020) and summer (June 2021) seasons were subjected to the in situ measurements of physico-chemical parameters (EC, TDS, pH, T) as well as the determination of major constituents (anions and cations) and trace elements.

3.1. Physico-Chemical Parameters

The results presented in Table 1 reveal relatively consistent temperature values, ranging from 24 to 27 °C, during January 2020 and June 2021. pH levels exhibited variability, with values ranging between 7.17 (Borehole 6) and 7.76 (Borehole 7) in January 2020 and between 7.02 (Borehole 6) and 7.83 (Borehole 2) in June 2021. In terms of electrical conductivity (EC), values ranged from 3700 $\mu s \cdot cm^{-1}$ (Borehole 6) to 7100 $\mu s \cdot cm^{-1}$ (Borehole 7) in January 2020 and from 4000 $\mu s \cdot cm^{-1}$ (Borehole 6) to 7200 $\mu s \cdot cm^{-1}$ (Borehole 7) in June 2021. Notably, EC does not appear to undergo significant seasonal fluctuations.

Regarding total dissolved solids (TDS), measured values spanned from 3986 $mg \cdot L^{-1}$ (Borehole 6) to 6184 $mg \cdot L^{-1}$ (Borehole 2) in January, and from 4394 $mg \cdot L^{-1}$ (Borehole 5) to 6603 $mg \cdot L^{-1}$ (Borehole 2) in June. This range suggests a prevalent brackish quality in the water throughout the measured period.

3.2. Major Constituents (Cations and Anions)

Table 1 summarizes the major ionic species (cations and anions) concentrations present in the studied waters. The ionic balance (IB) values for all samples during the two campaigns were lower than 5%.

Na^+ content in the water varies between 700 $mg \cdot L^{-1}$ (Borehole 6) and 1400 $mg \cdot L^{-1}$ (Borehole 7) for January 2020 and between 780 $mg \cdot L^{-1}$ (Borehole 6) and 1250 $mg \cdot L^{-1}$ (Borehole 7) for June 2021. The maximum Na^+ concentration is observed in the Plio-Quaternary aquifer of the Métlaoui area (1420 $mg \cdot L^{-1}$ in Borehole Bh2).

Ca^{++} contents vary between 325 $mg \cdot L^{-1}$ (Boreholes 6 and 7) and 479 $mg \cdot L^{-1}$ (Borehole2) for January 2020 and between 310 $mg \cdot L^{-1}$ (Borehole 7) and 488 $mg \cdot L^{-1}$ (Borehole 2) for June 2021. Mg^{++} contents fluctuate between 174 $mg \cdot L^{-1}$ (Borehole 5) and 250 $mg \cdot L^{-1}$ (Borehole 2) for January 2020 between 200 $mg \cdot L^{-1}$ (Borehole 5) and 264 $mg \cdot L^{-1}$ (Borehole 2) for June 2021. K^+ content varies between 3.9 $mg \cdot L^{-1}$ (Borehole 3) and 10.54 $mg \cdot L^{-1}$ (Borehole 7) in January 2020 and between 4.4 $mg \cdot L^{-1}$ (Borehole 3) and 9.7 $mg \cdot L^{-1}$ (Borehole 7) in June 2021. $SO_4{}^{2-}$ contents vary between 1500 $mg \cdot L^{-1}$ (Borehole 6) and 1899 $mg \cdot L^{-1}$ (Borehole 1) for January 2020 and between 1648 $mg \cdot L^{-1}$ (Borehole 6) and 2100 $mg \cdot L^{-1}$ (Borehole 2) for June 2021. Cl^- concentrations vary between 946 $mg \cdot L^{-1}$ (Borehole 5) and 1899 $mg \cdot L^{-1}$ (Borehole 2) for January 2020 and between 1040 $mg \cdot L^{-1}$ (Borehole 3) and 1900 $mg \cdot L^{-1}$ (Borehole 2) for June 2021. $HCO_3{}^-$ contents range from 210 $mg \cdot L^{-1}$ (Borehole 6) to 438 $mg \cdot L^{-1}$ (Borehole 2) for January 2020 and between 210 $mg \cdot L^{-1}$ (Borehole 7) and 394 $mg \cdot L^{-1}$ (Borehole 2) for June 2021. $NO_3{}^-$ contents vary between 10.1 $mg \cdot L^{-1}$ (Borehole 1) and 60 $mg \cdot L^{-1}$ (Borehole 6) for January 2020 and between 14.4 $mg \cdot L^{-1}$ (Borehole 5) and 48.3 $mg \cdot L^{-1}$ (Borehole 6) for June 2021.

The results of Table 1 show that these waters have the following general pattern: $[Na^+] > [Ca^{2+}] > [Mg^{2+}] >> [K^+]$ and $[SO_4{}^{2-}] > [Cl^-] >> [HCO_3{}^-] > [NO_3{}^-]$. Compared to the WHO standards, these results show that the investigated waters are enriched in sodium (>200 $mg \cdot L^{-1}$), calcium (>200 $mg \cdot L^{-1}$), magnesium (>50 $mg \cdot L^{-1}$), chlorine (>200 $mg \cdot L^{-1}$), sulfate (>500 $mg \cdot L^{-1}$), and bicarbonate (>15 $mg \cdot L^{-1}$) with a deficit in potassium (<20 $mg \cdot L^{-1}$) and nitrate (<50 $mg \cdot L^{-1}$), except the slight excess in nitrate (60 $mg \cdot L^{-1}$) in the borehole 6 of Redeyef in January 2020.

Table 1. Physico-chemical parameters and geochemical characteristics of groundwater in the Gafsa mining basin.

Locality	Borehole Reference	X_UTM	Y_UTM	Depth (m)	T °C	pH	EC (µs·cm^{-1})	TDS (mg·L^{-1})	Na^+ (mg·L^{-1})	K^+ (mg·L^{-1})	Ca^{2+} (mg·L^{-1})	Mg^{2+} (mg·L^{-1})	Cl^- (mg·L^{-1})	SO_4^{2-} (mg·L^{-1})	HCO_3^- (mg·L^{-1})	NO_3^- (mg·L^{-1})	IB%
January 2020 (a)																	
Metlaoui	Bh1	451,522	3,783,603	186	27.1	7.73	6000	5315	910	4.68	413	300	1431	1899	299	41	−4
	Bh2	374,046	3,798,126	249	27.1	7.35	7000	6184.5	1260	8.85	497	250	1899	1860	436.7	21.6	+0
Moulares	Bh3	432,239	3,813,085	197	26.1	7.27	4000	4282.4	779	3.9	375	184.3	967.6	1664	270.3	13.25	+1
	Bh4	437,821	3,816,611	210	26.8	7.39	4000	4880.6	709	4.55	455	270	1220	1890	291	41	−3
Rdeyef	Bh5	426,959	3,808,056	80	25.7	7.41	3900	4076	740	8.7	355	174	946	1612	230	10.1	+0
	Bh6	440,695	3,807,689	150	26.8	7.17	3700	3936	700	4.43	325	220	990	1500	231.6	60	+1
M'dhilla	Bh7	479,595	3,797,186	175	27.5	7.76	7100	5890.5	1400	10.54	325	234	1840	1830	210.6	40.5	+1
June 2021 (b)																	
Metlaoui	Bh1	451,522	3,783,603	186	25.01	7.76	6400	4828.7	970	9.7	360	220	1320	1650	278.2	20.3	−3
	Bh2	374,046	3,798,126	249	24.6	7.83	6520	6603.2	1420	8.85	488	264	1900	2100	394	28.6	+3
Moulares	Bh3	432,239	3,813,085	197	28.1	7.45	5900	4663.5	880	4.4	390	202.9	1040	1833	299.4	18.6	+2
	Bh4	437,821	210,611	210	26.7	7.61	5400	5062.8	879	6.1	440	240	1250	1900	299	30.7	+1
Rdeyef	Bh5	426,959	3,808,056	80	26.2	7.68	4100	4393.9	800	8.9	381	200	1110	1630	259	14.4	+1
	Bh6	440,695	3,807,689	150	26.2	7.2	4000	4544.3	780	5.41	360	256	1200	1648	255.6	48.3	+0
M'dhilla	Bh7	479,595	3,797,186	175	27.5	7.76	7200	5390.4	1250	9.7	310	220	1650	1720	210.4	40.5	+2
WHO Guidelines (2011)		-	-	-	-	6.5–8.5	-	1000	200	20	200	50	250	500	-	50	-

WHO: World Health Organization, EC: electrical conductivity, TDS: total dissolved solid.

3.3. Trace Elements

The results presented in Table 2 enable a comparison of the concentrations of heavy metals and fluoride in two distinct sampling campaigns, January 2020 and June 2021, and thus, to evaluate any significant variation in water quality. Table 2 shows that there is no significant variation in the concentration levels of heavy metals (Fe, Mn, Al, Zn, Cr, Pb) between the two sampling campaigns of January 2020 and June 2021. For example, the levels of iron vary between 0.041 $\mu g \cdot L^{-1}$ and 0.099 $\mu g \cdot L^{-1}$ (BH6) in January 2020 and between 0.011 $\mu g \cdot L^{-1}$ (BH1) and 0.092 $\mu g \cdot L^{-1}$ (BH7) in June 2021. The levels of fluoride do not show significant variation between the two campaigns. Compared to WHO standards, the results from (Table 2) indicate that the examined waters are not enriched in heavy metals. However, fluorine contents exceed the WHO standards in almost all samples.

Table 2. Trace metals and fluoride contents in the groundwater of the Gafsa mining basin.

Area	Borehole Reference	F^- ($mg \cdot L^{-1}$)	Fe ($\mu g/L^{-1}$)	Mn ($\mu g/L^{-1}$)	Al ($\mu g/L^{-1}$)	Zn ($\mu g/L^{-1}$)	Cr ($\mu g/L^{-1}$)	Pb ($\mu g/L^{-1}$)
January 2020 (a)								
Metlaoui	BH1	2.193	0.018	0.035	0.106	0.02	<0.01	0.001
	BH2	1.724	0.061	0.029	0.083	<0.01	0.036	0.002
Moulares	BH3	1.802	0.091	0.011	0.102	0.015	0.056	0.003
	BH4	1.611	0.085	<0.01	0.111	<0.01	0.061	0.004
Rdeyef	BH5	2.247	0.041	0.036	0.099	0.013	<0.01	0.002
	BH6	2.931	0.099	0.032	0.091	0.021	<0.01	0.005
M'dhilla	BH7	2.944	0.072	<0.01	0.102	0.013	<0.01	0.005
June 2021 (b)								
Métlaoui	BH1	2.432	0.011	0.033	0.098	0.02	<0.01	0.001
	BH2	1.955	0.053	0.036	0.074	<0.01	0.029	0.004
Moulares	BH3	2.012	0.072	0.016	0.089	0.012	0.048	0.002
	BH4	1.876	0.067	<0.01	0.101	<0.01	0.059	0.003
Rdeyef	BH5	2.675	0.033	0.022	0.078	0.015	<0.01	0.001
	BH6	2.981	0.077	0.028	0.086	0.029	<0.01	0.004
M'dhilla	BH7	2.991	0.092	<0.01	0.095	0.011	<0.01	0.003
WHO Guidelines (2011)		1.5	<0.3	<0.5	0.2	<0.5	0.05	<0.05

WHO: World Health Organization.

4. Discussion

The pH levels observed in the analyzed waters during both the winter (January 2020) and summer seasons (June 2021) demonstrate significant consistency, highlighting the marked hydrochemical stability of the Gafsa mining basin and consistently align with the circum-neutral range of 7.1 to 7.8 sensu Bocoum (2004), which is in agreement with previous studies conducted on groundwaters in the same basin [25,26,39]. It is worth noting, however, that such pH values indicate the intervention of carbonates as buffering minerals. In such an environment, buffering provides a notably stable pH level [39].

Electrical conductivity (EC) exhibits significant variability, ranging from 3700 to 7200 $\mu S \cdot cm^{-1}$ during winter (Figure 3a) and from 4100 to 7200 $\mu S \cdot cm^{-1}$ during summer seasons (Figure 3b). However, no significant variation is observed between the two seasons. Total dissolved solids (TDS) do not exhibit noteworthy variations, ranging from 3986 to 6184.5 $mg \cdot L^{-1}$ (Figure 3a) during winter seasons and from 4393.9 $mg \cdot L^{-1}$ to 6603.2 $mg \cdot L^{-1}$ (Figure 3b) during summer seasons. Nonetheless, no notable difference was observed between the two seasons. Figure 3a,b reveals a high concentration of total dissolved solids (TDS) during both winter and summer seasons in the areas of Métlaoui (Borehole 2) and M'dhillla (Borehole 7). In contrast, a lower concentration of (TDS) is observed in the Moulares area (Boreholes 3 and 4) and the Redyef area (Boreholes 5 and 6). (Borehole 2) in the Métlaoui area and (Borehole 7) in the M'dhilla area are near the Gafsa Phosphate Company, which could potentially explain these elevated TDS values attributed to the overexploitation of the aquifer by the phosphate industry.

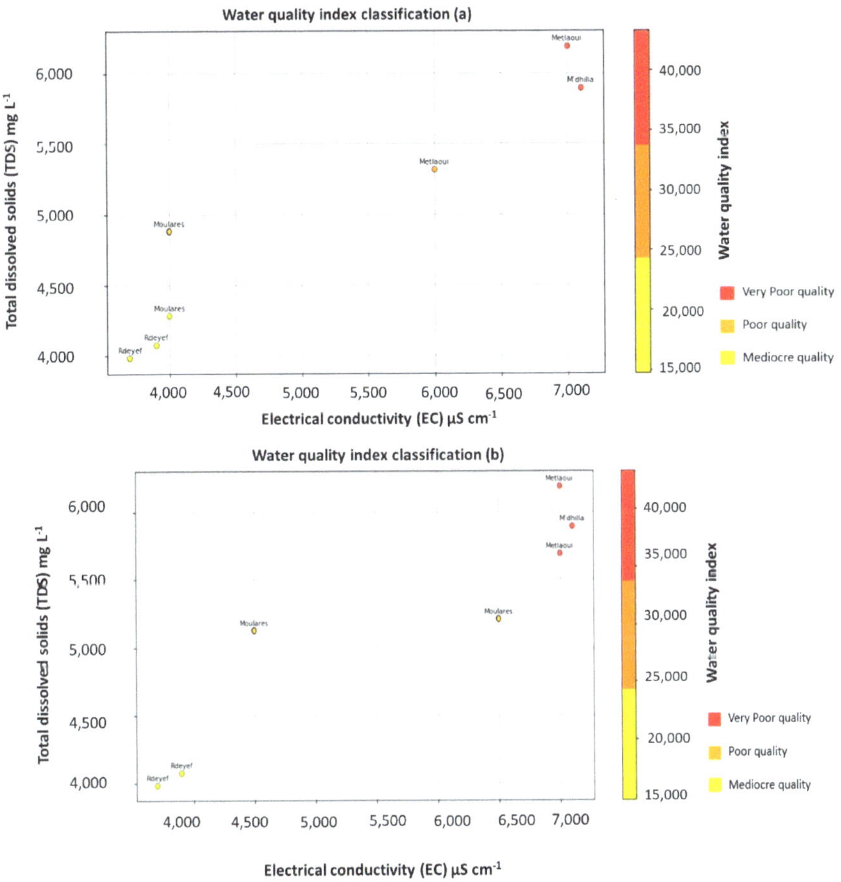

Figure 3. Index water quality of mining basin of Gafsa for ("January 2020 (**a**); June 2021 (**b**)").

These TDS values exceed the WHO standard (1000 mg·L^{-1}). However, falling within the 1000 to 10,000 mg·L^{-1} range, sense Rodier (1996) and [21,22,40–42], these waters may be classified as brackish. Similar values have been reported by [25].

Based on the measurements of electrical conductivity and dissolved solids concentrations (Table 1a,b), the classification of the water quality index in the Gafsa mining basin for two distinct seasons, winter and summer, is illustrated (Figure 3a,b). According to this graphical representation, the water quality in the mining basin is systematically categorized into three classes, each assigned a specific color (red, orange, yellow) (Figure 3a,b). Specifically, the observations disclose the following water quality assessments: (i) Mediocre water quality is evident in the Moulares and Redeyef (Figure 3a,b); (ii) poor water quality is identified in the Metlaoui zone (Figure 3a,b); (iii) inferior water quality is observed in the M'thilla and Metlaoui (Figure 3a,b). This classification presents the nuanced variations in water quality in the mining basin.

4.1. Hydrochemical Characteristics of the Investigated Waters

The concentrations of major elements observed in the two seasons exhibit significant elevations, particularly in sodium (700 to 1400 mg·L^{-1}) in January 2020 (Table 1a) and from 780 to 1420 mg·L^{-1} in June 2021 (Table 1b), chloride (946 to 1899 mg·L^{-1}) in January 2020 (Table 1a) and 1040 to 1900 mg·L^{-1} in June 2021 (Table 1b), sulfate (1500 to 1899 mg·L^{-1}) in January 2020 and 1630 to 2100 mg·L^{-1} in June 2021, calcium (325 to 497 mg·L^{-1}) in January 2020 (Table 1a), and 310 to 488 mg·L^{-1} in June 2021 (Table 1b), and magnesium (174 to 300 mg·L^{-1}) in January 2020 (Table 1a) and 200 to 264 mg·L^{-1} in June 2021 (Table 1b). However, it is essential to note that no significant variation is observed in these physicochemical parameters between the winter and summer seasons. These values consistently surpass their respective WHO standards (Table 1).

Heightened high sodium levels were observed in the study by [19] in the Moulares Redeyef mining area, specifically around the Tebedditt region. The ionic composition is prominently characterized by SO_4^{2-} (0.54–1.75 g/L), Cl^- (0.36–0.43 g/L), Na^+ (0.19–0.28 g/L), Ca^{2+} (0.18–0.35 g/L), and Mg^{2+} ranging between 0.11 and 0.24 g/L. This phenomenon can be attributed to various factors, including the hydrodynamic characteristics of the aquifer and intricate water–rock interactions involving processes such as dissolution and infiltration.

Moreover, discernible human influences, mainly from mining and agricultural activities, significantly contribute to the observed compositional trends. These anthropogenic factors amplify the impact on groundwater quality, necessitating careful consideration of environmental management strategies for the Gafsa mining basin.

The high concentrations of sulfate and chloride in this study are further supported by the study conducted by [25,26], which indicates that the concentrations of sodium, chloride, calcium, and sulfate ions surpass the standards set by the World Health Organization (WHO). This phenomenon was attributed to the dissolution of halite and gypsum.

The drawing of water samples onto the Piper diagram (Figure 4), of two seasons (January 2020 (a) and June 2021 (b)) shows that on the triangle of anions, water samples are located next to the SO_4^- Cl^- lines and far from the pole HCO_3· note however that for the samples Bh1, Bh3, Bh4, Bh5, and Bh6 the SO_4^{2-} content is slightly higher than that of Cl^-, and vice versa for Bh2 and Bh7 these two boreholes are richer in NaCl than the others, If we take into consideration the cation triangle, and assuming that potassium contents are accessory water samples are highly concentrated in Na, Ca and Mg, with Na > Ca > Mg. Thereby, the waters analyzed belong to the Na-Ca-Mg-SO_4-Cl chemical facies. The chemical particularities of (Bh1, Bh2, Bh3, Bh4, Bh5, Bh6, and Bh2, Bh7) depend on many natural factors, such as the lithology of the aquifer, the quality of recharge waters, and the types of interaction between water and aquifer.

The representation of the ion concentrations on the Scholler–Berkalov diagram (Figure 5) of two seasons (January 2020 (a) and June 2021 (b)) demonstrates that all the analyzed waters display identical profiles and that there is a cationic dominance of sodium, calcium, and magnesium in all examined samples and are highly loaded in Na^+, Ca^{2+}, Mg^{2+}, SO_4^{2-} and Cl^-. We can conclude that the groundwater in the mining basin is heavily loaded and oversaturated with calcite, halite, dolomite, and gypsum (Hamed et al., 2008) [25]. Figure 5a,b also show that all the aforementioned ions, in addition to bicarbonates (HCO_{3-}) in the analyzed groundwater, exceed their respective WHO guidelines. The only exception is nitrate (NO_{3-}) concentrations, which remain below the recommended limit, with the exception of sample Bh6, where the concentration (60 mg/L) slightly exceeds the WHO standard of 50 mg/L.

The sulfated, calcic, magnesium, sodium, and chlorinated facies result from the presence of sulfate, calcium, sodium, magnesium, and chloride in the soil due to the leaching of these ions by rainwater during infiltration and/or the dissolution of evaporates.

Comparing our results with other studies on water hydrochemistry in phosphate mining areas in various countries worldwide is paramount for a better understanding of the underlying factors contributing to the deterioration of water quality.

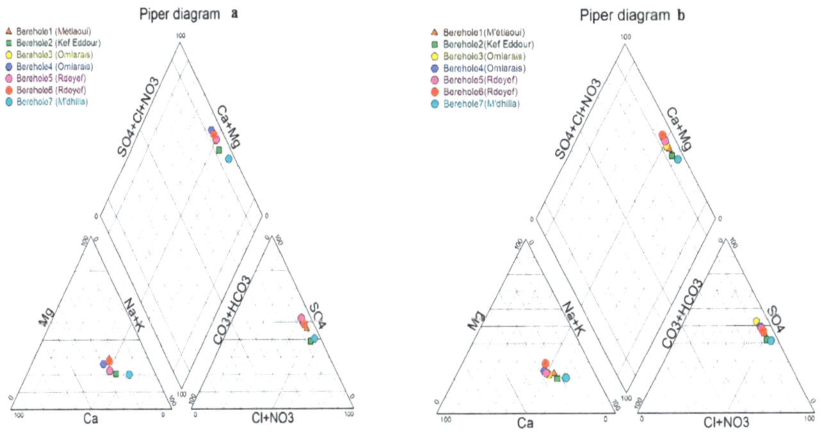

Figure 4. Plot of the groundwater samples on the Piper diagram: (**a**) January 2020; (**b**) June 2021.

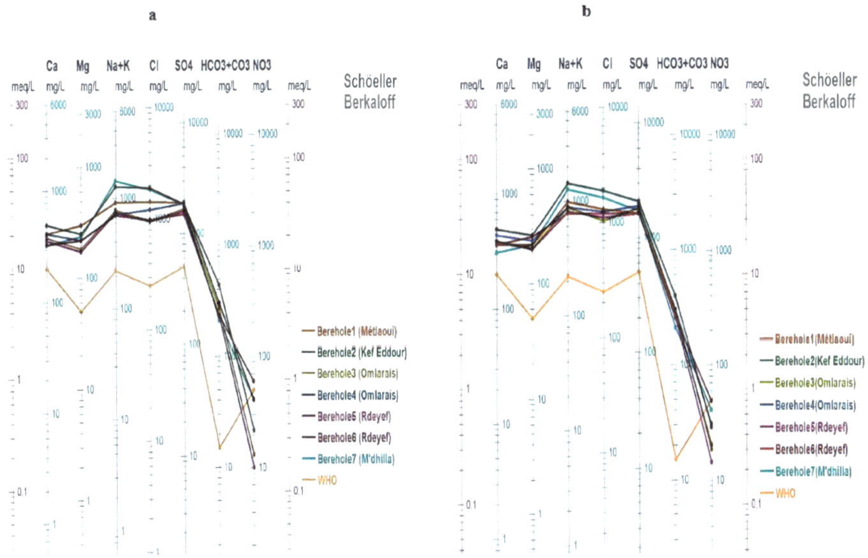

Figure 5. Plot of the characteristics of the studied groundwater samples on the Schoeller–Berkalov diagram, (**a**) January 2020), (**b**) June 2021).

Based on internal research and the findings of our study, it can be stated that the dynamic interaction between water and rock primarily influences the groundwater quality in the Moulares Redeyef basin. This complex relationship, integral to the hydrogeological dynamics of the aquifer, manifests through processes such as dissolution, and infiltration. The distinct geological characteristics of the study area play a pivotal role in driving these phenomena, exerting a substantial influence on the composition of groundwater resources. Notably, the etiology and geological attributes of the region function as foundational elements initiating shifts in water quality. This primary factor lays the groundwork for subsequent considerations, facilitating a comprehensive understanding of the geological intricacies that underscore the multifaceted challenges within the Moulares Redeyef basin. As we delve into the ensuing sections, we will expound upon additional contributing factors, including anthropogenic influences and intensified industrial activities, enriching the narrative with a holistic exploration of the diverse elements shaping the groundwater landscape in Phosphatic mining regions.

Phosphate extraction activities in the Gafsa mining basin pose significant risks to human health. The contamination of groundwater with harmful pollutants, as found by Hamed et al., 2014 [42], can lead to serious health conditions. For instance, fluoride presence in drinking water can lead to fluorosis, a condition characterized by dental and skeletal damage following prolonged exposure to elevated fluoride concentrations, as noted by Umer 2023 [27]. Additionally, elevated sulfate levels in water can cause gastrointestinal issues, such as diarrhea and intestinal irritation, as they disrupt the body's fluid balance and irritate intestinal tissues, as found by Backer 2000 [43].

The contamination of local water sources with these pollutants has a direct and adverse impact on the health of the regional population, who are exposed to fluoride and sulfates through their drinking water Guissouma and Tarhouni 2015, Guissouma et al. 2017 [44,45]. Therefore, it is essential to implement stringent management and monitoring strategies to safeguard water quality in the region. This includes optimizing extraction techniques and employing effective water treatment processes to mitigate these health risks.

To reduce the impact of the phosphate industry on groundwater contamination, several strategies should be adopted. First, improving waste management practices, particularly in the storage and treatment of phosphate residues, is crucial. Advanced containment and treatment technologies, such as filtration and neutralization systems, can significantly reduce the risk of contaminants leaching into groundwater. Additionally, establishing vegetated buffer zones around industrial sites can filter runoff water and reduce pollutant transfer to aquifers. Continuous monitoring of groundwater contamination levels and rapid response protocols for pollution incidents are also necessary. Finally, promoting research into more environmentally sustainable phosphate production methods and encouraging the adoption of eco-friendly practices can help minimize the industry's overall environmental impact.

4.2. Statistical Parameters and Correlation Matrix

The results of the PCA of two seasons, winter and summer, are summarized in Table 3. The binary relationship and the correlation coefficients between the different variables are given by the correlation matrix (Table 4).

Table 3. Statistical parameters of physico-chemical characteristics of groundwaters ((**a**) January 2020, (**b**) June 2021).

	EC ($\mu S \cdot cm^{-1}$)	TDS ($mg \cdot L^{-1}$)	Na^+ ($mg \cdot L^{-1}$)	K^+ ($mg \cdot L^{-1}$)	Ca^{2+} ($mg \cdot L^{-1}$)	Mg^{2+} ($mg \cdot L^{-1}$)	Cl^- ($mg \cdot L^{-1}$)	$SO_4{}^{2-}$ ($mg \cdot L^{-1}$)	$HCO_3{}^-$ ($mg \cdot L^{-1}$)	$NO_3{}^-$ ($mg \cdot L^{-1}$)
January 2020 (a)										
Average	5100	4933	928	6.5	386	226	1328	1751	282	32.5
SD	1541	885	286	2.73	61	42	408	158	76	18
Min	3700	3986	700	3.9	325	174	946	1500	211	10.1
Max	7100	6185	1400	10.54	497	300	1899	1899	437	60
CV%	30.2	18	31	42	15.77	18.62	30.75	9	27	55.56
June 2021 (b)										
Average	5646	5073	997	7.58	390	229	1353	1783	285	29
SD	1223	755	244	2.23	58	25	311	173	57	12.3
Min	4000	4394	780	4.4	310	200	1040	1630	211	14.4
Max	7200	6603	1420	9.7	488	264	1900	2100	394	48.3
CV%	21.66	14.88	24.47	29.18	14.95	10.93	23	9.71	19.94	43

CV: coefficient of variation, Min: minimum; Max: maximum, SD: standard deviation; EC: electrical conductivity; TDS: total dissolved solid.

Table 4. Correlation matrix of the physico-chemical characteristics of the groundwater in the mining basin of Gafsa ((**a**) January 2020; (**b**) June 2021).

(a)

	CE ($\mu S \cdot cm^{-1}$)	TDS ($mg \cdot L^{-1}$)	Na^+ ($mg \cdot L^{-1}$)	K^+ ($mg \cdot L^{-1}$)	Ca^{2+} ($mg \cdot L^{-1}$)	Mg^{2+} ($mg \cdot L^{-1}$)	Cl^- ($mg \cdot L^{-1}$)	SO_4^{2-} ($mg \cdot L^{-1}$)	HCO_3^- ($mg \cdot L^{-1}$)	NO_3^- ($mg \cdot L^{-1}$)
CE ($\mu s \cdot cm^{-1}$)	1									
TDS ($mg \cdot L^{-1}$)	0.960257	1								
Na^+ ($mg \cdot L^{-1}$)	0.948736	0.898261	1							
K^+ ($mg \cdot L^{-1}$)	0.639910	0.561176	0.765793	1						
Ca^{2+} ($mg \cdot L^{-1}$)	0.389599	0.552566	0.222530	0.001241	1					
Mg^{2+} ($mg \cdot L^{-1}$)	0.646318	0.644785	0.642620	−0.040302	0.419687	1				
Cl^- ($mg \cdot L^{-1}$)	0.966586	0.987053	0.935567	0.63150	0.459418	0.610774	1			
SO_4^{2-} ($mg \cdot L^{-1}$)	0.658052	0.791295	0.505068	0.183585	0.615489	0.645925	0.704453	1		
HCO_3^- ($mg \cdot L^{-1}$)	0.415895	0.565311	0.279600	0.024971	0.976418	0.404011	0.496045	0.512013	1	
NO_3^- ($mg \cdot L^{-1}$)	0.018844	0.018501	−0.033818	−0.275180	−0.306015	0.478721	0.091080	−0.013506	−0.237463	1

(b)

	CE ($\mu S \cdot cm^{-1}$)	TDS ($mg \cdot L^{-1}$)	Na^+ ($mg \cdot L^{-1}$)	K^+ ($mg \cdot L^{-1}$)	Ca^{2+} ($mg \cdot L^{-1}$)	Mg^{2+} ($mg \cdot L^{-1}$)	Cl^- ($mg \cdot L^{-1}$)	SO_4^{2-} ($mg \cdot L^{-1}$)	HCO_3^- ($mg \cdot L^{-1}$)	NO_3^- ($mg \cdot L^{-1}$)
CE ($\mu s \cdot cm^{-1}$)	1									
TDS ($mg \cdot L^{-1}$)	0.631522	1								
Na^+ ($mg \cdot L^{-1}$)	0.78406	0.937732	1							
K^+ ($mg \cdot L^{-1}$)	0.436903	0.368669	0.547816	1						
Ca^{2+} ($mg \cdot L^{-1}$)	−0.02407	0.574666	0.286034	−0.16743	1					
Mg^{2+} ($mg \cdot L^{-1}$)	0.004762	0.604856	0.392455	−0.07183	0.50503	1				
Cl^- ($mg \cdot L^{-1}$)	0.652598	0.936914	0.961069	0.585113	0.298669	0.572488	1			
SO_4^{2-} ($mg \cdot L^{-1}$)	0.410439	0.828729	0.637077	−0.11055	0.848166	0.524135	0.58587	1		
HCO_3^- ($mg \cdot L^{-1}$)	0.150898	0.661284	0.427053	−0.09422	0.937969	0.501414	0.400993	0.853222	1	
NO_3^- ($mg \cdot L^{-1}$)	−0.01437	0.167201	0.149081	−0.15447	−0.24711	0.64831	0.312676	−0.03033	−0.27571	1

The PCA was performed to analyze water sampled on the entire study area to highlight the possible associations between the different variables and the possible existence of the sub-populations of anions and cations. PCA is a dimensionality reduction method that allows the identification and visualization of underlying structures in multivariate data. It transforms the original variables into a set of new variables known as principal components, which explain the maximum possible variance in the data.

In our analysis, we applied PCA to simplify the interpretation of the relationships between the different studied variables, reducing data complexity while retaining essential characteristics. This approach enabled us to better understand variation patterns and evaluate the main sources of contamination and the chemical elements responsible for the mineralization of water in the phosphate basin.

The statistical characterizing of the distribution of the major elements was calculated for each of the physicochemical variables (Table 3)

The coefficient of variation (CV) and the standard deviation ratio divided by the arithmetic mean are parameters used to measure the dispersion of a series of observations of a variable around its arithmetic mean. If the value of the CV is less than 50%, the observed variable has a homogeneous distribution around its arithmetic mean, whereas if the value of the CV is greater than 50%, the observed variable has a heterogeneous distribution around its arithmetic mean [29].

The results indicate a homogeneous geographical distribution of physicochemical parameters (EC and TDS) and major ions, with a coefficient of variation (CV) below 50% (Table 3). However, the nitrate anion (NO_3^-) exhibits a heterogeneous geographical distribution, with a CV exceeding 50% (Table 3a) during the two sampling periods.

A correlation matrix in which 10 parameters characterizing the analyzed waters of two seasons, January 2020 and June 2021, are involved is presented in Table 4. Significant associations between the various parameters, reflecting the links between the studied variables, are given by this matrix. The PCA shows that the major ions are well correlated with EC and TDS. However, the strong correlation between the contents of major ions reveals that the total mineralization of water is mainly due to these elements that increase the salt load. On the other hand, the nitrate ion (NO_3^-) (Table 4 contents are the least correlated with the salt load.

Table 4 shows a strong positive correlation between sodium and chloride (R = 0.93 and 0.96; Tables 4a and 4b, respectively), in the mining basin. This correlation reflects a proportional evolution between these two major elements in all sampling points. Therefore, these elements appear to share a common origin of mineralization, likely stemming from the dissolution of halite [38].

The good correlation between SO_4^{2-} and Ca^{2+} (R = 0.61 and 0.84; Tables 4a and 4b, respectively), shows that the two chemical compounds have common origins, e.g., the dissolution of gypsum ($CaSO_4$, $2H_2O$), or anhydrite ($CaSO_4$).

A good correlation (R = 0.97 and 0.93; Tables 4a and 4b, respectively), is noticed between Ca^{2+} and HCO_3^-.

Calcium and magnesium exhibit an average correlation (R = 0.41 and 0.50; Tables 4a and 4b, respectively). Furthermore, the dominance of calcium over magnesium is evident at all sampling points (Table 1a,b).

Eigenvalues and Variance

Table 5 shows the eigenvalues, the expressed variances for each factor, and their accumulation. Factor F1, with an expressed variance of 57.5% for the winter season of January 2020 and 52% for the summer season of June 2021, is the most important of all, followed by factors F2 and F3, with, respectively, 18.75% and 16% for January 2020 of the expressed variance and 23% and 16.57% for June 2021. The cumulative variance expressed is 92.23% for the three factors for January 2020 and 91.51% for June 2021.

Table 5. Eigenvalues and variance expressed by the main axes.

	January 2020			June 2021		
	F1	F2	F3	F1	F2	F3
Eigenvalue	5.749703827	1.874768305	1.599105973	5.198406086	2.29595319	1.657048015
Variability%	57.49703827	18.74768305	15.99105973	51.98406086	22.9595319	16.57048015
Cumulative%	57.49703827	76.24472132	92.23578105	51.98406086	74.94359275	91.5140729

These factorial axes chosen for this statistical analysis are, therefore, representative of the variance of the whole data set. The factorial plans F1–F2 and F1–F3 represent cumulative variances equal to 76.1% and 74.8% and 74.8% and 68.5%, respectively. These two plans not only represent almost the same cumulative variance but also integrate enough representativeness to highlight the general trends. Since the different results are almost the same, only those of the F1–F2 design have been presented because this design more clearly distinguishes the trends.

The analysis of the results of the PCA of the space of the variables of the factorial plan F1–F2 shows that the factor F1, the most important (57.4% and 51.9%; Figures 6a and 6b, respectively), is determined by Ca^{2+}, Na^+, Mg^{2+}, SO_4^{2-}, HCO_3^-, TDS, CE, and Cl^-, (Figure 6a,b). The F1 factor, therefore, groups together the bulk of the chemical elements and the electrical conductivity. The proximity of these variables in the community circle shows that they control the phenomenon of putting the mineral elements into a solution. These variables highlight a slow process of dissolution of the ions. Thus, the F1 factor seems to account for the conditions of acquisition of the water chemistry. It explains the phenomenon of mineralization residence time. Most of the variables are positively correlated with this phenomenon showing that the main phenomenon at the origin of the mineralization of the groundwater in the region is dissolution. This dissolution is due to the contact during a sufficient residence time of the water with the surrounding geological formations. Factor F2 (Figure 6), accounting for 18.7% and 22.9% of the inertia of the sample point cloud, is influenced by the elements NO_3^- and K^+. These results show that the F2 factor is not related to the hydrolysis phenomenon but rather to the redox phenomenon.

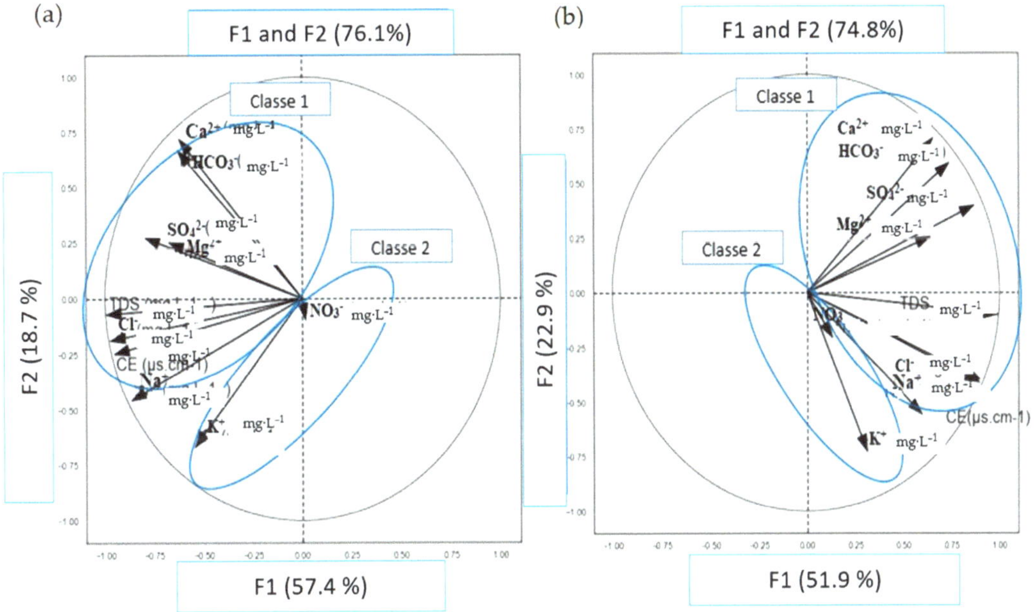

Figure 6. Variables space of the factorial plane ("F1–F2 ((**a**) January 2020; (**b**) June 2021").

The PCA community circle (Figure 6) serves as a graphical representation in the statistical unit space, depicting the distribution of water points based on the various F1–F2 factors, revealing two primary groupings of water points.

These two factors, F1–F2, can also be grouped into two classes. Class 1 encompasses the major elements that are well correlated with electrical conductivities (EC) and total dissolved solids (TDS) sulfate, chloride, sodium, magnesium, and calcium), as depicted in (Figure 6). Conversely, Class 2 includes major elements with a weaker correlation with (TDS) and (EC) (nitrate, potassium), as shown in (Figure 5).

The concentrations of Pb, Mn, Zn, Cr, Fe, and Al in the analyzed water samples fell for winter (January 2020) and summer (June 2021) within the regulatory limits established by the WHO (2011) during the two sampling campaigns. Conversely, fluoride (F^-) concentrations exhibited values ranging from (1.6 mg·L^{-1} to 2.9 mg·L^{-1}) in the winter of January 2020 and (1 to 2.9 mg·L^{-1}) in the summer of June 2021 (Table 2) surpassing its WHO limit (1.5 mg·L^{-1}) in almost all water samples. According to (Table 2), it is observed that the M'étlaoui zone (Borehole 1) and the M'dhilla zone (Borehole 7) are characterized by a high concentration of fluoride, with respective values of (2.1 and 2.9 mg·L^{-1}) for January 2020 and (2.4 to 2.9 mg·L^{-1}) for June 2021 season. On the other hand, a lower concentration is observed in the Moulares and Redeyef zones (Table 2). These elevated fluoride concentrations in these zones are attributed to high TDS values. These boreholes are near the phosphate industry, explaining the high TDS concentration resulting from overexploitation of the aquifer by the phosphate industry.

These findings align with [46], who investigated fluoride concentrations in drinking water and revealed that the fluoride concentration in drinking water in the Metlaoui and Redeyef regions ranged from 2.2 to 3.39 mg·L^{-1}. Furthermore, [44,45] conducted comprehensive fluoride health risk assessments and revealed that the fluoride concentration in groundwater in the Gafsa regions is (2.22; 1.27; 2.22 mg·L^{-1}) surpassing its WHO limit. They highlighted that groundwater in Southern Tunisia, notably in Gafsa, naturally contains elevated fluoride levels due to geological factors. In contrast, northern and central areas exhibit lower concentrations, falling short of meeting WHO water quality standards.

The elevated levels of fluoride in drinking water have become a focal point of extensive scientific investigation at both national and international levels. That is why a water purification program for the mining basin is strongly recommended, especially for domestic use and particularly as drinking water.

5. Conclusions

In conclusion, this study highlights the groundwater quality in the Mio-Plio-Quaternary aquifer of the Gafsa mining basin, using a hydrochemical approach and multivariate statistical methods based on PCA to assess the mineralization processes of these waters.

The hydrochemical study of the Gafsa basin groundwaters, conducted during both the winter and summer seasons, has shown that these waters are characterized all year round by the dominance of the Na-Ca-Mg-SO_4-Cl facies.

The interpretation of major ion analyses indicates that mineralization is influenced by the prolonged interaction of water with geological formations. This highlights the predominant role of dissolution (calcite, dolomite, gypsum, and halite) and infiltration in the high mineralization of groundwater at a regional scale.

The results of the PCA represent two correlation classes. Class 1 comprises major elements strongly correlated with electrical conductivity (EC) and total dissolved solids (TDS)—sulfate, chloride, sodium, magnesium, and calcium. Thus, these strong correlations among these major elements are attributed to high mineralization in the waters of the basin. Class 2 includes major elements with a weaker correlation with (TDS) and (EC)—nitrate and potassium. These two elements do not play a role in the mineralization phenomenon.

The levels of heavy metals in the groundwater of the Gafsa mining basin do not exceed the WHO limits. However, the presence of fluoride in this basin's groundwater exceeds the WHO limits [47]

These observations suggest a high degree of mineralization, rendering the investigated waters unsuitable for human consumption. These implications underscore the importance of carefully considering alternative water sources and implementing robust purification processes in the studied region. Furthermore, this study provides valuable insights into groundwater management in similar areas, highlighting the importance of implementing appropriate management strategies to ensure water quality and protect public health.

Author Contributions: Conceptualization, N.N., R.S., E.A.L.M. and F.S.; methodology, N.N., F.S., A.I., T.B.A., O.S., D.T., E.A.L.M. and R.S.; software, N.N., D.T., F.S., O.S. and A.I., validation, N.N., R.S., A.I. and T.B.A., formal analysis; N.N., F.S., D.T. and A.I.; investigation, N.N., F.S., T.B.A., A.I., O.S., D.T., E.A.L.M. and R.S.; resources, E.A.L.M. and R.S.; data curation, N.N., R.S., A.I. and T.B.A.; writing—original draft preparation, N.N., F.S., A.I., O.S., D.T., E.A.L.M. and R.S.; writing—review and editing, E.A.L.M. and R.S.; supervision, E.A.L.M. and R.S., project administration, N.N., E.A.L.M. and R.S.; funding acquisition, E.A.L.M. and R.S. All authors have read and agreed to the published version of the manuscript.

Funding: The authors are grateful to the institutions involved in developing this research project. They also appreciate the Consejo Nacional de Humanidades, Ciencias y Tecnologías (CONAHCyT) for the financial support (Grant No. A1-S-38139).

Data Availability Statement: The data sets generated during and/or analyzed during the current study are available from the corresponding author upon reasonable request.

Acknowledgments: The authors would like to express their gratitude to the International Center for Environmental Technologies of Tunis (CITET), for their collaboration and assistance.

Conflicts of Interest: Authors Nada Nasri, Fouad Souissi, Amina Ismailia, Olfa Smida and Radhia Souissi were employed by the company Technopark of Sidi Thabet. The remaining authors declare that the research was conducted in the absence of any commercial or financial relationships that could be construed as a potential conflict of interest.

Abbreviations

PCA: principal component analysis; WHO: World Health Organization; EC: electrical conductivity; TDS: total dissolved solids; GPC: Gafsa Phosphate Company; TGC: Tunisian Chemical Group.

References

1. Li, P.; Wu, J.; Tian, R.; He, S. Geochemistry, hydraulic connectivity and quality appraisal of multilayered groundwater in the Hongdunzi Coal Mine, Northwest China. *Environ. Earth Sci.* **2018**, *37*, 222–237. [CrossRef]
2. Pandey, A. The realm of microbial lipases in biotechnology. *Biotechnol. Appl. Biochem.* **1999**, *29*, 119–131. [CrossRef]
3. Chkir, N.; Trabelsi, R.; Bahir, M.; Hadj Ammar, F.; Zouari, K.; Chaamchati, H.; Manteito, J.P. Vulnérabilité des ressources en eaux des aquifères côtiers en zones semi-arides–Etude comparative entre les bassins d'Essaouira (Maroc) et de la Jeffara (Tunisie). *Comun. Geológicas* **2008**, *95*, 107–121.
4. Navarro, M.C.; Perez, C.; Martinez, M.J.; Vidal, J. Abandoned mine sites as a source of contamination by heavy metals: A case study in a semi-arid zone. *Environ. Pollut.* **2008**, *96*, 183–193. [CrossRef]
5. Qadir, M.; Charma, B.; Karajeh, F. Non-conventional water resources and opportunities for water augmentation to achieve food security in water scarce countries. *Agric. Water Manag.* **2007**, *87*, 2–22. [CrossRef]
6. Ramadam, E. Sustainable water resources management in arid environment: The case of Arabian Gulf. *Int. J. Waste Resour.* **2015**, *5*, 3–6.
7. Salameh, E. Over-exploitation of groundwater resources and their environmental and socio-economic implications: The case of Jordan. *Water Int.* **2008**, *33*, 55–68. [CrossRef]
8. Hamed, Y. The hydro geochemical characterization of groundwater in Gafsa-Sidi Boubaker region (Southwestern Tunisia). *Arab. J. Geosci.* **2011**, *6*, 697–710. [CrossRef]
9. Ncibi, K.; Chaar, H.; Hadji, R.; Baccari, N.; Sebei, A.; Khelifi, F.; Abbes, M.; Hamed, Y. A GIS-based statistical model for assessing groundwater susceptibility index in shallow aquifer in Central Tunisia (Sidi Bouzid basin). *Arab. J. Geosci.* **2020**, *13*, 98. [CrossRef]
10. Gasmi, N.; Bouissou, S.; Piriou, F. Comparison of Potential Dual Formulations Developed with Different Elements. In *Electric and Magnetic Fields*; Nicolet, A., Bellman, R., Eds.; Springer: Boston, MA, USA, 1995. [CrossRef]
11. Tarki, M.; Ben Hammadi, M.; El Mejri, H.; Dassi, L. Assessment of hydrochemical processes and groundwater hydrodynamics in a multilayer aquifer system under long-term irrigation condition: A case study of Nefzaoua basin, Southern Tunisia. *Appl. Radiat. Isot.* **2016**, *110*, 138–149. [CrossRef]
12. Dassi, L. Investigation by multivariate analysis of groundwater composition in a multilayer aquifer system from North Africa: A multi-tracer approach. *Appl. Geochem.* **2011**, *26*, 1386–1398. [CrossRef]
13. Naseem, S.; Erum, B.; Ahmed, P.; Rafique, T. Impact of seawater intrusion on the geochemistry of groundwater of Gwadar District, Balochistan and its appraisal for drinking water quality. *Environ. Monit. Assess.* **2018**, *13*, 281–293. [CrossRef]
14. Sharaf, M.; Amin, M. Major elements hydrochemistry and groundwater quality of Wadi Fatimah, West Central Arabian Shield, Saudi Arabia. *Arab. J. Geosci.* **2013**, *6*, 2633–2653. [CrossRef]
15. Benmarce, K.; Hadji, R.; Hamed, Y.; Zahri, F.; Zighmi, K.; Hamad, A.; Gentilucci, M.; Ncibi, K.; Besser, H. Hydrogeological and water quality analysis of thermal springs in the Guelma region of North- Eastern Algeria: A study using hydrochemical, statistical, and isotopic approaches. *J. Afr. Earth Sci.* **2023**, *205*, 105011. [CrossRef]
16. WHO. *Fluoride in Drinking-Water, Background Document for Development of Who Guidelines For Drinking-Water Quality*; WHO: Geneva, Switzerland, 1996.
17. WHO. *Fluor et Santé, Série de Monographie, Genève, 59*; WHO: Geneva, Switzerland, 1972.
18. Cidu, R.; Ridou, R.; Fanfani, L.; Luca, F. Impact of past mining activity on SW Sardinia (Italy) groundwater quality. *J. Geochem. Explor.* **2009**, *100*, 125–132. [CrossRef]
19. Hamdi, M.; Goïta, K.; Karaouli, F.; Zagrarni, M.F. Hydrodynamic groundwater modeling and hydro chemical conceptualization of the mining area of Moulares Redeyef (southwestern of Tunisia): New local insights. *Phys. Chem. Earth Parts A/B/C* **2021**, *121*, 102974. [CrossRef]
20. Mokadem, N.; Hamed, Y.; Ben Sâad, A.; Gargouri, I. Atmospheric pollution in North Africa (ecosystems–atmosphere interactions): A case study in the mining basin of El Guettar–M'Dilla (southwestern Tunisia). *Atmos. Pollut. Res.* **2014**, *7*, 2071–2079. [CrossRef]
21. Yermani, M.; Zouari, K.; Michelot, J.L.; Mamou, A.; Moumni, L. Approche géochimique du fonctionnement de la nappe profonde de Gafsa Nord (*Tunisie centrale*). *Hydrol. Sci. J.* **2003**, *48*, 95–108. [CrossRef]
22. Farhat, H.; Moumni, L. *Etude Hydrogéologique de la Nappe de Gafsa Nord*; DGRE: Tunis, Tunisia, 1989.
23. Malik, N.; Slim, N.; Shimi, N. Etude de la vulnérabilité des eaux souterraines de la ville de Gafsa (Sud-Ouest de la Tunisie): Effets anthropiques et conséquences. *Alger. J. Environ. Sci. Technol.* **2019**, *5*, 1127–1134.
24. Majdoub, R.; Dridi, L.; M'nasri, S. Caractérisation de la nappe profonde Gafsa nord suite à la surexploitation des eaux souterraines. *Larhyss J.* **2014**, *17*, 179–192.
25. Hamed, Y.; Dassi, L.; Ahmadi, R.; Dhia, H.B. Geochemical and isotopic study of the multilayer aquifer system in the Moulares-Redayef basin, southern Tunisia / Etude géochimique ET isotopique du système aquifère multicouche du bassin de Moulares-Redayef, sud tunisien. *Hydrol. Sci. J.* **2008**, *53*, 1241–1252. [CrossRef]

26. Hamed, Y.; Dassi, L.; Tarki, M.; Ahmadi, R.; Mehdi, K.; Dhia, H.B. Groundwater origins and mixing pattern in the multilayer aquifer system of the Gafsa-south mining district: A chemical and isotopic approach. *Environ. Earth Sci.* **2010**, *63*, 1355–1368. [CrossRef]
27. Umer, M.F. A Systematic Review on Water Fluoride Levels Causing Dental Fluorosis. *Sustainability* **2023**, *15*, 12227. [CrossRef]
28. El Hasnaoui, B.; Younsi, A.; Mountadar, M.; Garmes, H.; Mouhab, I. Impacts négatifs d'une zone industrielle sur les eaux souterraines et sur le cheptel (Cas du Jorf Lasfar, Maroc): Approches pluridisciplinaires. *Déchets Sci. Tech.* **2011**, *59*, 2945.
29. Jiries, A.; El-Hasan, T.; Al-Hweiti, M.; Seiler, K.-P. Evaluation of the Effluent Water Quality Produced at Phosphate Mines in Central Jordan. *Mine Water Environ.* **2004**, *23*, 133–137. [CrossRef]
30. Khelifi, F.; Besser, H.; Ayadi, Y.; Liu, G. Evaluation of potentially toxic elements' (PTEs) vertical distribution in sediments of Gafsa–Metlaoui mining basin (Southwestern Tunisia) using geochemical and multivariate statistical analysis approaches. *Environ. Earth Sci.* **2019**, *78*, 53. [CrossRef]
31. Burollet, P.F. Contribution à l'étude Stratigraphique de la Tunisie Centrale. *Ann. Mines Geol.* **1956**, *18*, 352.
32. Sassi, S. La Sedimentation Phosphatee au Paléocène dans le Sud et le Centre Ouest de la Tunisie. Ph.D. Thesis, Université de Paris Sud, Orsay, France, 1974.
33. Dlala, M.; Hfaiedh, M. Le séisme du 7 novembre 1989 à Metlaoui (*Tunisie méridionale*): Une tectonique active en compression. *Comptes Rendus L'Académie Sci. Série 2 HFAIEDH Mécanique Phys. Chim. Sci. L'Univers Sci. Terre* **1993**, *317*, 1297–1302.
34. Hamed, Y. Caractérisation Hydrogéologique, Hydrochimique et Isotopique du Système Aquifère de Moularès-Tamerza. Ph.D. Thesis, University of Sfax, Sfax, Tunisia, 2009.
35. Chraiti, R.; Raddaoui, M.; Hafiane, A. Effluent Water Quality at Phosphate Mines in M'Dhilla, Tunisia and its Potential Environmental Effects. *Mine Water Environ.* **2016**, *35*, 462–468. [CrossRef]
36. Salem, M.; Souissi, R.; Souissi, F.; Abbes, N.; Moutte, J. Phosphoric acid purification sludge: Potential in heavy metals and rare earth elements. *Waste Manag.* **2019**, *83*, 46–56. [CrossRef]
37. Smida, O.; Souissi, R.; Marzougui, S.; Souissi, F. Geochemical Assessment and Mobility of Undesired Elements in the Sludge of the Phosphate Industry of Gafsa-Metlaoui Basin, (Southern Tunisia). *Minerals* **2021**, *11*, 1075. [CrossRef]
38. Souissi, R.; Souissi, F.; Chakroun, H.K.; Bouchardon, J.L. Mineralogical and geochemical characterization of mine tailings and Pb, Zn, and Cd mobility in a carbonate setting (Northern Tunisia). *Environ. Earth Sci.* **2013**, *32*, 16–27. [CrossRef]
39. Bocoum, M. *Méthodes D'analyses des Sols. Document de Travail*; Institut National de Pédologie: Dakar, Sénégal, 2004; 55p.
40. Rodier, J. *L'Analyse de l'Eau, Eaux Naturelles, Eaux Résiduaires, Eaux de Mer*; Edition Dunod: Malakoff, France, 1996; 1434p.
41. Banton, O.; Bangoy, L.M.; Chevalier, S.; Houenou, P.; Lafrance, P.; Rivard, C. *Hydrogéologie: Multi Science Environnementale des Eaux Souterraines*; Presses de l'Université du Québec/AUPELF: Quebec, Canada, 1997; 460p.
42. Hamed, Y.; Gentilucci, M.; Mokadem, N.; Khalil, R.; Ayadi, Y.; Hadji, R.; Elaloui, E. Assessment and Mitigation of Groundwater Contamination from Phosphate Mining in Tunisia: Geochemical and Radiological Analysis. *Hydrology* **2024**, *11*, 84. [CrossRef]
43. Backer, L.C. Assessing the acute gastrointestinal effects of ingesting naturally occurring, high levels of sulfate in drinking water. *Crit. Rev. Clin. Lab. Sci.* **2000**, *37*, 389–400. [CrossRef] [PubMed]
44. Ben Nasr, K.; Walha, C.; Charcosset, R. Ben Amar, Removal of fluoride ions using cuttlefish bones. *J. Fluor. Chem.* **2011**, *132*, 57–62. [CrossRef]
45. Guissouma, W.; Tarhouni, J. Fluoride in Tunisian Drinking Tap Water. *J. Water Resour. Prot.* **2015**, *7*, 860. [CrossRef]
46. Essouli, O.F. Impact de la Décharge Publique du lac Mbeubeuss sur la Ressource en eau de L'aquifère des Sables Quaternaires de Thiaroye (Dakar, Sénégal). Ph.D. Thesis, Département de Géologie, Faculté des Sciences de Sénégal, Dakar, Senegal, 2005.
47. Guissouma, W.; Hakami, O.; Al-Rajab, A.J.; Tarhouni, J. Risk assessment of fluoride exposure in drinking water of Tunisia. *Chemosphere* **2017**, *177*, 102–108. [CrossRef]

Disclaimer/Publisher's Note: The statements, opinions and data contained in all publications are solely those of the individual author(s) and contributor(s) and not of MDPI and/or the editor(s). MDPI and/or the editor(s) disclaim responsibility for any injury to people or property resulting from any ideas, methods, instructions or products referred to in the content.

hydrology

MDPI

Article

Combining Hydro-Geochemistry and Environmental Isotope Methods to Evaluate Groundwater Quality and Health Risk (Middle Nile Delta, Egypt)

Atef M. Abu Khatita [1,2,*], Robert van Geldern [3], Abdullah O. Bamousa [1], Dimitrios E. Alexakis [4,*], Esam Ismail [1,5], Wael R. Abdellah [1,6] and Ibrahim A. A. Babikir [1]

1 Geology Department, College of Science, Taibah University, Medina 344, Saudi Arabia; abamousa@taibahu.edu.sa (A.O.B.); ebismail@taibahu.edu.sa (E.I.); wabdellah@taibahu.edu.sa (W.R.A.); ibabiker@taibahu.edu.sa (I.A.A.B.)
2 Geology Department, Faculty of Science, Al-Azhar University, Cairo 11651, Egypt
3 GeoZentrum Nordbayern, Department of Geography and Geosciences, University of Erlangen-Nuremberg, Schlossgarten 5, 91054 Erlangen, Germany; robert.van.geldern@fau.de
4 Laboratory of Geoenvironmental Science and Environmental Quality Assurance, Department of Civil Engineering, School of Engineering, University of West Attica, 250 Thivon & P. Ralli Str., 12241 Athens, Greece
5 Department of Geology, Faculty of Science, Minia University, Minia 61111, Egypt
6 Mining and Metallurgical Engineering Department Faculty of Engineering, University of Assiut, Assiut 71515, Egypt
* Correspondence: atefabukhatita@yahoo.com (A.M.A.K.); d.alexakis@uniwa.gr (D.E.A.); Tel.: +966-530298340 (A.M.A.K.)

Academic Editor: Meysam Vadiati

Received: 13 February 2025
Revised: 20 March 2025
Accepted: 24 March 2025
Published: 27 March 2025

Citation: Khatita, A.M.A.; van Geldern, R.; Bamousa, A.O.; Alexakis, D.E.; Ismail, E.; Abdellah, W.R.; Babikir, I.A.A. Combining Hydro-Geochemistry and Environmental Isotope Methods to Evaluate Groundwater Quality and Health Risk (Middle Nile Delta, Egypt). *Hydrology* **2025**, *12*, 72. https://doi.org/10.3390/hydrology12040072

Copyright: © 2025 by the authors. Licensee MDPI, Basel, Switzerland. This article is an open access article distributed under the terms and conditions of the Creative Commons Attribution (CC BY) license (https://creativecommons.org/licenses/by/4.0/).

Abstract: This study aims to assess the vulnerability of groundwater in the Nile Delta to contamination and evaluate its suitability for drinking and irrigation. A total of 28 groundwater wells (ranging from 23 to 120 m in depth) and two Nile surface water samples were analyzed for total dissolved solids (TDS), heavy metals, groundwater quality index (GWQI), and hazard quotient (HQ). The findings reveal that deep groundwater (60–120 m) displays paleo-water characteristics, with low TDS, total hardness, and minimal heavy metal contamination. In contrast, shallow groundwater (<60 m) is categorized into three groups: paleo-water-like, recent Nile water with elevated TDS and heavy metals, and mixed water. Most groundwater samples (64%) are of the Ca-HCO_3 type, while 28% are Na-HCO_3, and 8% are Na-Cl, the latter associated with sewage infiltration. Most groundwater samples were deemed suitable for irrigation, but drinking water quality varied significantly—4% were classified as "excellent", 64% as "good", and 32% as "poor". HQ analysis identified manganese as a significant health risk, with 56% of shallow groundwater samples exceeding safe levels. These findings highlight the varying groundwater quality in the Nile Delta, emphasizing concerns regarding health risks from heavy metals, particularly manganese, and the need for improved monitoring and management.

Keywords: Nile Delta aquifers; heavy metal pollution; groundwater quality index (GWQI); manganese contamination; paleo-groundwater

1. Introduction

Groundwater is a vital source of potable water, particularly in arid and semi-arid regions, where it serves as a critical resource for drinking, irrigation, and industrial purposes [1,2]. However, its vulnerability to contamination is increasing due to urbanization, limited recharge from surface water, and various anthropogenic activities. Common sources of groundwater degradation include the leakage of uncontrolled domestic sewage and sanitation systems [3–6], industrial and mining contamination [7–11], landfill leachates [5,12–15], and contamination from surface water bodies [5,16–20]. Additionally, agricultural practices, such as waterlogging, cause soil salts and agrochemicals to leach into groundwater, especially in flood irrigation systems [7,21–25]. The chemical composition of groundwater significantly affects its usability for drinking. Major ions such as Ca^{2+}, Na^{+}, and K^{+} influence its taste, but heavy metal contamination is one of the most pressing concerns for both environmental and public health [26]. While trace metals such as Fe, Mn, Cu, and Zn are essential nutrients at low concentrations, they can be toxic at high levels, whereas metals like As, Cr, and Pb are toxic even at low concentration [27].

The suitability of groundwater for various purposes (drinking, agriculture, and industrial supply) can be assessed using a variety of geochemical methods. Major ions, as well as stable isotopes of oxygen, hydrogen, and carbon, offer valuable insights into the water's origin, recharge zones, and factors influencing the aquifer system [17,25,28–35]. Other techniques, such as statistical processing and water quality indices, help analyze the interrelationships between water constituents and provide estimates of overall water quality [36–38].

In the Nile Delta, Egypt, groundwater is the primary drinking water source, particularly for rural inhabitants who rely on municipal boreholes or private hand-dug wells. Many households, especially in the Middle Nile Delta, use shallow wells (less than 30 m deep) to extract water for both drinking and irrigation, with the latter consuming substantial amounts of water. Groundwater contamination in the Nile Delta has become a significant issue which is not adequately addressed or controlled [22,25,39–43]. The main sources of contamination in this region are (1) waterlogged soil that affects shallow groundwater quality, (2) inadequate sanitation systems in some villages, and (3) untreated industrial waste discharges into rivers. These contaminants infiltrate the groundwater through downward percolation, affecting both shallow and deeper aquifers. Consequently, assessing groundwater quality and its vulnerability to contamination in this region is essential.

The novelty of this study lies in its comprehensive approach to evaluating the hydrogeochemical characteristics of groundwater in the Middle Nile Delta. By combining traditional hydrogeochemical analyses with advanced stable isotope techniques (oxygen and hydrogen), this study aims to provide a more thorough understanding of groundwater recharge, contamination sources, and health implications. This integrated approach also facilitates the identification of heavy metal contamination, specifically targeting arsenic (As), chromium (Cr), lead (Pb), and manganese (Mn), all of which pose serious environmental and health risks. This study is unique in its investigation of these contaminants in the Middle Nile Delta, an area with limited previous research on this specific issue [44–50].

The problem statement of this study focuses on the contamination of groundwater by heavy metals in the Middle Nile Delta. Arsenic, chromium, lead, and manganese were chosen as target pollutants due to their widespread occurrence in the region and their known health impacts. Arsenic, chromium, and lead are recognized carcinogens, while manganese, although an essential nutrient, can become toxic at higher concentrations.

The methodology used to assess these pollutants includes both hydrogeochemical analyses and stable isotope techniques. Hydrogeochemical analyses provide valuable data on the concentration and distribution of major ions and heavy metals, offering insights into contamination levels and potential sources. However, these analyses are limited in their ability to detect long-term contamination trends or to account for complex interactions between pollutants. Stable isotope techniques, specifically those involving oxygen and hydrogen, provide crucial information on groundwater recharge sources, flow paths, and age, which helps trace how contaminants enter the system. Despite their powerful insights, isotopic methods are more resource-intensive, requiring specialized equipment and expertise, which can limit their accessibility for routine analysis. Despite these challenges, combining both techniques offers a holistic understanding of groundwater quality in the Nile Delta, providing essential data for improved management strategies.

This study aims to bridge critical gaps in understanding groundwater quality in the Middle Nile Delta by assessing contamination levels and health risks using a combination of geochemical and isotopic methods. The findings will contribute significantly to water resource management, environmental protection, and public health efforts in the region.

Study Area

The Nile Delta covers an area of about 25,000 km^2 in Northern Egypt and is located ~20 km north of Cairo (Figure 1). The study wells are in the middle part of the Nile Delta between latitudes 30°45′ and 30°57′ N and longitudes 30°45′ and 31°05′ E in an area of about 720 km^2. This area is mainly cultivated and intersected by a web of sealed roads that interconnect between major industrial and commercial centers, e.g., the cities of Tanta and Kafr El Zayat, which are considered the heart of the Nile Delta with many surrounding villages and hamlets.

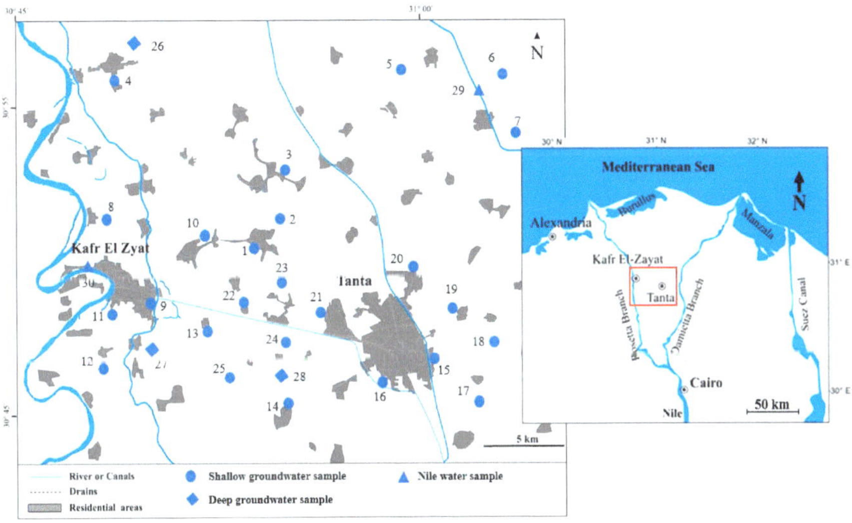

Figure 1. Study area in the Nile River Delta (Egypt) with sampling locations.

From a geological and hydrogeological point of view, the Nile Delta is covered by the soil of the Holocene alluvial plain. The sedimentary sequence of the Nile Delta basin is subdivided into three sedimentary cycles [51] of Miocene, Plio-Pleistocene, and Holocene age. Subsequently, the Plio-Pleistocene and Holocene sequences are subdivided into two rock units, the Bilqas Formation underlain by the Mit Ghamr Formation. These two formations constitute the primary aquifer of the Nile Delta [36,39].

The Bilqas Formation represents an organic-rich agricultural soil plain with predominantly fluvisols that consist of mixed clay and silty clay fluviatile sediments deposited by overbank flooding with an approximate accumulation rate of 5 mm per year [52] and a thickness in the study area that ranges from 10 to 23 m [36,53]. The thickness decreases with increasing distance from the main Nile branches and reaches its lowest thickness towards the middle part of the study area. Interbedded within this layer are silt and sand lenses with a thickness between 3 and 5 m.

The underlying Mit Ghamr Formation of Plio-Pleistocene age is the main water-bearing formation in the Middle Nile Delta aquifer [36,39], with thicknesses that range from 500 to 700 m [54]. It is composed of graded sand and gravel with thin interbedded clay layers. These clay layers divide the primary aquifer into several connected and disconnected minor aquifers [55]. The Mit Ghamr Formation is considered a leaky aquifer in the Middle Nile Delta area, which is overlain by and hydraulically connected to the aquitard of the semi-pervious Bilqas Formation [37,56].

The groundwater flows from the southeast to the northwest following the Nile Delta topographic slope, with an average hydraulic gradient of 11 cm/km [57]. Discharge occurs freely northward to the Mediterranean Sea and the Rosetta branch of the Nile River and artificially by municipal and private wells. After the Aswan High Dam construction between 1960 and 1970, the piezometric levels of the aquifer ranged from 2 to 8 m above sea level (m.a.s.l.) and 0.5 to 5.0 from the ground surface [58]. According to Salem [59] and Masoud [60], extensive recharge from unsewered villages and irrigation water combined with the limited permeability of the surface clay layer leads to saturated infiltration conditions and waterlogged soils in large parts of the study area.

2. Materials and Methods

2.1. Field Work and Laboratory Analyses

A 28 groundwater samples were collected from wells with depths ranging between 25 and 120 m. The groundwater sampling protocol involved collecting 28 groundwater samples and two surface water samples using pre-cleaned polyethylene bottles, which were rinsed with sample water before collection. Sub-samples for heavy metal analysis were acidified with nitric acid to pH < 2, while others were kept at 4 °C to preserve their integrity until analysis. Major anions were determined using ion chromatography (Thermo Dionex ICS 2000), Sunnyvale, CA, U.S.A and major cations and trace metals were analyzed by ICP-MS (Thermo iCAP-Q), Bremen, Germany. Calibration was performed using standard solutions, and quality control measures included the use of blanks, duplicates, and certified reference materials to ensure accuracy and precision.

Groundwater samples were selected across different land use types: cultivated, residential, and industrial areas. Additionally, two surface water samples from the Nile River were collected for comparative analysis. The coordinates of the sampling sites were recorded using a portable GPS device (Figure 1). The following physicochemical parameters were measured directly in the field using a multi-parameter portable instrument (YSI model 63): temperature (T), pH, electrical conductivity (EC), and total dissolved solids (TDS), the latter being calculated from EC. The wells were hand-pumped for 10–15 min before sampling to ensure that the measured parameters were stabilized. The total hardness (TH) of the groundwater was calculated according to the formula suggested by Sawyer et al. [61]:

$$\mathrm{TH(asCaCO_3)mg/L} = \left(\mathrm{Ca^{2+}} + \mathrm{Mg^{2+}}\right)\mathrm{meq/L} \times 50 \quad (1)$$

Major anions were analyzed using ion chromatography (Thermo Dionex ICS 2000), while major cations and trace metals were determined by inductively coupled plasma-mass spectrometry (ICP-MS) (Thermo iCAP-Q). For isotopic analysis, the oxygen and hydrogen isotope ratios in the water samples were measured using wavelength-scanned cavity ring-down infrared spectroscopy (CRDS), employing a Picarro L1102-i instrument, which was coupled with a vaporization module. Post-run data corrections were applied according to the methodology described by van Geldern and Barth [62]. Isotopic results are reported in the conventional delta (δ) notation, calculated with respect to Vienna Standard Mean Ocean Water (V-SMOW) and expressed in per mil (‰):

$$\delta = \left(\frac{Rsample}{Rreference} - 1\right) \quad (2)$$

where R represents the $^{18}O/^{16}O$ or the $^{2}H/^{1}H$ stable isotope ratios [63]. The external reproducibility for the isotope measurements was found to be better than 0.1‰ for $\delta^{18}OH_2O$ and $\delta^{2}HH_2O$, as determined by control standards.

All chemical analyses were performed at Geozentrum Nordbayern (GZN), Erlangen, Germany.

2.2. Groundwater Quality Assessment

The groundwater quality index (GWQI) was used to assess the suitability of the groundwater for drinking purposes. The GWQI was calculated using the following equation [64]:

$$\mathrm{GWQI} = \sum \mathrm{SI_i} = \sum (\mathrm{W_i} + \mathrm{q_i}) = \sum \left(\left(\frac{\mathrm{w_i}}{\sum_{i=1}^{n} \mathrm{w_i}} \right) \times \left(\frac{\mathrm{C_i}}{\mathrm{S_i}} \times 100 \right) \right) \quad (3)$$

where C_i is the concentration of each parameter, Si is the limit values concerning WHO standards [26], w_i is the assigned weight according to its relative importance in the overall quality of water for drinking purposes (Table 1), q_i is the water quality rating, W_i is the relative weight, and SI_i is the subindex of *ith* parameter. In this study, the parameters that have been considered are pH, TDS, Na^+, Mg^{2+}, Ca^{2+}, Cl^-, SO_4^{2-}, Mn, Fe, Cu, Zn, As, and Pb.

Table 1. Relative weight of chemical parameters in the groundwater quality index (GWQI). * WHO—World Health Organization [26].

Parameter	WHO * Standards	Weight, w_i	Relative Weight, W_i
pH	7	3	0.06
TDS (mg/L)	500	5	0.11
Na^+ (mg/L)	200	2	0.04
Mg^{2+} (mg/L)	30	3	0.06
Ca^{2+} (mg/L)	75	3	0.06
Cl^- (mg/L)	250	5	0.11
SO_4^{2-} (mg/L)	250	4	0.09
Fe (mg/L)	0.3	4	0.09
Mn (mg/L)	0.4	4	0.09
Cu (mg/L)	2	2	0.04
Zn (mg/L)	5	3	0.06
As (mg/L)	0.01	4	0.09
Pb (mg/L)	0.01	5	0.11
		$\sum w_i = 47$	$\sum W_i = 1.0$

* World Health Organization.

Suitability for irrigation. Assessment of the suitability of groundwater for agricultural irrigation purposes is vital for soil productivity and crop yield [65]. In this study, the sodium percentage (Na%) parameter is applied to assess the suitability of the groundwater for irrigation, which is expressed as follows [56]:

$$\mathrm{Na}\ \% = \frac{\mathrm{Na^+} + \mathrm{K^+}}{\mathrm{Na^+} + \mathrm{K^+} + \mathrm{Ca^{2+}} + \mathrm{Mg^{2+}}} \times 100 \quad (4)$$

2.3. Health Risk Estimation

The human health risk from heavy metals in drinking water was evaluated following established protocols [66–69]. Health risks from ingestion were considered the primary pathway, as inhalation and dermal contact were deemed negligible. The hazard quotient (*HQ*) for each heavy metal was calculated using the formula from the US EPA model [70–72]:

$$HQ = \frac{ADD}{RfD} \tag{5}$$

with

$$ADD = \frac{C \times IR \times EF \times ED}{BW \times AT} \tag{6}$$

where *ADD* is he average daily dose (mg/kg-day), *RfD* is the reference dose for different metals, USEPA, 2012 [73], *C* is the concentration of the heavy metal in groundwater (mg/L), IR is the intake rate (2 L/day), *EF* is the exposure frequency (365 day/year), *ED* is the exposure duration (in this study, 60 years), *BW* is the average body weight (70 kg), and *AT* is the averaging time for non-carcinogens (21,900 days). An excess risk exists when the *HQ* is greater than 1.

3. Results

Table 2 lists the physicochemical properties, major and trace element concentrations, and stable isotope values of the analyzed surface and groundwater in the studied area.

3.1. Physicochemical Parameters (pH, Eh, TDS, and TH)

The pH values of the groundwater wells ranged from 7.4 to 8.2 for different groundwater samples. These values were comparable to the Nile River water samples' pH, which was between 8.0 and 8.2. As the WHO prescribes, the allowable pH limit for drinking water is 6.5 to 8.5 [26].

Along with pH, Eh is the other physicochemical parameter affecting elemental mobility. In shallow groundwater wells, Eh values ranged from −14 mV to +23 mV, indicating the water is weakly reducing to weakly oxidizing. The low Eh values may be attributed to the waterlogging problem, where the reduction processes in waterlogged soils result from the anaerobic respiration of soil bacteria [73].

The total dissolved solids (TDS) ranged between 275 and 1274 mg/L, while the deep groundwater wells showed values from 288 to 608 mg/L. The high concentration values were observed in the southeastern and western sections of the study area. In contrast, the Nile River samples had lower TDS values (279 and 394 mg/L). The TDS values of 64% of the shallow groundwater wells were in critical condition according to the US EPA permissible limit (500 mg/L).

The values of total hardness (TH) of groundwater wells ranged from 126 to 496 mg/L $CaCO_3$ and indicate that the studied groundwater falls into the hard water category (>120 mg/L $CaCO_3$; WHO, 2011). TH was higher in the shallow (average 299 mg/L) than the deep groundwater (average 210 mg/L) and surface Nile water (140 and 177 mg/L). A range of epidemiological evidence has demonstrated the relationship between health risk and long-term consumption of hard water. Its effects on human health include cardiovascular disease, cerebrovascular mortality, central nervous system malformations, childhood atopic dermatitis, digestive health and constipation, bone mineral density, and urolithiasis and kidney stones [74–77].

Table 2. Geochemical and isotopic composition of the groundwater of the area studied.

S. No.	pH	Eh (mV)	EC (µS/cm)	TDS (mg/L)	TH (mg/L CaCO3)	K^+ (mg/L)	Na^+ (mg/L)	Ca^{2+} (mg/L)	Mg^{2+} (mg/L)	HCO_3^- (mg/L)	SO_4^{2-} (mg/L)	Cl^- (mg/L)	As (µg/L)	Cu (µg/L)	Cr (µg/L)	Fe (µg/L)	Mn (µg/L)	Pb (µg/L)	Sr (µg/L)	Zn (µg/L)	δ^2HH_2O (‰)	$\delta^{18}OH_2O$ (‰)
Shallow Groundwater																						
1	8.1	−14	624	411.5	254.3	4.6	31	70.8	19	273	1.2	73.9	0.5	1.24	0.52	270	670	2.4	490	170	8.61	0.46
2	7.9	3	710	451.2	315.8	3.6	43	82.4	27	273	1.1	136.1	0.5	0.54	1.13	360	690	0.5	690	2200	11.03	0.79
3	7.99	12	823	553	242.2	3.5	36	64.3	20	265	0.6	77.1	0.6	0.5	0.67	600	670	0.6	450	310	10.25	0.65
4	7.98	−5	1572	1002	353.8	5.4	200	96	28	525	38.6	195.0	0.5	0.5	0.82	410	1000	0.5	720	890	16.78	1.68
5	7.86	0	931	595	226.1	3.9	110	57.8	20	331	23.9	124.7	0.5	1.07	0.98	63	770	0.9	689	482	−5.44	−1.39
6	7.8	0	1004	651	262.4	3.5	110	65.8	24	299	43.3	154.8	0.5	0.5	0.96	240	410	0.5	810	260	8.94	0.39
7	7.78	−1	776	500	298.6	3.1	55	75.4	27	397	48.9	34.2	0.5	0.56	0.50	420	530	0.5	750	4600	20.68	2.28
8	7.93	−9	1161	752	339.1	5.0	130	95	25	516	50.7	109.5	0.5	1.3	0.50	19	1700	0.5	750	12.8	21.19	2.3
9	7.92	−7	800	523	311.0	3.9	36	83.7	25	292	0.7	111.1	0.7	0.5	2.18	990	460	0.5	540	520	7.38	0.23
10	8.21	20	961	614	211.7	4.1	140	57	17	479	51.8	54.2	0.5	57.6	0.70	240	530	9.4	440	630	27.61	3.29
11	8.12	−12	928	607	354.4	5.0	47	99.5	26	310	32.8	126	0.5	4.5	0.97	640	530	0.5	710	230	8.94	0.47
12	7.89	−4	1957	1274	442.0	41.0	260	110	41	678	203	209	0.5	2.5	0.50	57.3	1600	0.5	1300	600	19.36	2.2
13	8.11	−13	523	338.6	195.3	3.3	28	53.7	15	232	2.8	48.7	0.5	0.5	0.70	180	410	1.0	360	520	11.08	0.79
14	7.43	23	1057	681.6	466.6	3.8	48	110	47	511	54.9	66.3	0.5	0.9	0.50	100	1700	0.6	640	1600	23.45	2.71
15	7.97	−7	558	356.5	319.6	3.4	52	87.2	25	407	3.6	72.3	1.4	0.5	0.50	290	1300	0.5	690	949	11.51	0.74
16	7.52	15	1198	788.5	367.9	4.5	110	100	29	680	26	21.8	0.5	1.06	0.72	13.4	630.4	1.2	890	280	19.38	2.23
17	7.9	−5	419	274	132.1	2.8	36	36.5	10	169	19	39.9	0.5	0.65	1.55	340	360	0.5	330	150	12.62	0.95
18	7.58	12	1571	1104	496	4.2	180	130	42	342	280	230	0.5	1.07	1.25	450	1500	1.0	940	1400	12.74	1.16
19	7.88	0	571	378	192.4	2.2	53	52.5	15	340	8.4	17.5	0.63	0.91	1.34	150	520	0.7	450	409	10.16	0.65
20	7.97	9	654	424	142.1	2.1	60	35.6	13	245	120	250	0.5	2.17	0.50	16.8	220.3	0.5	380	390	16.36	1.79
21	7.96	−4	656	438.4	243.6	5.7	55	61.6	22	313	73.2	28	0.9	5.76	1.91	1100	1300	2.2	480	840	22.49	2.51
22	8.07	−7	795	518	322.5	4.0	50	86.7	26	454	1.7	46	0.5	0.5	0.50	18.7	970	0.5	590	400	9.6	0.54
23	8.1	−11	721	487	213.5	4.6	90	54.4	19	385	45.1	38.2	0.5	0.87	0.58	510	850	1.9	430	57.8	25.98	3.1
24	8.08	−11	804	531	345.5	4.1	43	95.9	26	448	1.6	55	0.5	0.5	0.50	369.8	1100	0.6	690	160	15.07	1.32
25	7.97	−5	981	641	401.0	4.5	49	110	31	551	0.7	45.6	0.5	0.5	0.50	21.4	1000	0.5	760	150	11.64	0.84
Min	7.43	−14	419	274	132	2.1	28	35.6	10	169.4	0.6	17.5	0.5	0.5	0.5	13.4	220.3	0.5	330	12.8	−5.44	−1.39
Max	8.21	23	1957	1274	496	41	260	130	47	680	280	250	1.4	57.6	2.2	1100	1700	9.4	1300	4600	27.61	3.29
avg.	7.9	−0.4	930	609	299	6.6	86.6	79.1	25	391.3	54.6	97.4	0.6	5.4	0.89	332.6	864	1.4	651	845	14.1	1.2
Deep Groundwater																						
26	7.8	3	936	608	324.8	4.29	53	84.3	28	226	15.0	175	0.5	0.5	0.5	170	790	0.07	730	11	8.74	0.42
27	7.87	0	446	288	147.5	6.22	35	39.4	12	241	0.6	24.7	0.5	5.4	0.67	160	370	0.5	270	5.9	1.4	−0.48
28	8	−8	351	233	126.3	3.25	16	34.4	9.9	163	0.8	22.7	0.5	0.5	0.5	210	340	0.5	240	5.8	8.89	0.49
Min	7.8	−8	351	232	126.2	3.2	16	34.4	9.9	162.6	0.6	22.7	0.5	0.5	0.5	160	340	0.07	240	5.8	1.4	−0.48
Max	8	3	936	608	324.7	6.2	53	84.3	28	240.6	15	175	0.5	5.4	0.67	210	790	0.5	730	11	8.89	0.49
avg.	7.8	−2	604	394	209.9	4.6	34.6	55.36	17.56	206.5	5.4	84.02	0.5	2.46	0.56	182	526	0.32	442	7.9	5.8	0.08
Nile Water																						
29	8.07	5	476	279	139.6	5.4	36	36.3	12	203	27	22.1	0.5	2.07	3.6	440	71.9	0.53	310	8.4	20.3	2.22
30	8.25	−18	658	394	176.6	7.5	56	44.6	16	240	44.7	47	0.6	1.05	0.8	100	98.3	0.50	389	4.8	20.5	2.36

3.2. Major Ions

Chloride concentration varied between 17.5 and 250 mg/L for the shallow groundwater wells, whereas the deeper wells (depth > 60 m) showed higher chloride concentrations (avg. 175 mg/L) compared to wells with depths of 100 m (23 mg/L) and 120 m (25 mg/L). In Nile River water, the average chloride concentration was 34.5 mg/L. Leaching of soil and domestic and municipal effluents are the potential main sources of chloride. For all studied wells, the chloride concentration is below the WHO and Egypt's national permissible limit of 250 mg/L [26].

Chloride and sulfate concentrations ranged widely from 0.6 to 280 mg/L in the shallow groundwater, whereas deeper wells generally displayed lower concentrations (0.6 to 15.0 mg/L).

Except for a single well (no. 18; Figure 1), all studied groundwater possesses SO_4^{2-} content below the permissible limit of 250 mg/L. The surface water of the Rosetta branch of the Nile River had a higher sulfate content (44.5 mg/L) than the Damietta branch (27 mg/L) (Figure 1). This probably reflects the impact of industrial wastewater discharging from the chemical industry factories along the Nile River Rosetta branch.

The HCO_3^- concentration varied between 163 and 680 mg/L in the studied groundwater. The shallow groundwater wells displayed relatively high concentrations (avg. 388 mg/L) compared to the deep wells (avg. 210 mg/L) and Nile water samples (avg. 222 mg/L).

The concentrations of Ca^{2+}, Na^+, K^+, and Mg^{2+} were in the range of 38–130, 28–260, 2.1–41, and 27.8–47 mg/L for the water in shallow wells and 34–48, 16–53, 3.2–6.2, and 10–28 mg/L for the water in deep wells, respectively. Except for Na^+ in shallow groundwater wells, the other major cations are within the permissible limits of Egypt's national standards. Concerning Na^+, two shallow groundwater wells (nos. 4 and 19) have critical values of 200 and 260 mg/L, respectively, which are above the acceptable limit of 200 mg/L. Lower sodium levels in drinking water (30–60 mg/L) are essential for human health, but high levels can cause hypertension [78].

The Piper diagram [79] has been frequently applied to investigate the hydro-chemical facies of groundwater wells, the evolution of phreatic water and understanding the hydro-chemical characteristics, as well as the formation mechanism of groundwater (Lu et al., 2010). The relative abundance of anions and cations is $HCO_3^- > SO_4^{2-} > Cl^-$ and $Ca^{2+} > Na^+ > Mg^{2+} > K^+$, respectively, for most of the studied wells. The Piper diagram (Figure 2) indicates that deep wells and most of the shallow wells (64%) are of the Ca-HCO_3 type, while the other samples (28%) are of the Na-HCO_3 type. Wells nos. 18 and 20 are of the Na-Cl type, which suggests that the shallow groundwater in the western part of the study area might be affected by sewage infiltration. The Na-HCO_3 water type may indicate an ion exchange between sodium-rich soil and the infiltrating surface water with higher calcium and magnesium concentrations [59].

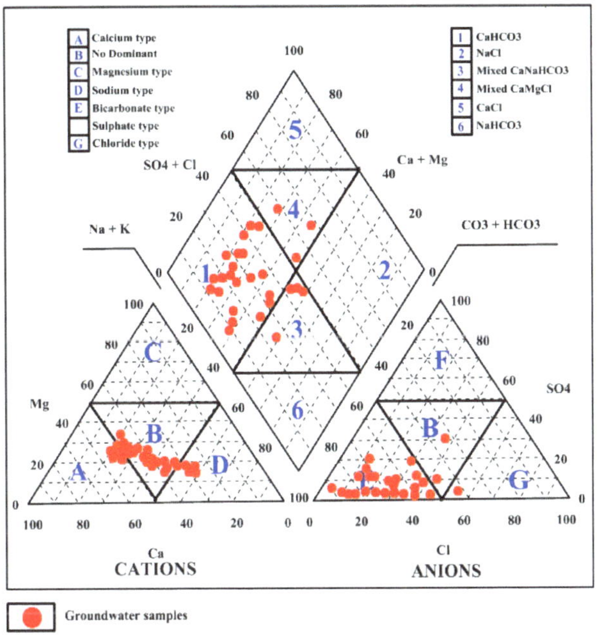

Figure 2. Piper plot of surface water and groundwater samples collected from the central area of the Nile Delta.

3.3. Stable Isotopes

The $\delta^{18}O$ and δ^2H values of the shallow hand-pumped water samples (Figure 3) correlate along a regression line with $\delta^2H = 6.6 \times \delta^{18}O + 5.6$. The oxygen and hydrogen isotopes of the water molecule can be used to categorize the sampled water into two main groups representing (1) surface water and (2) deep groundwater endmembers (Figure 3). The surface water was enriched in the heavy isotopes, showing δ^2H values of ~+20‰ and $\delta^{18}O$ values from +2.2 to +2.5‰. In contrast, the deep groundwater samples ranged from +1.4 to +8.9‰ for δ^2H and between −0.5 and +0.5‰ for $\delta^{18}O$.

The samples from the deep well plot in the paleo-water field represent the isotopic signature of the Nile River before the completion of the first stage of the Aswan High Dam in 1964 [32,59,80,81]. The broader range of isotope values for the deep wells might indicate that portions of paleo-groundwater, formed under cooler and more humid climatic conditions in the Pleistocene aquifer of the floodplain, are admixed here [82].

The higher isotope values of the shallow groundwater can be attributed to the extensive evaporation processes in Lake Nasser, the Aswan High Dam reservoir, with losses of about 19% of lake water [83], and through the circulation of different irrigation canals [32]. This process enriches the heavy stable isotopes ^2H and ^{18}O in the surface (recent Nile) water [83–85].

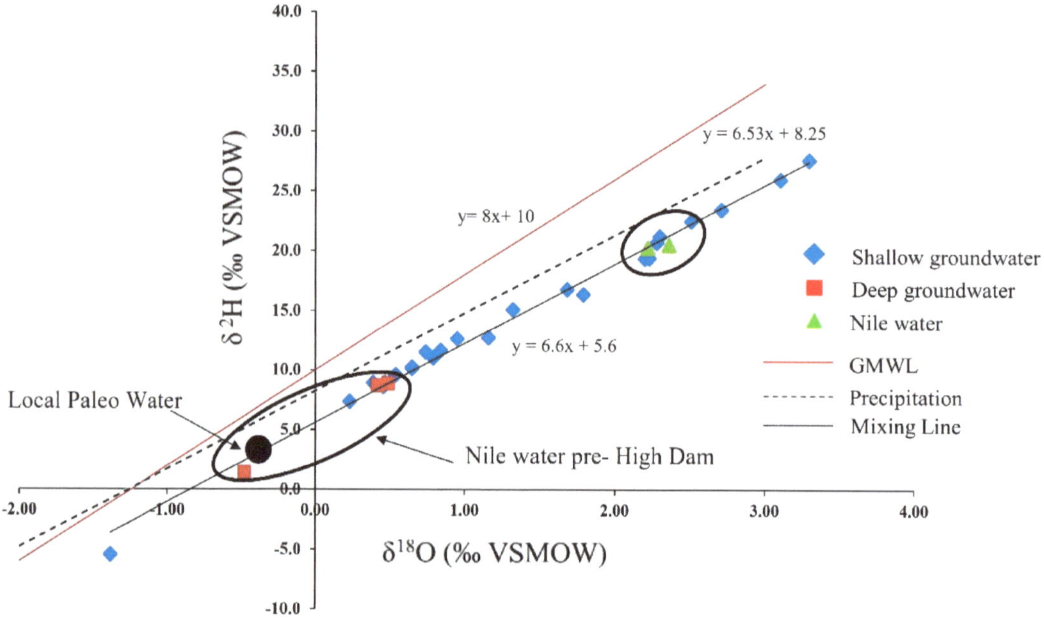

Figure 3. Plot of $\delta^{18}O$ and $\delta^{2}H$ for groundwater from the Nile Delta aquifer system with plots of the GMWL according to Rozanski et al. [86] and Nile LPW according to Awad et al. [87].

Figure 3 shows, however, that the shallow groundwater wells can be categorized into three categories: (1) wells with the same isotopic characteristics as deep groundwater (nos. 1, 8, 14, 17, and 36), (2) wells that displayed the same isotopic characteristics as recent Nile water (sample nos. 9, 10, 28, and 34), and (3) mixed water samples that fall along a mixing line between both endmembers. Water samples with isotope values higher than recent Nile water indicate direct recharge from surface irrigation water subject to evaporation before infiltration or from recycling drainage water [29,32,88]. After infiltration, further evaporation of soil water is suspected because of the rising water table during the last few decades, particularly after the completion of the first stage of the Aswan High Dam in 1964.

The spatial distribution pattern of the oxygen and hydrogen isotope ratios (Figure 4) indicates that the recent Nile water is distributed in the study area's western, southern, and eastern parts. On the other hand, the aquifers holding older water are in the northern, central, and southeastern parts, while aquifers with mixed recharged water are located in between. This pattern most probably reflects the spatial distribution of impermeable clay lenses within the Mit Ghamr Fm [36]. The presence or absence of such clay lenses tends to subdivide the main aquifer into sub-restricted or perched aquifers with variable portions of recent Nile water. Samples of paleo-water and recent Nile water are both Ca-HCO_3 type, while most mixed water samples belong to the Na-HCO_3 and Na-Cl water types.

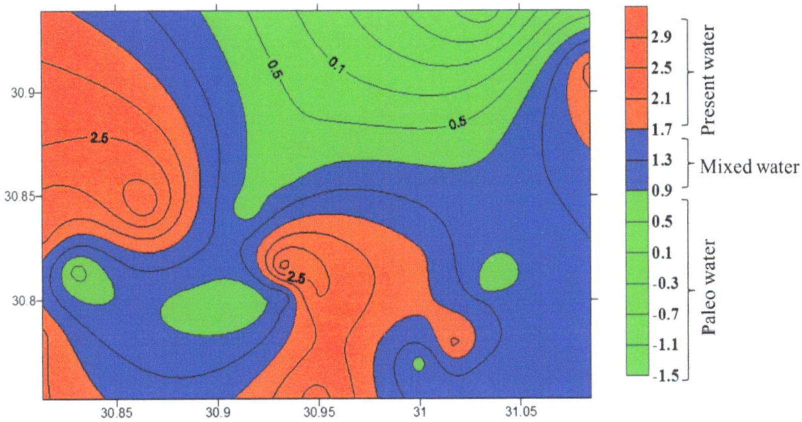

Figure 4. Spatial distribution map of $\delta^{18}O$ for shallow groundwater from the Nile Delta aquifer system.

3.4. Heavy Metal Contents

The total concentrations of potentially toxic metals in the studied groundwater wells are tabulated in Table 2.

Among the trace elements, Mn and Fe represent critical concentrations compared to the acceptable limits. The manganese concentration in 80% of shallow groundwater samples exceeds the WHO standard limit of 400 μg/L [26]. In shallow groundwater, the Mn concentration ranges from 220 to 1700 μg/L, while in deep groundwater, it ranges from 340 to 790 μg/L, whereas lower values of 72 to 98 μg/L were measured in Nile River water. The spatial distribution of high Mn concentrations is related to the old and highly populated residential areas in the study area's western and southern parts suffering from waterlogging problems.

Forty-four percent of the shallow groundwater exceeded the permissible limit of 300 μg/L. Fe concentrations ranged from 13 to 1100 μg/L in shallow groundwater, 160 to 210 μg/L in deep groundwater, and 100 to 400 μg/L in Nile River water. The spatial distribution of the iron concentration suggests that the pattern is related to the land use type, where the high Fe contents are related to cultivated lands, small unsewered villages, and industrial areas such as Kafr El Zayat city.

The dissimilarity of the spatial distribution of Fe and Mn and the reverse correlation between them (Table 3) indicates the different sources of these elements. Slow groundwater runoff in the poorly drained cultivated lands and waterlogged soil of unsewered villages provides appropriate conditions for the accumulation of soluble Fe and Mn. Reducing Fe and Mn is one of the most critical chemical transformations in waterlogged soils [73]. The uppermost part of cultivated soil is enriched in Fe-organic content compared to urban soil [89]. In addition, industrial activities may contribute in part to the excessive Fe.

On the other hand, the frequently occurring Mn concentration may indicate a pedogenic source rather than anthropogenic. The correlation between Mn and Ca^{2+}, Mg^{2+}, and Sr suggests that Mn is strongly associated with carbonate minerals of the water-bearing layer (Table 3). Under acidic and reducing conditions, the dissolution of Mn-carbonates from poorly drained soils close to water channels and drains can be the primary source of Mn.

Table 3. Correlation matrix of studied physicochemical parameters and major and trace elements of the groundwater in the study area.

	pH	Eh	EC	TDS	TH	K	Na	Ca	Mg	HCO_3	SO_4	Cl	Cr	Mn	Fe	Cu	Zn	As	Sr	Pb
pH	1.00																			
Eh	−0.57	1.00																		
EC	−0.37	0.23	1.00																	
TDS	−0.38	0.24	1.00	1.00																
TH	−0.48	0.16	0.77	0.78	1.00															
K	−0.02	−0.08	0.60	0.58	0.33	1.00														
Na	−0.20	0.18	0.89	0.89	0.48	0.63	1.00													
Ca	−0.41	0.09	0.75	0.76	0.99	0.28	0.46	1.00												
Mg	−0.59	0.28	0.77	0.78	0.97	0.38	0.49	0.92	1.00											
HCO_3	−0.27	0.18	0.72	0.70	0.68	0.45	0.63	0.68	0.65	1.00										
SO_4	−0.36	0.30	0.70	0.73	0.46	0.48	0.75	0.41	0.52	0.29	1.00									
Cl	−0.18	0.13	0.62	0.62	0.39	0.32	0.57	0.36	0.40	0.07	0.66	1.00								
Cr	0.06	0.06	−0.22	−0.23	−0.27	−0.09	−0.15	−0.28	−0.25	−0.37	0.00	−0.15	1.00							
Mn	−0.34	0.04	0.61	0.62	0.76	0.37	0.49	0.74	0.77	0.62	0.46	0.24	−0.32	1.00						
Fe	0.19	−0.17	−0.15	−0.13	−0.05	−0.14	−0.18	−0.05	−0.06	−0.33	0.00	−0.08	0.52	−0.07	1.00					
Cu	0.31	0.40	0.04	0.03	−0.15	0.00	0.21	−0.15	−0.15	0.14	0.05	−0.11	−0.03	−0.11	0.00	1.00				
Zn	−0.34	0.21	0.12	0.11	0.28	−0.04	0.05	0.23	0.36	0.12	0.21	0.04	−0.09	0.13	0.14	−0.02	1.00			
As	0.06	−0.14	−0.22	−0.21	0.02	−0.07	−0.15	0.03	−0.01	−0.01	−0.12	−0.13	0.09	0.24	0.32	−0.05	0.06	1.00		
Sr	−0.43	0.07	0.86	0.85	0.84	0.58	0.70	0.82	0.82	0.70	0.53	0.52	−0.27	0.62	−0.19	−0.15	0.26	−0.02	1.00	
Pb	0.34	0.33	0.01	0.01	−0.16	−0.05	0.19	−0.15	−0.16	0.15	0.04	−0.18	−0.04	−0.07	0.06	0.95	−0.02	−0.03	−0.18	1.00

3.5. Water Quality Assessment

The calculated GWQI indicated that deep groundwater samples are categorized as "excellent" to "good" quality for drinking water (Table 4). The GWQI values categorized the samples into three categories for shallow groundwater samples, ranging from "excellent" to "poor" water quality. Only 4% of the total wells show "excellent" quality, most of the samples (64%) are categorized as "good" quality, and the rest (32%) exhibit "poor" quality [64]. Spatially, there is no circumscribed distribution of "excellent"- and "good"-quality samples for shallow wells (Figure 5). The eastern and southern parts of the study area, which have waterlogged soils, are in critical condition. Poor-quality water wells roughly correlated with the spatial distribution of high Mn concentration and enriched stable isotope values. However, it is noted that a weak correlation ($r = 0.56$) exists between GWQI values and isotope values. Water samples with higher, evaporative-enriched isotope values display higher levels of substances found in samples with undesirable GWQI values.

Table 4. GWQI classification of the groundwater in the area studied.

S. No	GWQI	Water Type
1	77.2	Good water
2	85.4	Good water
3	82.2	Good water
4	118.9	Poor water
5	83.7	Good water
6	64.1	Good water
7	78.0	Good water
8	147.9	Poor water
9	83.0	Good water
10	169.2	Poor water
11	89.0	Good water
12	162.2	Poor water
13	54.0	Good water
14	155.4	Poor water
15	120.6	Poor water
16	76.1	Good water
17	52.9	Good water
18	165.9	Poor water
19	62.2	Good water
20	48.4	Excellent water
21	154.4	Poor water
22	90.6	Good water
23	94.8	Good water
24	110.0	Poor water
25	96.1	Good water
26	87.1	Good water
27	55.2	Good water
28	44.2	Excellent water

The methods used in Figures 4 and 5 likely involve spatial interpolation techniques such as Kriging or inverse distance weighting (IDW) to visualize the distribution of $\delta^{18}O$ and groundwater quality index (GWQI) values across the study area. These techniques are commonly used in hydrogeological studies to estimate values at unsampled locations based on nearby measured data points.

For Figure 4, the distribution of $\delta^{18}O$ values suggests that recent Nile water and older groundwater are separated due to geological barriers, particularly impermeable clay lenses within the Mit Ghamr Formation. The isotope analysis, typically conducted using mass spectrometry, helps differentiate water sources by identifying evaporation effects and recharge origins [88].

For Figure 5, the groundwater quality index (GWQI) was calculated using a weighted sum of key water quality parameters such as pH, total dissolved solids (TDS), Mn concentration, and major ions. The spatial mapping of GWQI values likely employed geographic information system (GIS) tools to identify areas with poor water quality, which correlated with high Mn concentrations and isotope enrichment.

These findings align with previous hydrogeological studies that emphasize the role of aquifer heterogeneity in water quality variations [89,90].

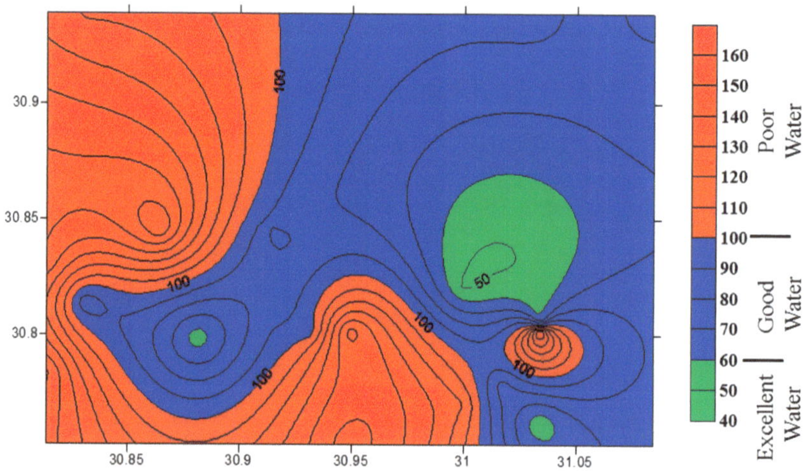

Figure 5. Spatial distribution map of GWQI in the Central Nile Delta groundwater.

3.6. Irrigation Suitability

Wilcox [56] suggested a graphical method based on the sodium percentage parameter (Na%) for irrigation purposes. Sodium percentage values classify the water into five classes: 'excellent' (<20 Na%), 'good' (20–40 Na%), 'permissible' (40–60 Na%), 'doubtful' (60–80 Na%), and 'unsuitable' (>80 Na%). The Wilcox diagram (Figure 6) revealed that 40% of the samples belong to the excellent-to-good class and 50% are good to permissible, while 10% of the groundwater samples are categorized as permissible-to-doubtful irrigation water.

3.7. Health Risk Assessment

The calculated values of health risk are listed in Table 5. HQ values ranked the heavy metals in the order of Mn > Cr > Zn >As > Fe > Pb > Cu based on their detriments. Except for Mn, the HQ values for the evaluated heavy metals were less than one, which indicates little or no risk to residents. On the other hand, Mn shows HQ values indicating an unacceptable non-carcinogenic health risk for 14 wells, representing 56% of the total shallow groundwater wells, mainly located in the southern and western areas of the study region. Manganese is an essential microelement in nutrition, and Mn deficiency (less than 0.1 mg/day) can cause health problems (e.g., weight gain, glucose intolerance, blood clotting, skin problems, lowered cholesterol levels, skeleton disorders, and congenital disabilities). It contributes to maintaining healthy nerves, the immune system, and helps reduce blood sugar regulation symptoms [91].

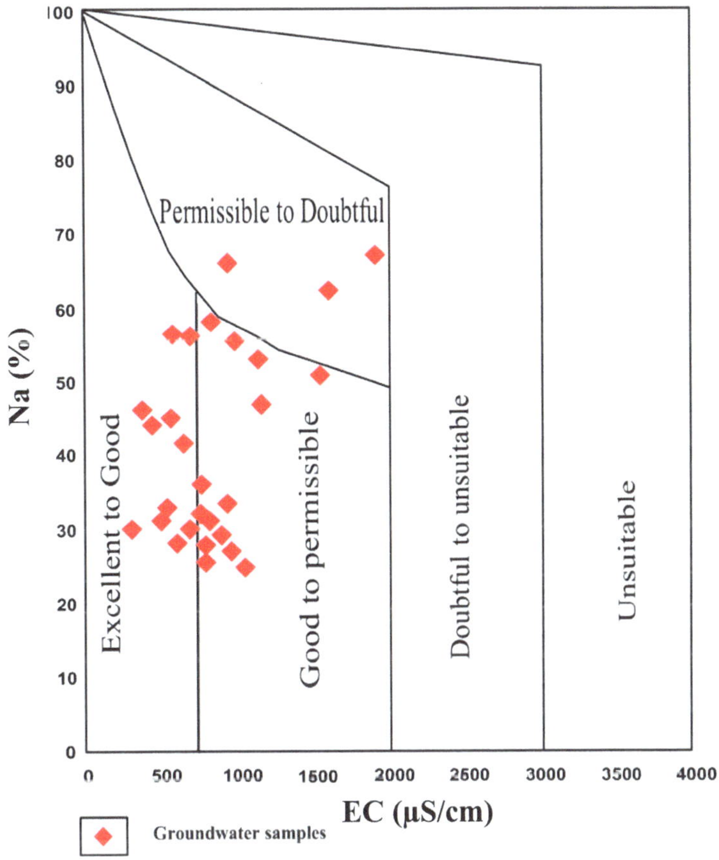

Figure 6. Wilcox's diagram illustrates the suitability of groundwater for irrigation based on electrical conductivity and sodium percentage.

Table 5. Non-cancer health risk (HQ) from heavy metals in the water at each surface and groundwater site (HQ values > 1.0 are shown in bold).

S No.	Mn	Fe	Cu	As	Pb	Cr	Zn
1	9.57×10^{-1}	2.57×10^{-2}	8.91×10^{-4}	4.76×10^{-2}	1.91×10^{-2}	5.00×10^{-2}	1.62×10^{-2}
2	9.86×10^{-1}	3.43×10^{-2}	3.90×10^{-4}	4.76×10^{-2}	3.97×10^{-3}	1.07×10^{-1}	2.10×10^{-1}
3	9.57×10^{-1}	5.72×10^{-2}	3.57×10^{-4}	5.71×10^{-2}	4.45×10^{-3}	6.38×10^{-2}	2.95×10^{-2}
4	1.43	3.90×10^{-2}	3.57×10^{-4}	4.76×10^{-2}	3.97×10^{-3}	7.82×10^{-2}	8.48×10^{-2}
5	1.10	6.07×10^{-3}	7.67×10^{-4}	4.76×10^{-2}	6.80×10^{-3}	9.36×10^{-2}	4.60×10^{-2}
6	5.86×10^{-1}	2.29×10^{-2}	3.57×10^{-4}	4.76×10^{-2}	3.97×10^{-3}	9.11×10^{-2}	2.48×10^{-2}
7	7.57×10^{-1}	4.00×10^{-2}	4.00×10^{-4}	4.76×10^{-2}	3.97×10^{-3}	4.76×10^{-2}	4.38×10^{-1}
8	2.43	1.81×10^{-3}	9.14×10^{-4}	4.76×10^{-2}	3.97×10^{-3}	4.76×10^{-2}	1.23×10^{-3}
9	6.57×10^{-1}	9.43×10^{-2}	3.57×10^{-4}	6.88×10^{-2}	4.29×10^{-3}	2.08×10^{-1}	4.95×10^{-2}
10	7.57×10^{-1}	2.29×10^{-2}	4.12×10^{-2}	4.76×10^{-2}	7.49×10^{-2}	6.65×10^{-2}	6.00×10^{-2}
11	7.57×10^{-1}	6.10×10^{-2}	3.28×10^{-3}	4.76×10^{-2}	3.97×10^{-3}	9.25×10^{-2}	2.19×10^{-2}
12	2.29	5.46×10^{-3}	1.83×10^{-3}	4.76×10^{-2}	3.97×10^{-3}	4.76×10^{-2}	5.71×10^{-2}
13	5.86×10^{-1}	1.71×10^{-2}	3.57×10^{-4}	4.76×10^{-2}	8.28×10^{-3}	6.67×10^{-2}	4.95×10^{-2}
14	2.43	9.52×10^{-3}	6.38×10^{-4}	4.76×10^{-2}	5.01×10^{-3}	4.76×10^{-2}	1.52×10^{-1}
15	1.86	2.76×10^{-2}	3.57×10^{-4}	1.37×10^{-1}	3.97×10^{-3}	4.76×10^{-2}	9.04×10^{-2}

Table 5. *Cont.*

S No.	Mn	Fe	Cu	As	Pb	Cr	Zn
16	9.01×10^{-1}	1.28×10^{-3}	7.63×10^{-4}	4.76×10^{-2}	9.32×10^{-3}	6.88×10^{-2}	2.67×10^{-2}
17	5.14×10^{-1}	3.24×10^{-2}	4.65×10^{-4}	4.76×10^{-2}	3.97×10^{-3}	1.48×10^{-1}	1.43×10^{-2}
18	2.14	4.29×10^{-2}	7.65×10^{-4}	4.99×10^{-2}	7.91×10^{-3}	1.19×10^{-1}	1.33×10^{-1}
19	7.43×10^{-1}	1.43×10^{-2}	6.51×10^{-4}	5.99×10^{-2}	5.86×10^{-3}	1.27×10^{-1}	3.90×10^{-2}
20	3.15×10^{-1}	1.61×10^{-3}	1.56×10^{-3}	5.08×10^{-2}	3.97×10^{-3}	4.76×10^{-2}	3.71×10^{-2}
21	1.86	1.05×10^{-1}	4.12×10^{-3}	9.30×10^{-2}	1.73×10^{-2}	1.82×10^{-1}	8.00×10^{-2}
22	1.39	1.79×10^{-3}	3.57×10^{-4}	4.76×10^{-2}	3.97×10^{-3}	4.76×10^{-2}	3.81×10^{-2}
23	1.21	4.86×10^{-2}	6.22×10^{-4}	4.89×10^{-2}	1.49×10^{-2}	5.54×10^{-2}	5.50×10^{-3}
24	1.57	3.52×10^{-2}	3.57×10^{-4}	4.76×10^{-2}	5.03×10^{-3}	4.76×10^{-2}	1.53×10^{-2}
25	1.43	2.04×10^{-3}	3.57×10^{-4}	4.76×10^{-2}	3.97×10^{-3}	4.76×10^{-2}	1.43×10^{-2}
Min.	3.15×10^{-1}	1.28×10^{-3}	3.57×10^{-4}	4.76×10^{-2}	3.97×10^{-3}	4.76×10^{-2}	1.23×10^{-3}
Max.	2.43	1.05×10^{-1}	4.12×10^{-2}	1.37×10^{-1}	7.49×10^{-2}	2.08×10^{-1}	4.38×10^{-1}
Avg.	1.23	3.17×10^{-2}	3.85×10^{-3}	5.78×10^{-2}	1.15×10^{-2}	8.53×10^{-2}	8.05×10^{-2}
26	1.13	1.62×10^{-2}	3.57×10^{-4}	4.76×10^{-2}	5.64×10^{-4}	5.16×10^{-2}	1.05×10^{-3}
27	5.29×10^{-1}	1.52×10^{-2}	3.91×10^{-3}	4.76×10^{-2}	3.97×10^{-3}	6.45×10^{-2}	5.67×10^{-4}
28	4.86×10^{-1}	2.00×10^{-2}	3.57×10^{-4}	4.76×10^{-2}	3.97×10^{-3}	4.87×10^{-2}	5.55×10^{-4}
Min.	4.86×10^{-1}	1.52×10^{-2}	3.57×10^{-4}	4.76×10^{-2}	5.64×10^{-4}	4.87×10^{-2}	5.55×10^{-4}
Max.	1.13	2.00×10^{-2}	3.91×10^{-3}	4.76×10^{-2}	3.97×10^{-3}	6.45×10^{-2}	1.05×10^{-3}
Avg.	7.51×10^{-1}	1.73×10^{-2}	1.78×10^{-3}	4.76×10^{-2}	2.61×10^{-3}	5.56×10^{-2}	7.54×10^{-4}
R*f*D Mg/kg/day	2.00×10^{-2}	3.00×10^{-1}	4.00×10^{-2}	3.00×10^{-4}	3.60×10^{-3}	3.00×10^{-3}	3.00×10^{-1}

On the other side, it is toxic at higher concentrations (>300 µg/L, USEPA [92]), causing Parkinson's disease, hallucinations, forgetfulness, lung embolism and bronchitis, and nerve damage [93], affects fertility in mammals, and is toxic to the embryo and fetus [94]. Manganese neurotoxicity results from an accumulation of metal in brain tissue [95,96].

It is worth mentioning that the intake of Mn by direct water consumption is not the only source of ingestion. Residents could have deleterious health effects from Mn bioaccumulation through the food chain since shallow groundwater is the primary drinking water source for all domestic animals, including cattle, poultry, and other livestock.

The study acknowledges the importance of broadening the scope of health risk assessment beyond manganese (Mn) to encompass other heavy metals. Future research will aim to conduct a more comprehensive evaluation of cumulative exposure and potential synergistic effects among multiple contaminants. Additionally, while the current analysis primarily focuses on direct water consumption as the main exposure pathway, other potential routes such as inhalation of aerosols, dermal absorption, and consumption of contaminated agricultural products will be explored to provide a more holistic understanding of exposure risks.

Moreover, the study recognizes the need for demographic-specific analysis to account for variations in susceptibility among different population groups, including children, pregnant women, and the elderly. Incorporating these factors into future assessments will enhance the accuracy of risk estimations. The temporal aspect of heavy metal concentrations is another area for further investigation. Examining seasonal or temporal variations could provide deeper insights into how exposure levels and associated health risks fluctuate over time.

Lastly, the study highlights the need to propose practical mitigation measures alongside identifying health risks. Future work will focus on recommending strategies to reduce exposure, improve water quality, and safeguard public health, thereby enhancing the study's practical implications. These considerations will be prioritized in subsequent research to address the limitations and strengthen the overall assessment.

4. Discussion

The results of this study on the physicochemical characteristics of groundwater provide valuable insights into the water quality of the region, and a comparison with the existing literature enhances our understanding of regional trends. The pH values of groundwater in this study ranged from 7.4 to 8.2, which is within the acceptable range for drinking water as set by the World Health Organization (WHO), i.e., 6.5–8.5. This slight alkalinity is typical of regions with limestone-dominated geology, where groundwater often interacts with mineral deposits, leading to higher pH values. Similar pH values have been reported in other groundwater studies in the Middle East and North Africa, with pH values ranging from 7.2 to 8.5 in limestone-dominated aquifers [25]. The consistency in pH levels across these studies suggests that the groundwater in the study area remains within a safe range for consumption, with no immediate concerns for acidity or alkalinity that could affect water quality.

The redox potential (Eh) in this study ranged from −14 mV to +23 mV, reflecting a transition from reducing to weakly oxidizing conditions. These values suggest that the shallow groundwater is influenced by organic matter and microbial activity, leading to reducing conditions, while deeper groundwater exhibits weakly oxidizing conditions, indicative of a more stable geochemical environment. Similar findings have been reported in agricultural regions with organic-rich soils, where Eh values range from −50 mV to +50 mV in shallow aquifers, and more oxidizing conditions are found in deeper aquifers [27]. The variation in Eh values highlights the dynamic nature of groundwater systems, where microbial and organic processes in shallow aquifers significantly affect the redox conditions.

The total dissolved solids (TDS) concentrations in shallow groundwater ranged from 500 to 1200 mg/L, while deeper wells had lower concentrations, ranging from 200 to 400 mg/L. These elevated TDS levels in shallow groundwater are typical of regions with high evaporation rates and agricultural activities, where salts from irrigation and evaporation contribute to increased TDS. This observation aligns with other studies from arid regions, such as those in North Africa and the Middle East, where TDS levels in shallow aquifers are often elevated, ranging from 500 to 1500 mg/L, due to similar agricultural practices [29]. In the study area, the highest TDS values were observed in the southeastern and western zones, which are also heavily influenced by surface evaporation and irrigation, consistent with findings from other arid and semi-arid regions.

Water hardness in this study ranged from 126 to 496 mg/L as $CaCO_3$, categorizing the water as hard, with higher hardness in shallow wells (299 mg/L) compared to deeper wells (210 mg/L). This is in line with studies conducted in areas with carbonate aquifers, where the dissolution of calcium and magnesium from limestone formations results in elevated hardness levels. For example, a study in Saudi Arabia reported hardness values ranging from 150 to 500 mg/L in carbonate aquifers [31]. The higher hardness in shallow groundwater samples in our study suggests that groundwater chemistry is influenced by the dissolution of minerals from the surrounding limestone geology.

Chloride concentrations were generally higher in deeper groundwater, ranging from 100 to 220 mg/L, compared to 50 to 150 mg/L in shallow wells. This pattern suggests the presence of saline intrusion, which is a common phenomenon in coastal and semi-arid regions. Similar patterns of elevated chloride concentrations in deeper groundwater have been observed in areas like coastal Mediterranean aquifers, where saline water from surrounding sources intrudes into deeper wells, with chloride concentrations often exceeding 200 mg/L [33]. Elevated sulfate concentrations, particularly in shallow groundwater, were observed in the study area, ranging from 40 to 150 mg/L. This is consistent with studies in agricultural and industrial regions, where sulfate contamination from agricultural runoff and industrial effluents is common. In contrast, sulfate concentrations in deeper groundwater in the study area ranged from 20 to 80 mg/L, indicating less contamination from surface activities.

Stable isotopic analysis revealed that shallow groundwater samples were enriched in heavier isotopes, with $\delta^{18}O$ values ranging from $-3.5‰$ to $-2.0‰$ and $\delta^{2}H$ values from $-25‰$ to $-18‰$. These isotopic signatures indicate the influence of evaporation, particularly from Lake Nasser, which is located nearby. This finding is consistent with other studies in arid regions, where evaporation leads to the enrichment of heavier isotopes in shallow groundwater, with $\delta^{18}O$ values typically ranging from $-3‰$ to $-1‰$ in similar settings [35]. In contrast, deeper groundwater samples displayed isotopic signatures consistent with paleo-recharge from wetter climatic periods, with δ18O values ranging from $-7.0‰$ to $-5.0‰$, which is typical of recharge that occurred during periods of higher precipitation. This pattern is also seen in other studies of deep aquifers in arid regions, where deep groundwater is recharged during wetter periods that occurred centuries or even millennia ago [36].

Iron and manganese concentrations were elevated in some shallow groundwater samples, particularly in areas with reduced conditions. Iron concentrations ranged from 0.3 to 1.2 mg/L, while manganese concentrations ranged from 0.2 to 0.8 mg/L, exceeding the WHO guidelines for manganese in 80% of shallow wells. These elevated concentrations are typical of groundwater in areas with reducing conditions, where iron and manganese are released from insoluble forms into the dissolved state. Similar findings have been reported in other agricultural regions, where iron and manganese levels are elevated due to reduction processes in shallow aquifers [37]. The elevated manganese concentrations in our study raise concerns about potential health risks, as prolonged exposure to high levels of manganese can lead to neurological disorders, as noted in studies in rural areas with high manganese levels in drinking water [38].

The groundwater quality index (GWQI) classification in this study showed that shallow groundwater was generally classified as "fair" to "poor", with values ranging from 40 to 60, while deeper groundwater samples were classified as "excellent" to "good", with values ranging from 75 to 90. These classifications align with findings from other agricultural regions, where shallow aquifers tend to have lower water quality due to contamination from agricultural practices, while deeper groundwater remains of higher quality. For instance, in India, similar GWQI classifications have been reported for shallow and deep groundwater in agricultural zones, with shallow wells classified as "poor" and deep wells as "excellent" due to contamination from agricultural runoff and fertilizers [39].

Consequently, the results of this study provide a detailed comparison of the physicochemical characteristics of groundwater in the region with findings from other studies conducted in similar geological and climatic settings. The study highlights that shallow groundwater in the area is more vulnerable to contamination from anthropogenic activities, with elevated levels of TDS, hardness, and trace metals, particularly iron and manganese, posing potential risks to water quality. In contrast, deeper groundwater remains of higher quality, with less contamination. These findings underscore the importance of ongoing groundwater monitoring and the implementation of sustainable water management practices, especially in agricultural regions, to ensure safe water resources for the future.

5. Conclusions

Groundwater contamination in the Nile Delta region is a highly significant problem that is currently not adequately controlled. The deterioration of the groundwater is attributed to agricultural practices, sanitation systems, and industrial activities. The deep (>60–120 m) groundwater wells are characterized by slightly alkaline (pH = 7.8–8.0), fresh (TDS = 232–608 mg/L), hard water (TH = 126–325 mg/L) that belongs to the Ca-HCO_3

water type. The shallow (<60 m) groundwater wells are slightly alkaline (pH = 7.4–8.2), mostly (92% of the total samples) fresh water (TDS = 274–1274 mg/L), hard-to-very-hard water (TH = 299–496 mg/L), and the majority (64%) are of the Ca-HCO_3 type, while 28% are of the Na-HCO_3 type and 8% of the Na-Cl water type.

Based on isotope geochemistry, the deep groundwater wells display paleo-water (pre-high-dam-construction), while the shallow groundwater differentiates into paleo-water, recent Nile water (post-high-dam-construction), or mixed between both categories. The presence or absence of clay lenses within the aquifer is the main factor which controls the water type.

Most of the groundwater samples (90%) studied are classified as excellent to good for irrigation purposes, while 10% of the samples are categorized as permissible-to-doubtful irrigation water. Manganese proved to be the most significant element in the Middle Nile Delta groundwater, where its concentration exceeds the permissible limits in up to 95% of the groundwater samples studied. HQ values indicate an unacceptable non-carcinogenic health risk for 56% of the shallow groundwater samples concerning Mn. Detailed studies about the geochemical characteristics of manganese in Nile Delta soil are needed to control the high Mn concentration in underlying groundwater.

This study only provides a thorough description, assessment, and interpretation of inorganic contaminants. However, additional studies are required to assess the organic contaminants in the Nile Delta groundwater. Rural communities that consume shallow groundwater should be warned about the potential adverse effects of such water.

6. Recommendations and Way Forward

Future research should focus on applying advanced isotopic techniques to trace groundwater recharge rates and contamination sources. Long-term monitoring is crucial to detect quality changes and emerging pollutants. Interdisciplinary collaboration among hydrogeologists, geochemists, health experts, and policymakers is essential for effective risk assessment and sustainable management. Additionally, developing accessible databases and cost effective methods will support broader implementation, especially in resource-limited regions. These efforts will enhance sustainable management of groundwater resources and safeguard public health in arid areas.

Author Contributions: Conceptualization: A.M.A.K.; methodology: A.M.A.K., R.v.G., and E.I.; formal analysis and investigation: A.M.A.K., R.v.G., W.R.A., and I.A.A.B.; data curation; A.M.A.K., E.I., I.A.A.B., and D.E.A.; writing—original draft preparation: I.A.A.B., R.v.G., and A.O.B.; writing—review and editing: A.M.A.K., A.O.B., D.E.A., and W.R.A. All authors have read and agreed to the published version of the manuscript.

Funding: This research received no external funding.

Data Availability Statement: The data that support the findings of this study are available from the corresponding author upon reasonable request.

Acknowledgments: The first author gratefully acknowledges a grant from the Egyptian mission (Higher Education Ministry) as well as R. Koch and H. de Wall for hosting him during his stay at the FAU, Germany. Thanks also to M. Hassaan and M. A. Hekal (Geology Dep., Faculty of Science, Al Azhar University, Egypt) for reviewing an early version of the manuscript. Sincere thanks to J. Barth and Ch. Hanke (FAU) for their help with chemical and stable isotope analyses.

Conflicts of Interest: The authors declare no conflicts of interest.

References

1. Aeschbach-Hertig, W.; Gleeson, T. Regional strategies for the accelerating global problem of groundwater depletion. *Nat. Geosci.* **2012**, *5*, 853–861. [CrossRef]
2. Wang, X.; Xiao, C.; Yang, W.; Liang, X.; Zhang, L.; Zhang, J. Analysis of the quality, source identification and apportionment of the groundwater in a typical arid and semi-arid region. *J. Hydrol.* **2023**, *625*, 130169. [CrossRef]
3. Pastén-Zapata, E.; Ledesma-Ruiz, R.; Harter, T.; Ramírez, A.I.; Mahlknecht, J. Assessment of sources and fate of nitrate in shallow groundwater of an agricultural area by using a multi-tracer approach. *Sci. Total. Environ.* **2014**, *470–471*, 855–864. [CrossRef]
4. Wang, S.; Tang, C.; Song, X.; Wang, Q.; Zhang, Y.; Yuan, R. The impacts of a linear wastewater reservoir on groundwater recharge and geochemical evolution in a semi-arid area of the Lake Baiyangdian watershed, North China Plain. *Sci. Total. Environ.* **2014**, *482*, 325–335. [CrossRef]
5. Akhtar, N.; Syakir Ishak, M.I.; Bhawani, S.A.; Umar, K. Various natural and anthropogenic factors responsible for water quality degradation: A review. *Water* **2021**, *13*, 2660. [CrossRef]
6. Zeydalinejad, N.; Javadi, A.A.; Webber, J.L. Global perspectives on groundwater infiltration to sewer networks: A threat to urban sustainability. *Water Res.* **2024**, *262*, 122098. [CrossRef] [PubMed]
7. Arumugam, K.; Elangovan, K. Hydrochemical characteristics and groundwater quality assessment in Tirupur Region, Coimbatore District, Tamil Nadu, India. *Environ. Geol.* **2009**, *58*, 1509–1520. [CrossRef]
8. Nagaraju, A.; Sunil Kumar, K.; Thejaswi, A. Assessment of groundwater quality for irrigation: A case study from Bandalamottu lead mining area, Guntur District, Andhra Pradesh, South India. *Appl. Water Sci.* **2014**, *4*, 385–396. [CrossRef]
9. Lghoula, M.; Maqsoudb, A.; Hakkouc, R.; Kchikacha, A. Hydrogeochemical behavior around the abandoned Kettara mine site, Morocco. *J. Geochem. Expl.* **2014**, *144*, 456–467. [CrossRef]
10. Abu Khatita, A.M.; de Wall, H.; Koch, R. Anthropogenic particle dispersions in topsoils of the Middle Nile Delta: A preliminary study on the contamination around industrial and commercial areas in Egypt. *Environ. Earth. Sci.* **2016**, *75*, 264. [CrossRef]
11. Dinis, M.D.L.; Fiúza, A. Mitigation of uranium mining impacts—a review on groundwater remediation technologies. *Geosciences* **2021**, *11*, 250. [CrossRef]
12. Mor, S.; Ravindra, K.; Dahiya, R.P.; Chandra, A. Leachate characterization and assessment of groundwater pollution near municipal solid waste landfill site. *Environ. Monit. Assess.* **2006**, *118*, 435–456. [CrossRef] [PubMed]
13. Parameswari, K.; Mudgal, B.V. Geochemical investigation of groundwater contamination in Perungudi dumpsite, South India. *Arab. J. Geosci.* **2013**, *7*, 1363–1371. [CrossRef]
14. Gómez-Puentes, F.J.; Reyes-López, J.A.; López, D.L.; Carreón-Diazconti, C.; Belmonte-Jiménez, S. Geochemical processes controlling the groundwater transport of contaminants released by a dump in an arid region of México. *Environ. Earth. Sci.* **2014**, *7*, 1609–1621. [CrossRef]
15. Parvin, F.; Tareq, S.M. Impact of landfill leachate contamination on surface and groundwater of Bangladesh: A systematic review and possible public health risks assessment. *Appl. Water Sci.* **2021**, *11*, 100. [CrossRef] [PubMed]
16. Diaw, M.; Faye, S.; Stichler, W.; Maloszewski, P. Isotopic and geochemical characteristics of groundwater in the Senegal River delta aquifer: Implication of recharge and flow regime. *Environ. Earth. Sci.* **2012**, *66*, 1011–1020. [CrossRef]
17. Chandrajith, R.; Chaturangani, D.; Abeykoon, S.; Barth, J.A.C.; Geldern, R.; Edirisinghe, E.A.N.V.; Dissanayake, C.B. Quantification of groundwater–seawater interaction in a coastal sandy aquifer system: A study from Panama, Sri Lanka. *Environ. Earth. Sci.* **2014**, *72*, 867–877. [CrossRef]
18. Wang, F.; Song, K.; He, X.; Peng, Y.; Liu, D.; Liu, J. Identification of groundwater pollution characteristics and health risk assessment of a landfill in a low permeability area. *Int. J. Environ. Res. Public Health* **2021**, *18*, 7690. [CrossRef]
19. Elazhary, M.A.E.; Sabet, H.S.; Abu Khatita, A.M.; Shehata, A.F.; El-Hemamy, S.T. Using hydrogeochemical and geographic information system methods as an environmental approach for evaluating coastal aquifers in the west northern coast of Egypt. *Egypt. J. Geol.* **2023**, *67*, 95–110. [CrossRef]
20. Fida, M.; Li, P.; Alam, S.K.; Wang, Y.; Nsabimana, A.; Shrestha, P.S. Review of Groundwater Nitrate Pollution from Municipal Landfill Leachates: Implications for Environmental and Human Health and Leachate Treatment Technologies. *Expo. Health* **2024**, *16*, 1225–1249. [CrossRef]
21. Schullehner, J.; Hansen, B.; Thygesen, M.; Pedersen, C.B.; Sigsgaard, T. Nitrate in drinking water and colorectal cancer risk: A nationwide population-based cohort study. *Int. J. Cancer* **2018**, *143*, 73–79. [CrossRef] [PubMed]
22. Wongsasuluk, P.; Chotpantarat, S.; Siriwong, W.; Robson, M. Heavy metal contamination and human health risk assessment in drinking water from shallow groundwater wells in an agricultural area in Ubon Ratchathani province, Thailand. *Environ. Geochem. Health* **2013**, *36*, 169–182. [CrossRef] [PubMed]
23. Abuzaid, A.S.; Jahin, H.S. Implications of irrigation water quality on shallow groundwater in the Nile Delta of Egypt: A human health risk prospective. *Environ. Technol. Innov.* **2021**, *22*, 101383. [CrossRef]

24. Hossain, M.E.; Shahrukh, S.; Hossain, S.A. Chemical Fertilizers and Pesticides: Impacts on Soil Degradation, Groundwater, and Human Health in Bangladesh. In *Environmental Degradation: Challenges and Strategies for Mitigation*, 2nd ed.; Singh, V.P., Yadav, S., Yadav, K.K., Yadava, R.N., Eds.; Springer International Publishing: Cham, Switzerland, 2022; Volume 104, pp. 63–92. [CrossRef]
25. Dugan, S.T.; Muhammetoglu, A.; Uslu, A. A combined approach for the estimation of groundwater leaching potential and environmental impacts of pesticides for agricultural lands. *Sci. Total Environ.* **2023**, *901*, 165892. [CrossRef]
26. Kasem, A.M.; Xu, Z.; Jiang, H.; Liu, W.; Zhang, J.; Nosair, A.M. Nitrate source and transformation in groundwater under urban and agricultural arid environment in the southeastern Nile Delta, Egypt. *Water* **2024**, *16*, 22. [CrossRef]
27. Edition, F. Guidelines for drinking-water quality. *WHO Chron.* **2011**, *38*, 104–108.
28. Siegel, F.R. *Environmental Geochemistry of Potentially Toxic Metals*, 1st ed.; Springer: Berlin, Germany, 2002; p. 218. [CrossRef]
29. Gomes, M.; Ralph, T.J.; Humphries, M.S.; Graves, B.P.; Kobayashi, T.; Gore, D.B. Waterborne contaminants in high intensity agriculture and plant production: A review of on-site and downstream impacts. *Sci. Total Environ.* **2025**, *958*, 178084. [CrossRef]
30. Chen, F.; Yao, L.; Mei, G.; Shang, Y.; Xiong, F.; Ding, Z. Groundwater quality and potential human health risk assessment for drinking and irrigation purposes: A case study in the semiarid region of north China. *Water* **2021**, *13*, 783. [CrossRef]
31. Prapanchan, V.N.; Subramani, T.; Karunanidhi, D.; Gopinathan, P. Groundwater quality assessment for drinking and irrigation purposes and its human health risks in the Sevathur mine region, south India. *Desalination Water Treat.* **2024**, *320*, 100883. [CrossRef]
32. Schulte, P.; van Geldern, R.; Freitag, H.; Karim, A.; Négrel, P.; Petelet-Giraud, E.; Probst, A.; Probst, J.L.; Telmer, K.; Veizer, J.; et al. Applications of stable water and carbon isotopes in watershed research: Weathering, carbon cycling, and water balances. *Earth. Sci. Rev.* **2011**, *109*, 20–31. [CrossRef]
33. Yousif, M.; van Geldern, R.; Bubenzer, O. Hydrogeological investigation of shallow aquifers in an arid data-scarce coastal region (El Daba'a, northwestern Egypt). *Hydrogeol. J.* **2016**, *24*, 159–179. [CrossRef]
34. Liao, F.; Wang, G.; Yi, L.; Shi, Z.; Cheng, G.; Kong, Q.; Liu, C. Identifying locations and sources of groundwater discharge into Poyang Lake (eastern China) using radium and stable isotopes (deuterium and oxygen-18). *Sci. Total Environ.* **2020**, *740*, 140163. [CrossRef] [PubMed]
35. Mao, H.; Wang, G.; Shi, Z.; Liao, F.; Xue, Y. Spatiotemporal variation of groundwater recharge in the lower reaches of the Poyang Lake Basin, China: Insights from stable hydrogen and oxygen isotopes. *J. Geophys. Res. Atmos.* **2021**, *126*, e2020JD033760. [CrossRef]
36. Ahmed, M.; Chen, Y.; Khalil, M.M. Isotopic composition of groundwater resources in arid environments. *J. Hydrol.* **2022**, *609*, 127773. [CrossRef]
37. Dhanasekarapandian, M.; Chandran, S.; Devi, D.S.; Kumar, V. Spatial and temporal variation of groundwater quality and its suitability for irrigation and drinking purpose using GIS and WQI in an urban fringe. *J. Afr. Earth. Sci.* **2016**, *124*, 270–288. [CrossRef]
38. Ambrosino, M.; El-Saadani, Z.; Khatita, A.A.; Mingqi, W.; Palarea-Albaladejo, J.; Cicchella, D. Geochemical speciation, ecological risk and assessment of main sources of potentially toxic elements (PTEs) in stream sediments from Nile River in Egypt. *Water* **2023**, *15*, 2308. [CrossRef]
39. Egbueri, J.C.; Unigwe, C.O.; Omeka, M.E.; Ayejoto, D.A. Urban groundwater quality assessment using pollution indicators and multivariate statistical tools: A case study in southeast Nigeria. *Int. J. Environ. Anal. Chem.* **2023**, *103*, 3324–3350. [CrossRef]
40. Atwia, M.G.; Abou Heleika, M.M.; El-Shishtawy, A.M.; Sharp, J.M. Hydrostratigraphy of the Central Nile Delta, Egypt, Using Geoelectric Measurements. *EGS AGU EUG Jt. Assem.* **2003**, *5*, 6–11.
41. Gemail, K.S.; El-Shishtawy, A.M.; El-Alfy, M.; Ghoneim, M.F.; Abd El-Bary, M.H. Assessment of aquifer vulnerability to industrial waste water using resistivity measurements: A case study along El-Gharbyia main drain, Nile Delta, Egypt. *J. Appl. Geophys.* **2011**, *75*, 140–150. [CrossRef]
42. Bennett, P.C.; El Shishtawy, A.M.; Sharp, J.M.; Atwia, M.G. Source and migration of dissolved manganese in the Central Nile Delta Aquifer, Egypt. *J Afr. Earth. Sci.* **2014**, *96*, 8–20. [CrossRef]
43. Negm, A.M.; Sakr, S.; Abd-Elaty, I.; Abd-Elhamid, H.F. An Overview of Groundwater Resources in Nile Delta Aquifer. In *The Handbook of Environmental Chemistry*; Negm, A., Ed.; Groundwater in the Nile Delta; Springer: Cham, Switzerland, 2018; Volume 73, pp. 3–44. [CrossRef]
44. Arafa, N.A.; Salem, Z.E.S.; Ghorab, M.A.; Soliman, S.A.; Abdeldayem, A.L.; Moustafa, Y.M.; Ghazala, H.H. Evaluation of groundwater sensitivity to pollution using GIS-based modified DRASTIC-LU model for sustainable development in the Nile Delta Region. *Sustainability* **2022**, *14*, 14699. [CrossRef]
45. Shi, Q.; Gao, Z.; Guo, H.; Zeng, X.; Sandanayake, S.; Vithanage, M. Hydrogeochemical factors controlling the occurrence of chronic kidney disease of unknown etiology (CKDu). *Environ. Geochem. Health* **2023**, *45*, 2611–2627. [CrossRef] [PubMed]
46. Ahmad, A.; van der Wens, P.; Baken, K.; de Waal, L.; Bhattacharya, P.; Stuyfzand, P. Arsenic reduction to< 1 µg/L in Dutch drinking water. *Environ. Int.* **2020**, *134*, 105253. [CrossRef]

47. Chen, R.; Chai, X.; Zhang, Y.; Zhou, T.; Xia, Y.; Jiang, X.; Chen, C. Novel role of FTO in regulation of gut–brain communication via Desulfovibrio fairfieldensis-produced hydrogen sulfide under arsenic exposure. *Gut Microbes* **2025**, *17*, 2438471. [CrossRef]
48. Richter, F.; Kloster, S.; Wodschow, K.; Hansen, B.; Schullehner, J.; Kristiansen, S.M.; Petersen, M.M.; Strandberg-Larsen, K.; Ersbøll, A.K. Maternal exposure to arsenic in drinking water and risk of congenital heart disease in the offspring. *Environ. Int.* **2022**, *160*, 107051. [CrossRef]
49. Maity, J.P.; Vithanage, M.; Kumar, M.; Ghosh, A.; Mohan, D.; Ahmad, A.; Bhattacharya, P. Seven 21st century challenges of arsenic-fluoride contamination and remediation. In *Groundwater for Sustainable Development*; Elsevier B.V.: Amsterdam, The Netherlands, 2021; Volume 12. [CrossRef]
50. Nshimiyimana, C.; Thakur, N.; Alalawy, A.I.; Zheng, Y.; Kareri, T.; Salama, E.S. Low-water quality and non-communicable disease nexus: Health impacts, mechanisms, and advanced water treatment approaches. *J. Environ. Chem. Eng.* **2025**, *13*, 115744. [CrossRef]
51. Rizzini, A.; Vessani, F.; Coccetta, V.; Milad, G. Stratigraphy and Sedimentation of a Neogene—Quaternary Section in the Nile Delta Area (A.R.E.). *Mar. Geol.* **1978**, *27*, 327–348. [CrossRef]
52. Coutellier, V.; Stanley, D.J. Late Quaternary stratigraphy of the Eastern Nile Delta, Egypt. *Mar. Geol.* **1987**, *77*, 257–275. [CrossRef]
53. Stanley, D.J.; Warne, A.G. Nile delta: Recent geological evolution and human impact. *Science.* **1993**, *260*, 628–634. [CrossRef]
54. Zaghloul, Z.M.; Taha, A.A.; Hegab, O.; El Fawal, F. The Neogene—Quaternary sedimentary basins of the Nile Delta. *Egyptian J. Geol.* **1977**, *21*, 1–19.
55. Kashef, A.I. Salt-water intrusion in the Nile Delta. *Ground Water* **1983**, *21*, 160–167. [CrossRef]
56. Salem, Z.E.; Hasan, S.S.; Sefelnasr, A.M. Health risk assessment of groundwater use for drinking in West Nile Delta, Egypt. *Sci. Rep.* **2025**, *15*, 7414. [CrossRef]
57. Khan, M.A.; Akhtar, R.; Ahmad, I. Assessing Groundwater Contamination, Health Risk And Health Cost: A Regression Analysis And Fuzzy AHP Approach. *Spectr. Eng. Sci.* **2025**, *3*, 452–477.
58. RIGW. *Hydrological Map of Egypt, SCAL 1: 500,000. Nile Delta*; Research Institute for the Groundwater: El Kanater Ek Khairia, Egypt, 1992; p. 39.
59. Salem, Z.E. Natural and human impacts on the groundwater under an Egyptian village, central Nile Delta: A case study of Mehallet Menouf. In Proceedings of the Thirteenth International Water Technology Conference (IWTC 13), Hurghada, Egypt, 12 March 2009; pp. 1397–1414.
60. Bandara, S.; Rajapaksha, A.U.; Kandasamy, A.; Hettithanthri, O.; Magana-Arachchi, D.; Wanigatunge, R.; Vithanage, M. Food-mediated exposure of Hofmeister ions in Oryza sativa (Rice) from selected CKDu endemic regions in Sri Lanka. *Environ. Geochem. Health* **2023**, *45*, 8417–8432. [CrossRef]
61. Masoud, A.A. Geotechnical evaluation of the alluvial soils for urban land management zonation in Gharbiya governorate, Egypt. *J. Afr. Earth. Sci.* **2015**, *101*, 360–374. [CrossRef]
62. Sawyer, G.N.; McCartly, D.L.; Parkin, G. *Chemistry for Environmental Engineering and Science*, 5th ed.; McGraw Hill: New York, NY, USA, 2003; p. 752.
63. Van Geldern, R.; Barth, J.A.C. Optimization of instrument setup and post-run corrections for oxygen and hydrogen stable isotope measurements of water by isotope ratio infrared spectroscopy (IRIS). *Limnol. Oceanogr: Methods.* **2012**, *10*, 1024–1036. [CrossRef]
64. Clark, I.D.; Fritz, P. *Environmental Isotopes in Hydrogeology*; CRC Press/Lewis Publishers: Boca Raton, FL, USA, 1997; p. 328.
65. Vasanthavigar, M.; Srinivasamoorthy, K.; Vijayaragavan, K.; Ganthi, R.R.; Chidambaram, S.; Anandhan, P.; Manivannan, R.; Vasudevan, S. Application of water quality index for groundwater quality assessment: Thirumanimuttar sub-basin, Tamilnadu, India. *Environ. Monit. Assess.* **2010**, *171*, 595–609. [CrossRef]
66. Singh, V.; Singh, U.C. Assessment of groundwater quality of parts of Gwalior (India) for agricultural purposes. Indian. *Indian. J. Sci. Technol.* **2008**, *1*, 1–5. [CrossRef]
67. Wilcox, L.V. *Classification and Use of Irrigation Waters*; Circular 969; USDA: Washington, DC, USA, 1955.
68. Peplow, D.; Edmonds, R. Health risks associated with contamination of groundwater by abandoned mines near Twisp in Okanogan County, Washington, USA. *Environ. Geochem. Health* **2004**, *26*, 69–79. [CrossRef]
69. Dissanayake, C.B.; Chandrajith, R. *Introduction to Medical Geology*; Springer: Berlin, Germany, 2009; p. 297. [CrossRef]
70. Li, P.; Wu, J.; Qian, H.; Lyu, X.; Liu, H. Origin and assessment of groundwater pollution and associated health risk: A case study in an industrial park, northwest China. *Environ. Geochem. Health* **2014**, *36*, 693–712. [CrossRef]
71. Fajčíková, K.; Cvečková, V.; Stewart, A.; Rapant, S. Health risk estimates for groundwater and soil contamination in the Slovak Republic: A convenient tool for identification and mapping of risk areas. *Environ. Geochem. Health* **2014**, *36*, 973–986. [CrossRef] [PubMed]
72. USEPA (US Environmental Protection Agency). *Risk Assessment Guidance for Superfund, 1989, Volume I: Human Health Evaluation Manual (Part A)*; EPA 540/1-89/002; US Environmental Protection Agency, Office of Emergency and Remedial Response: Washington, DC, USA, 1989.

73. USEPA (US Environmental Protection Agency). *Guidelines for Carcinogen Risk Assessment*; EPA/630/P-03/001F; United States Environmental Protection Agency, Risk Assessment Forum: Washington, DC, USA, 2005.
74. Lu, S.G.; Tang, C.; Rengel, Z. Combined effects of waterlogging and salinity on electrochemistry, water-soluble cations and water dispersible clay in soils with various salinity levels. *Plant Soil* **2004**, *264*, 231–245. [CrossRef]
75. Durvey, V.S.; Sharma, L.L.; Saini, V.P.; Sharma, B.K. *Handbook on the Methodology of Water Quality Assessment Rajasthan*; Agriculture University: Jaipur City, India, 1991.
76. Agrawal, V.; Jagetai, M. Hydrochemical assessment of groundwater quality in Udaipur city, Rajasthan, India. In *Proceedings of the National Conference on Dimensions of Environmental Stress in India*; Department of Geology, MS University: Baroda, India, 1997; pp. 151–154.
77. Sengupta, P. Potential health impacts of hard water. *Int. J. Prev. Med.* **2013**, *4*, 866–875. [PubMed Central]
78. Kozisek, F. Regulations for calcium, magnesium or hardness in drinking water in the European Union member states. *Regul. Toxicol. Pharmacol.* **2020**, *112*, 104589. [CrossRef]
79. *USEP Drinking Water Advisory: Consumer Acceptability Advice and Health Effects Analysis on Sodium EPA 822-R-03-006*; US Environmental Protection Agency Office of Water (4304T), Health and Ecological Criteria Division: Washington, DC, USA, 2003.
80. Piper, A.M. A graphical interpretation of water analysis. *Trans. Am. Geophys. Union* **1944**, *25*, 914–928. [CrossRef]
81. Sultan, M.; Sturchio, N.C.; Gheith, H.; Hady, Y.A.; Anbeawy, M. Chemical and Isotopic Constraints on the Origin of Wadi El-Tarfa Ground Water, Eastern Desert, Egypt. *Groundwater* **2000**, *38*, 743–751. [CrossRef]
82. Sadek, M.A.; Samie, S.G. Pollution vulnerability of the quaternary aquifer near Cairo, Egypt, as indicated by isotopes and hydrochemistry. *J. Hydrogeol.* **2001**, *9*, 273–281. [CrossRef]
83. Geirnart, W.; Leaven, M.P. Composition and history of groundwater in the western Nile delta. *J. Hydrol.* **1992**, *138*, 169–189. [CrossRef]
84. Aly, A.I.M.; Froehlich, K.; Nada, A.; Awad, M.; Hamza, M.; Salem, W.M. Study of environmental isotope distribution in the Aswan High Dam Lake (Egypt) for estimation of evaporation of lake water and its recharge to adjacent groundwater. *Environ. Geochem. Health* **1993**, *15*, 37–49. [CrossRef]
85. Chandrajith, R.; Diyabalanage, S.; Premathilake, K.M.; Hanke, C.; van Geldern, R.; Barth, J.A.C. Controls of evaporative irrigation return flows in comparison to seawater intrusion in coastal karstic aquifers in northern Sri Lanka: Evidence from solutes and stable isotopes. *Sci. Total. Environ.* **2016**, *548–549*, 421–428. [CrossRef]
86. Rozanski, K.; Araguás-Araguás, L.; Gonfiantini, R. Isotopic patterns in modern global precipitation. In *Climate Change in Continental Isotopic Records*; Swart, P.K., Lohmann, K.L., McKenzie, J., Savin, S., Eds.; American Geophysical Union: Washington, DC, USA, 1993; pp. 1–37.
87. Awad, M.A.; Hamza, M.S.; Atwa, S.M.; Sallouma, M.K. Isotopic and hydrogeochemical evaluation of groundwater at Qusier-Safaga area, eastern desert, Egypt. *Environ. Geochem. Health* **1996**, *18*, 47–54. [CrossRef]
88. Abu Khatita, A.M.; Koch, R.; Bamousa, A.O. Sources identification and contamination assessment of heavy metals in soil of Middle Nile Delta, Egypt. *J. Taibah Univ. Sci.* **2020**, *14*, 750–761. [CrossRef]
89. Thomas, J.; Rose, T. Environmental isotopes in hydrogeology. *Environ. Earth Sci.* **2003**, *43*, 532. [CrossRef]
90. Edmunds, W.M.; Guendouz, A.H.; Mamou, A.; Moulla, A.; Shand, P.; Zouari, K. Groundwater evolution in the Continental Intercalaire aquifer of southern Algeria and Tunisia: Trace element and isotopic indicators. *Appl. Geochem.* **2003**, *18*, 805–822. [CrossRef]
91. Kendall, C.; McDonnell, J.J. (Eds.) *Isotope Tracers in Catchment Hydrology*; Elsevier: Amsterdam, The Netherlands, 2012.
92. Apostoli, P.; Cornelis, R.; Duffus, J.; Hoet, P.; Lison, D.; Templeton, D.M. *Elemental Speciation in Human Health Risk Assessment. Environment Health Criteria 234*; WHO: Geneva, Switzerland, 2006.
93. USEPA (US Environmental Protection Agency). Ground Water and Drinking Water. 2012. Available online: https://www.epa.gov/ground-water-and-drinking-water (accessed on 27 January 2025).
94. Wasserman, G.A.; Liu, X.; Parvez, F.; Ahsan, H.; Levy, D.; Factor-Litvak, P.; Kline, J.; Van Geen, A.; Slavkovich, V.; LoIacono, N.J.; et al. Water manganese exposure and children's intellectual function in Araihazar, Bangladesh. *Environ. Health. Persp.* **2006**, *114*, 124–129. [CrossRef] [PubMed]
95. Nádaská, G.; Lesny, J.; Michalik, I. Environmental aspect of manganese chemistry. *Health. Environ. J.* **2010**, *100702-A*, 1–16.
96. Crossgrove, J.; Zheng, W. Manganese toxicity upon overexposure. *NMR Biomed.* **2004**, *17*, 544–553. [CrossRef]

Disclaimer/Publisher's Note: The statements, opinions and data contained in all publications are solely those of the individual author(s) and contributor(s) and not of MDPI and/or the editor(s). MDPI and/or the editor(s) disclaim responsibility for any injury to people or property resulting from any ideas, methods, instructions or products referred to in the content.

MDPI AG
Grosspeteranlage 5
4052 Basel
Switzerland
Tel.: +41 61 683 77 34

Hydrology Editorial Office
E-mail: hydrology@mdpi.com
www.mdpi.com/journal/hydrology

Disclaimer/Publisher's Note: The title and front matter of this reprint are at the discretion of the Guest Editor. The publisher is not responsible for their content or any associated concerns. The statements, opinions and data contained in all individual articles are solely those of the individual Editor and contributors and not of MDPI. MDPI disclaims responsibility for any injury to people or property resulting from any ideas, methods, instructions or products referred to in the content.

www.ingramcontent.com/pod-product-compliance
Lightning Source LLC
Chambersburg PA
CBHW041409210926
51850CB00013B/3217